Climate Change Signals and Response

Chandra Venkataraman · Trupti Mishra
Subimal Ghosh · Subhankar Karmakar
Editors

Climate Change Signals and Response

A Strategic Knowledge Compendium for India

Springer

Editors
Chandra Venkataraman
Interdisciplinary Programme in Climate Studies, and Department of Chemical Engineering
Indian Institute of Technology Bombay
Mumbai, Maharashtra
India

Subimal Ghosh
Interdisciplinary Programme in Climate Studies, and Department of Civil Engineering
Indian Institute of Technology Bombay
Mumbai, Maharashtra
India

Trupti Mishra
Interdisciplinary Programme in Climate Studies, and Shailesh J. Mehta School of Management
Indian Institute of Technology Bombay
Mumbai, Maharashtra
India

Subhankar Karmakar
Interdisciplinary Programme in Climate Studies, and Centre for Environmental Science and Engineering
Indian Institute of Technology Bombay
Mumbai, Maharashtra
India

ISBN 978-981-13-0279-4 ISBN 978-981-13-0280-0 (eBook)
https://doi.org/10.1007/978-981-13-0280-0

Library of Congress Control Number: 2018939941

Printed on acid-free paper

This Springer imprint is published by the registered company Springer Nature Singapore Pte Ltd.
The registered company address is: 152 Beach Road, #21-01/04 Gateway East, Singapore 189721, Singapore

Prof. D. R. Sikka (1932–2017)

For his generosity in bestowing his science and his conviction on the DST-Centre of Excellence in Climate Studies Indian Institute of Technology, Bombay

Prof. Dev Raj Sikka, known as the "Monsoon Guru", was born in 1932 in Jhang Mighiana, in then undivided India, now Pakistan. Prof. Sikka joined the India Meteorological Department (IMD) in 1954 before moving to the Indian Institute of Tropical Meteorology (IITM) in Pune in 1962, becoming its Director in 1986. His contribution to the field of Indian Monsoon is beyond comparison. His passion for monsoon research and his breakthroughs in monsoon science merged the contributions of pioneers of Monsoon Meteorology—Blanford, Elliot, Walker, Riehl, Malkus, Anantha Krishnan, Bjerknes and Charneyto—to generate our current understanding of monsoon science and mechanisms. He employed traditional and space-borne observations over India within the context of well-conceived theoretical frameworks. This led to his discovery of coherent northward propagations of precipitation band over the equatorial Indian Ocean at intervals of 2–6 weeks, throughout the summer monsoon season of June to September, leading to improved forecasting of intra-seasonal variations. Prof. Sikka's work on the link between the Indian summer monsoon rainfall and ENSO is the basis of monsoon prediction by most of the state-of art dynamical models, even today. Prof. Sikka was elected a Fellow of the Indian Academy of Sciences in 1984, received the first Lifetime Achievement

Award in Atmospheric Science and Technology from the Ministry of Earth Sciences in 2007, the Sir Gilbert Thomas Walker Gold Medal in 2012 and Lifetime Achievement Award from the India Meteorological Society (IMS) in 2017. The "Monsoon Guru," Prof. Dev Raj Sikka, passed away on 18 March 2017 and is survived by a granddaughter, two daughters and the community of grateful monsoon and climate scientists.

Foreword I

As part of the National Action Plan on Climate Change (NAPCC), the National Mission on Strategic Knowledge for Climate Change (NMSKCC) is one of the eight national missions on climate change coordinated by Department of Science and Technology, through its Climate Change Programme, Strategic Programmes, Large Initiatives and Coordinated Action Enabler (SPLICE) Division. The broad objective of NMSKCC is to build a vibrant and dynamic knowledge system that would inform and support national action on climate change and sustainable development.

The "DST-Centre of Excellence in Climate Studies (DST-CoECS)" at IIT Bombay was initiated and supported by DST under the NMSKCC in January 2012 with the aim to undertake interdisciplinary, problem-driven research for end-to-end solutions covering the causes and consequences of and responses to climate change; to build long-term scientific capacity and systems for study of regional climate change and climate futures; to enable the creation of a pool of multidisciplinary researchers to serve the growing need for climate change scientists and professionals to serve R&D and policy needs in private, public and governmental institutions; and to provide critical assessments to support policy and governmental decision-making on air and water resources and climate mitigation and adaptation measures.

The DST-CoECS has participation of over 24 faculty members from 9 departments of IIT Bombay, who apply their expertise to interdisciplinary problems that cross traditional academic boundaries to address climate change not only through climate and environmental science but also through economics and engineering. Co-advised training of over 30 Ph.D. scholars has been achieved, along with significant enrichment through lecture series and short courses, from invited national and international experts. It is heartening that the Centre has produced significant research output in terms of high-impact international journal publications.

I am happy to note that the Centre is bringing out a book entitled *Climate Change Signals and Response: A Strategic Knowledge Compendium for India* which is a synthesis of the recent research findings in climate change from the

DST-CoECS. The general scope of the book is to chronicle strategic knowledge to inform, guide and support national- and state-level action on climate change.

I am confident that this book will serve as an invaluable tool to disseminate scientific findings to various stakeholders including students, academicians, industry professionals and policy makers as well as to inform general public at large.

I would like to congratulate the DST-CoECS for having brought out such a useful publication and thank the Editorial team and authors for their contributions to the book.

(Akhilesh Gupta)

New Delhi, India
March 2018

Dr. Akhilesh Gupta
Advisor and Head, Strategic Programmes
Large Initiatives and Coordinated Action Enabler (SPLICE)
Division and Climate Change Programme
Department of Science and Technology
Ministry of Science and Technology
Government of India

Foreword II

Climate change signals indicate increased climate variability over the Indian region during the last 50 years, which could accelerate during coming decades. We have witnessed extreme temperatures, persistent droughts and frequent floods. Urbanisation, though a worldwide phenomenon driving climate change, is starkly visible in India. Economic growth and development impact emissions of greenhouse gases and short-lived climate pollutants, which drive climate change.

India's Nationally Determined Contributions (NDCs) submitted to United Nations Framework Convention on Climate Change (UNFCCC) is an attestation of our commitment to global environmental concerns. The government is committed to enhancing investments in development programmes in sectors vulnerable to climate change, particularly agriculture and water resources, in ecologically sensitive regions.

The Department of Science and Technology is coordinating and implementing a National Mission on Strategic Knowledge for Climate Change (NMSKCC) under National Action Plan on Climate Change (NAPCC) with a broad objective of building S&T capacity and training of scholars and researchers to generate new knowledge to underpin government decision-making. Under this mission, the DST-Centre of Excellence in Climate Studies, IIT Bombay, was established in January 2012, with an objective to "develop a scientific understanding of regional climate change and connect it to impacts (socio-economic, environment, resources) and effective response (technology and adaptation)". I am happy to see that the Centre has performed well towards meeting these objectives and have shown good research outcomes.

I am pleased to note that some of the key findings of the Centre of Excellence are being brought out as a book entitled *Climate Change Signals and Response: A Strategic Knowledge Compendium for India*. The book attempts to highlight actionable outcomes from the research undertaken, encompassing emergence of regional climate signals, their regional impacts and mitigation and adaptation response, relevant in the Indian context.

I compliment the editors and contributors of the book for their sincere efforts to synthesise high-impact research findings into actionable recommendations and policy prescriptions at national and sub-national levels.

I am confident the book will be of great use to not only researchers working in the area of climate change but also the policy makers who would find some of the findings useful for policy formulation purposes.

New Delhi, India
March 2018

Prof. Ashutosh Sharma
Secretary, Department of Science and Technology
Ministry of Science and Technology
Government of India

Preface and Acknowledgements

The notion for a book synthesising strategic knowledge for climate change action in India arose towards the end of the first phase of the Department of Science and Technology-sponsored Centre of Excellence in Climate Studies (DST-CoECS, 2012–2017) at IIT Bombay. A group of colleagues at IIT Bombay, with diverse expertise and backgrounds, had endeavoured for 5 years to step out of their research comfort zones, cement working relationships with collaborators from disciplines distinct from their own, in keeping with the governing philosophy of the Centre, and engage with rookie Ph.D. scholars to begin to create multidisciplinary, strategic knowledge on climate change in India. In the first phase of the Centre, a substantial body of scientific outcomes was achieved, leading to a large number of high-impact research publications. We then were encouraged by our chief mentor at DST, Dr. Akhilesh Gupta, Adviser/Scientist-G and Head, SPLICE and CCP to go beyond, to try and distil actionable recommendations from the underlying science.

Periodic engagement with the DST Expert Review Committee led alternately to dismay and inspiration, in regard to the enormous task we had set ourselves. The DST Expert Review Committee, especially Prof. D. R. Sikka, to whom we dedicate this volume, injected realism into our endeavours, while providing a strong base of conviction in our vision and work. We sincerely acknowledge Strategic Programmes, Large Initiatives and Coordinated Action Enabler (SPLICE) and Climate Change Programme (CCP), of the Department of Science and Technology, both for foresight in envisaging the Centre and for a crucial seed grant, to enable its set-up and functioning. Importantly, the grant enabled fellowships for over 20 doctoral students, who cemented academic exchange between their faculty mentors. We convey our sincere gratitude to Dr. Akhilesh Gupta, Adviser/Scientist-G and Head, SPLICE and CCP, for his continuous support and encouragement of the research at the Centre. Professor Ashutosh Sharma, Honourable Secretary of DST, has displayed a keen interest in understanding both climate change and air pollution over India. Professor Sharma has provided encouragement through informal discussions at DST workshops and was kind enough to provide a foreword to this book.

At IIT Bombay, the new Interdisciplinary Programme in Climate Studies is home to the DST-CoECS. Institute functionaries at IIT Bombay provided far-sighted and perceptive leadership in setting up one of the first doctoral programmes in climate studies, in India, as an independent academic unit. We thank Director, Prof. Devang Khakhar, and then Dean, Research and Development, Prof. Rangan Banerjee, for their mentoring of the Interdisciplinary Programme in Climate Studies. The editors are grateful to all authors, for their contributions, and for making this a wholesome, cooperative venture. We thank the team of student assistants, Ms. Anjana Devanand, Ms. Piyali Chowdhury, Ms. Krishna Malakar and Mr. Kushal Tibrewal, for their coordination efforts. Ms. Archismita Banerjee, Manager of the DST-CoECS at IIT Bombay, has played a central role at every stage of evolution of the book from coordinating early meetings, to helping with the concept note, to final collation of materials. We appreciate her cordial and patient coordination efforts, which have been central to completion of the work under the required time frame.

The process adopted in developing this book involved consultation meetings of the author group and development of a concept note which, along with sample chapters, was subject to review by our publishers, Springer Nature Scientific Publishing Services (P) Ltd. The completed book manuscript was then subjected to peer review, leading to significant and important inputs towards improvement of the book. At Springer, our sincere thanks are due to their staff, especially Sagarika Ghosh and Nupoor Singh, for their keen interest and time, towards discussions to frame the scope and content of the book. The editors wholeheartedly appreciate the professional support of Springer staff, for a rapid and smooth journey in bringing this book to press.

The structure and framework for any such collaborative outcome are necessarily enriched by broader conversations with peers and mentors. Among these colleagues, CV thanks Dr. Veena Joshi, Prof. Milind Kandlikar and Prof. Ambuj Sagar; TM thanks Prof. Anand Patwardhan; SG and SK both thank Prof. Raghu Murtugudde and Prof. Dev Niyogi; and SK thanks Prof. Dilip Swain and Dr. Naresh Soora. A project of this nature necessarily involves a state of preoccupation, on part of the editors, for a period of time. We thank our families for tolerating any unwitting distraction or inattention, on our parts, during the course of writing.

Mumbai, India

Chandra Venkataraman
Trupti Mishra
Subimal Ghosh
Subhankar Karmakar

Contents

Part I Regional Climate Change Signals

Land-Surface Feedback and Impacts of Land-Use Change to Indian Monsoon Rainfall 3
Amey Pathak, Supantha Paul and Subimal Ghosh

Aerosol Influences on Cloud Modification and Rainfall Suppression in the South Asian Monsoon Region 21
Prashant Dave, Nitin Patil, Mani Bhushan and Chandra Venkataraman

Future Hydrologic Scenarios in India Under Climate Change 39
Tarul Sharma, Aditya Gusain and Subhankar Karmakar

Links Between Global Climate Teleconnections and Indian Monsoon Rainfall 61
S. Adarsh and M. Janga Reddy

Urbanisation and Surface Urban Heat Island Intensity (SUHII) 73
Hiteshri Shastri and Subimal Ghosh

Future Heat Wave Projections and Impacts 91
Kamal Kumar Murari and Subimal Ghosh

Decadal Estimates of Surface Mass Balance for Glaciers in Chandra Basin, Western Himalayas, India—A Geodetic Approach 109
RAAJ Ramsankaran, Ankur Pandit and Avinash Parla

An Experimental Approach Toward Modeling Atmosphere and Ocean Mixing Processes 127
Sridhar Balasubramanian

Nearshore Sediment Transport in a Changing Climate 147
Piyali Chowdhury and Manasa Ranjan Behera

Part II National and Sub-national Responses to Climate Change

A Comprehensive Social Vulnerability Analysis at a National Scale 163
H. Vittal and Subhankar Karmakar

Review of Indian Low Carbon Scenarios 177
Pankaj Kumar, Trupti Mishra and Rangan Banerjee

Greenhouse Gas Emissions in India's Road Transport Sector 197
Namita Singh, Trupti Mishra and Rangan Banerjee

A State-Level Methodology for Estimating Present-Day Emissions of Short-Lived Climate Pollutants from Fired Brick Production in India 211
Kushal Tibrewal, Sameer Maithel and Chandra Venkataraman

Adaptation and its Socioeconomic Facilitators in the Marine Fishing Community of Maharashtra, India 231
Krishna Malakar and Trupti Mishra

Biodiesel Blending in India—Analysis of National Biofuel Policy 247
Sreenivas Chigullapalli, Anand B. Rao and Anand Patwardhan

Cross-Scale Institutional Linkages in Climate Change Responses: An Indian Perspective 255
D. Parthasarathy, Yash Shethia and N. C. Narayanan

Economic Assessment of Climate Mitigation Pathways (2015–2050) for the Brick Sector in India 273
Priyanka Jajal, Kushal Tibrewal, Trupti Mishra and Chandra Venkataraman

Energy Efficiency Policies in India: Implications for Climate Change Mitigation 289
Manisha Jain, Anand B. Rao and Anand Patwardhan

Editors and Contributors

About the Editors

Chandra Venkataraman is a Professor of Chemical Engineering and Associate Faculty at the Interdisciplinary Programme in Climate Studies, IIT Bombay. She works on environmental and climate science and aerosol nanoparticle engineering. Her research has contributed influential scientific knowledge on black carbon aerosols, climate change and air quality degradation over South Asia. She has built research networks for climate and air quality studies. She is the National Coordinator of the NCAP-COALESCE project leading 17 institutions investigating regional climate impacts of carbonaceous aerosols (MoEFCC). At IIT Bombay, she was instrumental in the establishment of Interdisciplinary Programme in Climate Studies and was Principal Investigator of the DST-supported Centre of Excellence in Climate Studies. She is the recipient of a Fulbright-Nehru Research Fellowship (2012), Vikram Sarabhai Award (2005), Fellowship of the Indian National Academy of Engineering (2016), the National Academy of Science, India (2017), and the Indian Academy of Science (2017).

Trupti Mishra is an Associate Professor at Shailesh J. Mehta School of Management and Associate Faculty at Interdisciplinary Programme in Climate Studies, IIT Bombay. She holds a Ph.D. in Economics from IIT Kharagpur. Her research interest is focused on economics of pollution, economic assessment of climate change vulnerability, adaptation and mitigation options and corporate environmental sustainability. Her research involves developing different low carbon scenario in sectors, identifying cost-effective mitigation option for industry, suggesting environmental compliances measures to achieve low carbon growth. She has published her research findings in various chapters and journal papers. She is also working on climate change impact and adaptation at city and community level. Her co-authored 2017 publication in *Climate and Development* (Taylor & Francis) on assessing the socio-economic vulnerability of cities in India received wide mainstream media coverage.

Subimal Ghosh is an Associate Professor at the Interdisciplinary Programme in Climate Studies and a Faculty at the Department of Civil Engineering, IIT Bombay. He works in the area of hydro-meteorology. His research interests include climate change impact assessment with statistical downscaling, understanding land surface feedback to monsoon, hydrological simulation and urban extreme rainfall forecasts. He has received several awards in recognition of his research, a few among them being INAE Young Engineer Award, INSA Young Scientist Award, NASI Young Scientist Award and IEI Young Engineer Award. He has published articles in reputed journals such as *Nature Climate Change*, *Nature Communications*, *Scientific Reports*, *Geophysical Research Letters*, *Journal of Climate*, *Water Resources Research*. He has been selected as one of the lead authors for IPCC Assessment Report 6 for a chapter on extremes by Working Group-I.

Subhankar Karmakar is an Associate Professor at the Centre for Environmental Science and Engineering, and an associated faculty member at the Interdisciplinary Programme in Climate Studies, IIT Bombay. He obtained his Ph.D. on "Uncertainty Modelling in Water Resources Systems" from the Indian Institute of Science, Bangalore, India, and was a Postdoctoral Fellow at the University of Western Ontario, Canada. Further, he received a BOYSCAST Fellowship from the Government of India to pursue research on Ecological Modelling at Duke University, North Carolina, USA. His primary research interests are environmental systems analysis, uncertainty modelling and risk vulnerability analysis to climate-induced natural hazards. Some of his recent research contributions include understanding the dependence of Indian summer monsoon rainfall extremes on temperature, mapping disaster vulnerability for a densely populated coastal urban area, and developing lifecycle-based decision support tools for selecting wastewater treatment alternatives. He has published over 40 international journal papers, 5 chapters and several international conference proceedings.

Contributors

S. Adarsh Department of Civil Engineering, Indian Institute of Technology Bombay, Mumbai, Maharashtra, India; Department of Civil Engineering, TKM College of Engineering, Kollam, Kerala, India

Sridhar Balasubramanian Department of Mechanical Engineering and Interdisciplinary Program in Climate Studies, Indian Institute of Technology Bombay, Mumbai, India

Rangan Banerjee Interdisciplinary Program (IDP) in Climate Studies, Indian Institute of Technology Bombay, Mumbai, Maharashtra, India; Department of Energy Science and Engineering, Indian Institute of Technology Bombay, Mumbai, India

Manasa Ranjan Behera Interdisciplinary Program in Climate Studies, Indian Institute of Technology Bombay, Mumbai, India; Department of Civil Engineering, Indian Institute of Technology Bombay, Mumbai, India

Mani Bhushan Interdisciplinary Program (IDP) in Climate Studies, Indian Institute of Technology Bombay, Mumbai, Maharashtra, India; Department of Chemical Engineering, Indian Institute of Technology Bombay, Mumbai, Maharashtra, India

Sreenivas Chigullapalli Interdisciplinary Program (IDP) in Climate Studies, Indian Institute of Technology Bombay, Mumbai, India

Piyali Chowdhury Interdisciplinary Program in Climate Studies, Indian Institute of Technology Bombay, Mumbai, India

Prashant Dave Interdisciplinary Program (IDP) in Climate Studies, Indian Institute of Technology Bombay, Mumbai, Maharashtra, India

Subimal Ghosh Interdisciplinary Program in Climate Studies, Indian Institute of Technology Bombay, Mumbai, India; Department of Civil Engineering, Indian Institute of Technology Bombay, Mumbai, Maharashtra, India

Aditya Gusain Centre for Environmental Science and Engineering, Indian Institute of Technology Bombay, Mumbai, India

Manisha Jain Indira Gandhi Institute of Development Research, Mumbai, India

Priyanka Jajal Interdisciplinary Program (IDP) in Climate Studies, Indian Institute of Technology Bombay, Mumbai, Maharashtra, India

M. Janga Reddy Department of Civil Engineering, Indian Institute of Technology Bombay, Mumbai, Maharashtra, India; Interdisciplinary Program (IDP) in Climate Studies, Indian Institute of Technology Bombay, Mumbai, Maharashtra, India

Subhankar Karmakar Interdisciplinary Programme in Climate Studies, Indian Institute of Technology Bombay, Mumbai, India; Centre for Environmental Science and Engineering, Indian Institute of Technology Bombay, Mumbai, India; Centre for Urban Science and Engineering, Indian Institute of Technology Bombay, Mumbai, India

Pankaj Kumar Interdisciplinary Program in Climate Studies, Indian Institute of Technology Bombay, Mumbai, India

Sameer Maithel Greentech Knowledge Solutions Pvt. Ltd., New Delhi, India

Krishna Malakar Interdisciplinary Program (IDP) in Climate Studies, Indian Institute of Technology Bombay, Mumbai, Maharashtra, India

Trupti Mishra Interdisciplinary Program (IDP) in Climate Studies, Indian Institute of Technology Bombay, Mumbai, Maharashtra, India; Shailesh J. Mehta School of Management, Indian Institute of Technology Bombay, Mumbai, Maharashtra, India

Kamal Kumar Murari Centre for Climate Change and Sustainability Studies, School of Habitat Studies, Tata Institute of Social Science, Mumbai, Maharashtra, India

N. C. Narayanan Interdisciplinary Programme in Climate Studies, Indian Institute of Technology Bombay, Mumbai, India

Ankur Pandit Interdisciplinary Programme in Climate Studies, Indian Institute of Technology Bombay, Mumbai, India

Avinash Parla Water Resources Division, Department of Civil Engineering, Indian Institute of Technology Bombay, Mumbai, India

D. Parthasarathy Interdisciplinary Programme in Climate Studies, Indian Institute of Technology Bombay, Mumbai, India

Amey Pathak Department of Civil Engineering, Indian Institute of Technology Bombay, Mumbai, Maharashtra, India

Nitin Patil Interdisciplinary Program (IDP) in Climate Studies, Indian Institute of Technology Bombay, Mumbai, Maharashtra, India

Anand Patwardhan School of Public Policy, University of Maryland, College Park, USA

Supantha Paul Interdisciplinary Program in Climate Studies, Indian Institute of Technology Bombay, Mumbai, Maharashtra, India

RAAJ Ramsankaran Department of Civil Engineering, Indian Institute of Technology Bombay, Mumbai, India; Interdisciplinary Programme in Climate Studies, Indian Institute of Technology Bombay, Mumbai, India

Anand B. Rao Interdisciplinary Program (IDP) in Climate Studies, Indian Institute of Technology Bombay, Mumbai, India; Centre for Technology Alternatives for Rural Areas (CTARA), Indian Institute of Technology Bombay, Mumbai, India

Tarul Sharma Interdisciplinary Programme in Climate Studies, Indian Institute of Technology Bombay, Mumbai, India

Hiteshri Shastri M. S. Patel Department of Civil Engineering, C.S. Patel Institute of Technology, Charotar University of Science and Technology, Changa, Anand, Gujarat, India

Yash Shethia World Wildlife Fund India, New Delhi, India

Namita Singh Interdisciplinary Program (IDP) in Climate Studies, Indian Institute of Technology Bombay, Mumbai, Maharashtra, India

Kushal Tibrewal Interdisciplinary Program (IDP) in Climate Studies, Indian Institute of Technology Bombay, Mumbai, Maharashtra, India

Chandra Venkataraman Interdisciplinary Program (IDP) in Climate Studies, Indian Institute of Technology Bombay, Mumbai, Maharashtra, India; Department of Chemical Engineering, Indian Institute of Technology Bombay, Mumbai, Maharashtra, India

H. Vittal Centre for Environmental Science and Engineering, Indian Institute of Technology Bombay, Mumbai, India; Department of Civil Engineering, Manipal Institute of Technology, Manipal, India

Abbreviations

AAC	Autoclaved aerated concrete
AAI	Absorbing Aerosol Index
AIM	Asia-Pacific Integrated Model
AISMR	All India summer monsoon rainfall
ALOS	Advanced Land Observing Satellite
AMO	Atlantic Multidecadal Oscillation
AOD	Aerosol optical depth
ASTER	Advanced Spaceborne Thermal Emission and Reflection Radiometer
AVHRR	Advanced very-high-resolution radiometer
BAU	Business as usual
BC	Black carbon
BEE	Bureau of Energy Efficiency
BLY	Bachat Lamp Yojana
BTKs	Bull's trench kilns
BUs	Billion units
CAGR	Compounded annual growth rate
CBA	Cost–benefit analysis
CCAC	Climate and Clean Air Coalition
CCS	Carbon capture and storage
CDER	Cloud droplet effective radius
CEA	Cost-effective analysis
CEEMDAN	Complete ensemble empirical mode decomposition with adaptive noise
CF	Cloud fraction
CFLs	Compact fluorescent lamps
CGE	Compound general equilibrium
CH_4	Methane
CLM	Community Land Model
CMIP	Coupled Model Intercomparison Project

CNG	Compressed natural gas
CO	Carbon monoxide
CO_2	Carbon dioxide
CP	Carbon price
CPCB	Central Pollution Control Board
CT	Carbon tax
CTP	Cloud top pressure
CTT	Cloud top temperature
CWT	Continous wavelet transform
CWV	Column water vapour
DEM	Digital Elevation Model
DRM	Dynamic recycling model
DSM	Demand-side management
EBP	Ethanol blending programme
EC	Energy conservation
ECMWF	European Centre for Medium-Range Weather Forecasts
EER	Energy efficiency ratio
EFOM	Energy Flow Optimisation Model
EMD	Empirical mode decomposition
ENSO	El Niño–Southern Oscillation
EOS	Earth Observing System
EQUINOO	Equatorial Indian Ocean Oscillation
ERB	Edmonds–Reilly–Barns model
ERFaci	Effective radiative forcing from aerosol–cloud interactions
ERFari	Effective radiative forcing from aerosol–radiation interactions
FAME	Faster Adoption and Manufacturing of Electric Vehicles
FCBTKs	Fixed chimney Bull's trench kilns
GCAM	Global Change Assessment Model
GCMs	General circulation models
GDP	Gross domestic product
GHGs	Greenhouse gases
GoI	Government of India
GT	Gigatonne
GW	Gigawatt
GWP	Global warming potential
HCVs	Heavy commercial vehicles
HFCs	Hydrofluorocarbons
HHT	Hilbert–Huang transform
HKH	Hindu Kush Himalayan
HL	High AOD-low precipitation
HT	Hilbert transform
IAM	Integrated assessment model
IEPR	Integrated Energy Policy Report
IESM	Integrated Energy System Model
IGP	Indo-Gangetic Plains

IITM	Indian Institute of Tropical Meteorology
IMD	India Meteorological Department
IMF	Intrinsic mode decomposition
INDCs	Intended Nationally Determined Contributions
INR	Indian rupees
IOD	Indian Ocean Dipole
IPCC	Intergovernmental Panel on Climate Change
ISM	Indian summer monsoon
ISMR	Indian summer monsoon rainfall
ISRO	Indian Space Research Organisation
ITCZ	Intertropical Convergence Zone
IWP	Ice water path
JJAS	June–July–August–September
KPSS	Kwiatkowski–Phillips–Schmidt–Shin
LCHs	Light commercial vehicles
LL	Low AOD-low precipitation
LLGHGs	Long-lived greenhouse gases
LPG	Liquefied petroleum gas
LST	Longshore sediment transport
LULC	Land use/land cover
LWP	Liquid water path
MARKAL	MARKet Allocation model
MEPS	Minimum Energy Performance Standard
MISO	Monsoon intra-seasonal oscillation
MJO	Madden–Julian Oscillations
MODIS	Moderate Resolution Imaging Spectroradiometer
MOP&NG	Ministry of Petroleum and Natural Gas
MORTH	Ministry of Road Transport and Highways
MRI	Meteorological Research Institute
MT	Million tonnes
N_2O	Nitrous oxide
NAO	North Atlantic Oscillation
NAPCC	National Action Plan on Climate Change
NASA	National Aeronautics and Space Administration
NCEP	National Centre for Environmental Prediction
NCR	National Capital Region
NCT	National Capital Territory
NEMMP	National Electric Mobility Mission Plan
NEP	National Electricity Policy
NMEEE	National Mission for Enhanced Energy Efficiency
NMVOCs	Non-methane volatile organic compounds
NOAA	National Oceanic and Atmospheric Administration
NOx	Nitrogen oxides
O_3	Ozone
OC	Organic carbon

OMI	Ozone monitoring instrument
ONI	Oceanic Niño Index
PAT	Perform, Achieve and Trade
PBL	Planetary boundary layer
PFCs	Perfluorocarbons
Pkm	Passenger travel kilometre demand
PM	Particulate matters
$PM_{2.5}$	Particulate matter with diameter less than 2.5 μm
PRECIP	Precipitation
PRISM	Panchromatic Remote-sensing Instrument for Stereo Mapping
PV	Photovoltaic
QBO	Quasi-biennial oscillation
RCP	Representative Concentration Pathway
S&L	Standards and labelling
SAM	South Asian monsoon
SDB	AIM Strategic Database
SF_6	Sulphur hexafluoride
SLCPs	Short-lived climate pollutants
SMB	Surface mass balance
SO_2	Sulphur dioxide
SoV	Social vulnerability
SPOT	Satellite Pour l'Observation de la Terre
SRES	Special Report on Emission Scenarios
SRTM	Shuttle Radar Topography Mission
SSP	Shared Socioeconomic Pathways
SST	Sea surface temperature
T&D	Transmission and distribution losses
TanDEM-X	TerraSAR-X add-on for Digital Elevation Measurement
TBO	Tropospheric biennial oscillation
TDIC	Time-dependent intrinsic correlation
Tg	Teragram
TIAM	TIMES Integrated Assessment Model
TIMES	The Integrated MARKAL-EFOM System
TOMS	Total Ozone Mapping Spectrometers
UAV	Unmanned aerial vehicle
UJALA	Unnat Jyoti by Affordable LEDs for All
UNFCCC	United Nations Framework Convention on Climate Change
UVs	Utility vehicles
VSBK	Vertical shaft brick kilns
WITCH	World Induced Technological Change Model
WRF	Weather Research and Forecasting
WTP	Willingness to pay

Introduction

Climate change has emerged as one of the defining global environmental problems of this century, requiring an unprecedented global response, shaped by substantive and coordinated national responses of world nations. In early October 2016, India ratified the Paris Agreement on climate change, signalling its commitment to the "global cause of environmental protection and climate justice". At the national level, India has initiated an ambitious and sustained response through the National Action Plan on Climate Change (NAPCC), and its eight missions covering different sectors and systems. Gaps in strategic choices can create industrial and social structures that have significant adverse impact on environment and climate. India is at a critical juncture in development journey and facing choices regarding technology that could have huge long-term socio-economic and competitiveness implications. Many sectors and regions in India are highly vulnerable to climate change impacts. These include agriculture, water resources, ecosystems, as well as urban and rural settlements. Of particular importance is the fact that India is strongly exposed to the risk of a number of natural hazards of climatic and hydro-meteorological origin including, for example, extremes of temperature and rainfall perturbation.

With this background, under the auspices of the National Mission on Strategic Knowledge for Climate Change (NMSKCC) coordinated by the Department of Science and Technology through its Climate Change Programme (CCP) Division, a Centre of Excellence in Climate Studies (DST-CoECS) was established at IIT Bombay in January 2012, with a mission to "develop a scientific understanding of regional climate change and connect it to impacts (socio-economic, environment, resources) and effective response (technology and adaptation)". The core activity of the DST-CoECS is to undertake interdisciplinary, problem-driven research for end-to-end solutions addressing causes, consequences of and responses to climate change towards providing critical assessments to reduce scientific uncertainty and to support governmental decision-making on climate mitigation and adaptation measures.

In regard to India, recent influential books have addressed contemporary Indian climate politics and policy and long-term climate variability and climate change over India using observational data and analyses. A need was thus felt to bring into focus a scientific understanding of recent climate change over India and its connection to impacts (socio-economic, environment, resources) and effective response (technology and adaptation). The objective of this book is to provide a synthesis of research findings, in terms of strategic knowledge outcomes, in regard to the emergence of recent regional climate signals, implications for impacts assessment and for mitigation and adaptation response, relevant in the Indian context. A concise interpretation is attempted to distil actionable recommendations and policy prescriptions at national and sub-national levels. The book contains 18 chapters written by the subject experts, participating in the DST-CoECS at IIT Bombay, from different academic disciplines and with distinct research expertise.

Underlying the book is a broad framework for the integration of regional climate perturbation to impacts (specifically on socio-economic sectors) and responses (adaptation and mitigation through technology response) as laid out in Fig. 1. This integration was made towards the development of a more robust response to regional climate change, to eventually contribute to mainstreaming the socio-economic context of climate change in India's development strategy. For example, linking scientific understanding and response relates to the assessment of

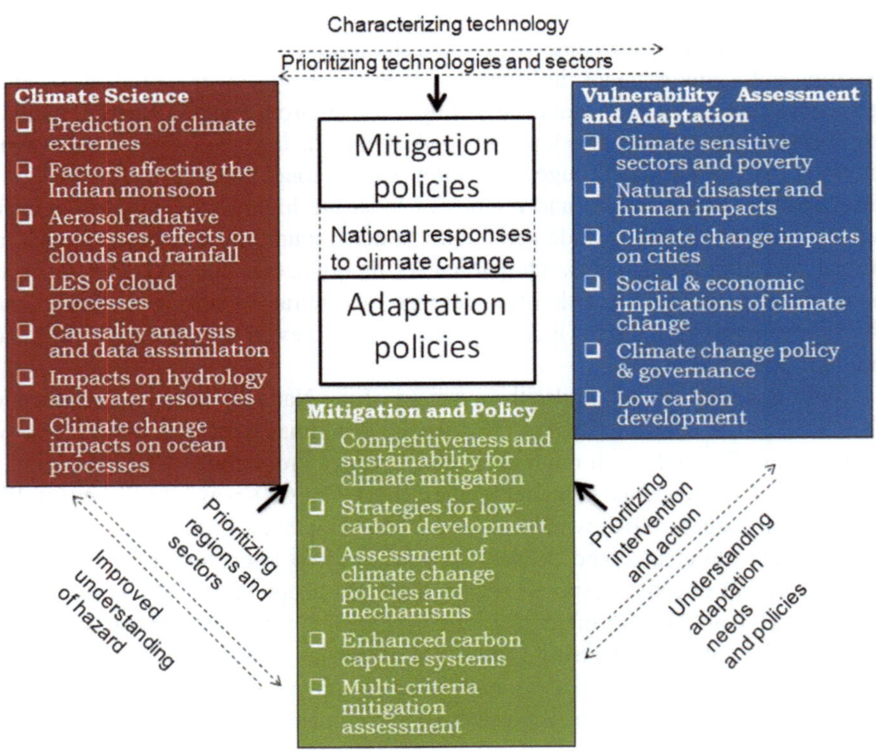

Fig. 1 Integration of understanding of regional climate perturbation, impacts and responses towards informing national policy

vulnerability and adaptation, where projections of climate and related variables (such as hydrological variables) need to be connected to impact models and used to evaluate potential adaptation responses.

Part I of the book is entitled "**Regional Climate Change Signals**".

Chapters in this part describe studies which provide evidence and scientific underpinning for the emergence of recent regional climate change signals over India and some aspects of climate response to the underlying atmosphere–ocean–land processes. Among important signals addressed are factors affecting changes in Indian monsoon rainfall and its extremes (floods and droughts), the effects of local processes and factors, like land-use changes and emission of pollution particles, on changes in summer monsoon rainfall, and hydrological scenarios, developed from the heat waves, cold waves, prolonged fog and snowfall.

Water resources are inextricably linked with climate change, and India's agricultural output, economy and societal well-being are dependent upon the stability of monsoon rainfall, its variability and its extremes. Changes in land use/land cover and direct aerosol radiative forcing significantly affect monsoon rainfall in India. This book examines the uniqueness of a computationally efficient framework, which involves an integration of global and local factors for precipitation projections through a conjugal statistical–dynamical approach and analysis based on synoptic scale fast responses acting over timescales of days to a month. Continental aerosols over India alter cloud properties in diametrically opposite ways during different monsoon years. The multidecadal gridded daily rainfall data over entire India indicates that the trend in spatial variability of mean monsoon rainfall is decreasing, on the contrary to that of extremes. There is a significant decline in the monsoon rainfall over major water surplus river basins in India. It is also revealed that the water yield in surplus river basins is decreasing but it is increasing in deficit basins.

Future projections of climate variables highlighted in the book, based on multiple climate models and scenarios for CMIP5 data, found that a sizable part of India will experience heat stress conditions in future. In northern India, the average number of days with extreme heat stress condition during the pre-monsoon hot season will reach 30. The intensification of heat waves might lead to severe heat stress and increased mortality. The information on the differential influence of various climate oscillations on ISMR at different scales is extracted and integrated into the development of improved rainfall forecasting model. From the HHT-based TDIC analysis, it is found that the strength and the nature of the association between ISMR and large-scale climatic oscillations (such as QBO, ENSO, AMO, EQINOO) vary with the timescales.

While global climate change in the long term (centuries) will be attributable to greenhouse gases, a multitude of short-lived emissions, with lifetimes of a few days to a few years, affect present-day atmosphere and climate on timescales of decades. Changes in energy consumption and development patterns will alter the regional profile of these emissions. The impacts of such changes on environmental resources will be experienced on local to regional scales, on the timescale of decades. Research which is relevant to furthering our understanding of these phenomena lies at the interfaces between energy use, climate modelling and the estimation of impacts on multiple scales. The influence of short-lived climate pollutants, specifically aerosols, is analysed on cloud modification and short-term rainfall inhibition over India. Aerosols were shown to modulate clouds in opposing ways, suggestive of cloud inhibition in years of deficient, but invigoration in years of abundant monsoon rainfall. Enhancement in aerosol levels was linked to repeated intra-seasonal short-term suppression of daily precipitation, with increases in the frequency and length of monsoon breaks, with implications for rainfall deficit and food grain production. Linkages borne out between enhanced air pollution and short-term suppression of regional monsoon rainfall indicate the need for greater synergy in policies addressing air pollution and climate change.

Throughout the Earth's history, there have been periods of glaciation followed by warming trends, when the glaciers retreated towards higher altitudes and latitudes. The glacier ice thickness is one of the cardinal input parameters for the estimation of the total glacier volume and future glacier evolution studies. Ice thickness distribution and total glacier volume are estimated using spatially distributed GlapTop-2 model over the Chhota Shigri Glacier.

Oceans have large heat capacity; hence, ocean heat storage mainly controls the timescales of variability to changes in the ocean–atmosphere system, including the timescales of adjustment to anthropogenic radiative forcing. Energy budgets in the surface layers, which depend on the exchange with deeper layers in the ocean, indicate the need to consider the processes which affect the circulation and water mass distribution in the deep ocean. The response of the climate system at decadal and longer timescales is considered in particular. The changing climate has altered the nearshore wave characteristics and has impacted the nearshore processes like sediment transport dynamics. This may directly impact the coastal response and the economic activities along the coastal belt. Links are established between the meteorological marine climate and global dynamics with coastal processes along the Indian coastline.

As delineated in this part, the primary signals of changing climate increase in average global land and sea surface temperature and changes in characteristics of cloud cover and precipitation, along with changes in coastal and glacial ecosystems.

Part II of the book is entitled "**National and Sub-national Responses to Climate Change**".

Chapters in this part describe studies related to responses to climate change in the form of adaptation or mitigation strategies. It is essential to undertake a careful assessment of technological solutions and adaptation measures to climate change for effective planning and implementation, given limited resources available in India. There are certain socio-economic, cultural and environmental factors in the country which make it imperative for us to undertake technology assessment studies and develop capabilities/tools for the same. Such local/regional factors, ground realities and public perceptions could play an important role in making technology ranking/choice. Choices regarding these technology options need to be informed by assessments of technology characteristics, cost, performance and expected future learning. Empirical and model-based assessments are essential for this purpose.

Adaptation refers to making adjustments including those in lifestyles, technology and behaviour, in order to avoid the adverse impacts of climate change, whereas mitigation seeks to reduce greenhouse gas (GHG) and short-lived climate pollutant (SLCP) emissions, which drive climate change. As the change in climate is inevitable, adaptation is necessary for sustaining life on the planet. Mitigation is crucial to avoid future over the edge climate change scenarios which might make adaptation strategies redundant. The National Action Plan on Climate Change (NAPCC) envisages both adaptation to and mitigation of climate change.

Adaptation in marine fishing has rarely been examined in the literature and seldom receives media and policy attention. Adaptation in marginalised marine fishing communities in Maharashtra is studied, using a state-level data set inclusive of all coastal districts/villages, to understand adaptation strategies and their socio-economic determinants in the communities of concern. It is found that the community has intensified their efforts to increase yields and have also diversified into other sources of livelihood. The findings suggest the need of interventions that provide education, cooperative and financial support, through specialised credit schemes, to the community. Socio-economic attributes crucial for adaptation are identified, which can thereby assist in developing state-level adaptation programmes for the community.

The transport sector is a significant contributor of GHG emissions. Trends in GHG emission from the Indian transport sector are analysed vis-à-vis transport characteristics such as vehicle population, transport demand, fuel share and mode share. Both top-down and bottom-up methodologies are applied for GHG emission estimation. Fleet models which have been used in the Indian context to estimate the age and technology of the vehicles on road are examined. Uncertainty in CO_2 emission from different studies for Indian transport sector has been addressed and attributed to the variation in the assumptions for input variables and emission factors. Policies undertaken for the reduction in emission from the transport sector are discussed, and gaps in emission estimation study are examined.

Adoption of biofuels and energy-efficient technologies are important mitigation strategies. Two chapters of the book examine the policies promoting these mitigation strategies. Biodiesel blending in India is examined, and salient features of India's national biofuel policy are highlighted. Biodiesel blending in India received a lot of attention over the past decade. India's national biofuel policy aimed at the promotion of biofuel use made from indigenous feedstock. The current status of biodiesel blending in India is discussed, looking at the issue of economic viability. Implications are examined vis-à-vis rural development needs, complexity of the value chain and grass-roots realities. Challenges in production and commercialisation fronts to meet the country's biodiesel blending policy requirements are examined.

Further, the role of policies in realising the energy efficiency potential in India is addressed. Bottom-up techno-economic analyses of technologies in different end-uses show the presence of cost-effective energy efficiency potential. The potential of energy savings directly translates into the potential of CO_2 emission reduction. Several barriers to adoption of energy-efficient measures have been identified, and policies have been implemented to address these barriers. Some of the important policies implemented in India include the standards and labelling (S&L) programme in appliances, and the Perform Achieve Trade (PAT) Scheme in industries and other demand-side management measures. Most of the studies report that the impact of these policies on energy and environment and the impact on consumer decisions have been largely overlooked. In this chapter, the impact of the S&L programme on both energy saving and emission reduction and on consumer decisions is presented. The cumulative emission reduction is calculated from the rate of efficiency improvements since the launch of the programme. The impact of the programme on consumer decisions is reported from a discrete choice experiment. Measures to improve the effectiveness of the programme and to enhance its contribution to energy saving and emission reduction are discussed.

The brick sector in India is of importance from its significant capacity, with an annual production of over 250 million fired bricks. The Indian brick sector is dominated by traditional kilns with inefficient combustion technology, which largely use coal, and sometimes combines with biomass and a host of other waste fuels, thus providing a unique mitigation opportunity for both GHGs and short-lived climate pollutants (SLCPs), through shifts towards more efficient technologies. Two chapters dealing with the brick sector estimate present-day emissions of SLCPs from brick production, and their evolution from 2015 to 2050, under two pathways with different levels of cleaner technology evolution, thus estimating SLCP mitigation potential. In terms of technology assessment, Bull's trench kilns contributed most to black carbon emissions while clamp kilns to emissions of organic carbon, carbon monoxide and methane. Significant achievable mitigation of SLCPs in 2050 was found possible under an ambitious prospective policies scenario, which assumes a significant shift to non-fired brick walling materials and large penetration of cleaner technologies like zigzag firing and vertical shaft brick kilns. A cost–benefit analysis shows that SLCP mitigation cost is significantly lower than those estimated for GHG mitigation in India.

Future climate action is embodied in Nationally Determined Contributions (NDCs), or national pledges to reduce emissions and undertake climate adaptation actions, given each country's unique circumstances. Under India's NDC, within a framework of common but differentiated responsibilities and respective capabilities, a broader engagement could be made, with diverse actions addressing climate change and sustainable development. While the focus of this book, on new knowledge arising from recent research in a single institution, may be seen as a modest compass of scholarship, it is hoped that its utility would lie in the synthesis of recent research findings, into strategic knowledge outcomes, relevant to policy making in the Indian context. It is envisaged that this book would serve as a reference point for understanding scientific advances and persisting uncertainty, future vulnerability and response capacity of interlinked human and natural systems, as pertaining to India.

Chandra Venkataraman

Part I
Regional Climate Change Signals

Land-Surface Feedback and Impacts of Land-Use Change to Indian Monsoon Rainfall

Amey Pathak, Supantha Paul and Subimal Ghosh

Abstract Indian summer monsoon rainfall (ISMR) being mainly controlled by the large-scale oceanic processes is also influenced by the land–atmosphere interactions. At the regional scale, land–atmosphere interactions govern the fate of precipitation by modulating regional hydro-climate variables such as evapotranspiration, snow cover, soil moisture, and surface temperature. The contribution of local evapotranspiration to the precipitation in the same region is known as precipitation recycling. The quantitative estimates of recycling in terms of regional ratio highlight the importance of land-surface feedback in the regional precipitation. In the present chapter, we discuss the role of land-surface feedback (through evapotranspiration) over the India subcontinent during summer monsoon rainfall. We also discuss how a significant change in land-surface condition due to changes in land use–land cover (LULC) could affect the summer monsoon rainfall in India with the help of a sensitivity experiment.

Keywords Indian monsoon rainfall · Precipitation recycling
Impact of changes in LULC type · Land-surface feedbacks

A. Pathak
Department of Civil Engineering, Indian Institute of Technology Bombay,
Mumbai 400076, Maharashtra, India
e-mail: pathak.amey1@gmail.com

S. Paul
Interdisciplinary Program in Climate Studies, Indian Institute of Technology Bombay,
Mumbai 400076, Maharashtra, India
e-mail: supanthapaul@gmail.com

S. Ghosh (✉)
Interdisciplinary Program in Climate Studies and Department of Civil Engineering, Indian
Institute of Technology Bombay, Mumbai 400076, Maharashtra, India
e-mail: subimal@civil.iitb.ac.in; subimal.ghosh@gmail.com

C. Venkataraman et al. (eds.), *Climate Change Signals and Response*,
https://doi.org/10.1007/978-981-13-0280-0_1

1 Introduction

Precipitation as a system over a region is basically a response to the coupled ocean—land–atmosphere interactions between the wide ranges of time-varying geophysical mechanisms. The mechanisms which influence the precipitation system over a region are mostly governed by different large-scale processes and altered by the regional land-surface feedbacks. However, the degree of land-surface feedback and its role in the precipitation generation process varies considerably with the geographic location and time. For example, globally, the subtropical and the tropical regions appear to have a strong land–atmosphere coupling (Koster et al. 2004), of which the Indian monsoon region is particularly more sensitive to the land-surface feedbacks. Furthermore, the summer monsoon rainfall in India is known for its complexity, and a considerable part of its complexity is governed by the land-surface feedbacks. Therefore, any change in the land-surface processes due to either natural climate variability or due to anthropogenic forcing such as changes in LULC type can significantly influence the monsoon variability. Hence, understanding the different land-surface feedbacks that influence the summer monsoon rainfall in India is of utmost importance in order to accurately model and predict the variations in the monsoon precipitation. Therefore, in the present chapter, we emphasize mainly on the role of land-surface feedbacks in summer monsoon rainfall in India.

2 Summer Monsoon Rainfall in India

A seasonal migration of intertropical convergence zone (ITCZ) to the northern latitudes, owing to land–sea thermal contrast, brings the copious amount of atmospheric moisture from the neighboring ocean to the Indian subcontinent. The rainfall generated by these moistures carrying winds during summer months of June, July, August, and September (JJAS) is known as the Indian summer monsoon rainfall (ISMR).

The rainfall during the monsoon is highly variable on many temporal (viz. intraseasonal to interannual, and multidecadal) and spatial scales, due to its dependence on multiple geophysical processes and severely affects the gross domestic product (GDP) of India. During recent past, our understanding of the summer monsoon rainfall in India is considerably improved. The monsoon rainfall was earlier believed to be affected only by the large-scale processes. However, now, with the recent advancement in the monsoon research, we can say that the summer monsoon rainfall in India is influenced by both the large-scale circulations as well as the regional land-surface feedback mechanisms.

The different geophysical processes which altogether alter the monsoonal circulations can be broadly categorized into the (i) large-scale processes and (ii) the local-/regional-scale feedbacks. The large-scale processes include: El-Niño Southern Oscillations (ENSO), Tropospheric Biennial Oscillation (TBO), Monsoon Intraseasonal Oscillations (MISO), Madden Julian Oscillation (MJO), Indian Ocean

Dipole (IOD), Atlantic Multidecadal Oscillation (AMO), whereas the main regional features are: Indian Ocean Dipole (IOD), geometry of the subcontinent, orography, tropospheric temperature variations, vegetation, soil moisture, and Eurasian snow cover, etc.

In a broader sense, we can say that the characteristics of the rainfall received during the summer monsoon months are actually a response to coupled ocean—land–atmosphere interactions between to the wide range of time-varying physical mechanisms. Furthermore, various land-surface feedbacks such as snow–precipitation feedback, vegetation–evapotranspiration–precipitation feedback, soil moisture feedback, and regional topography are also associated with the monsoon rainfall variability. Therefore, in the present chapter, we mainly focus on the role of land–atmosphere interactions during the summer monsoon rainfall in India. So, let us have a look at the different land–atmosphere interaction mechanisms that exist in the Indian monsoon system.

3 Impact of Land-Surface Feedback in ISMR

The interaction between earth's surface and adjoining atmosphere results into mass, momentum, and energy exchanges (Dominguez and Kumar 2008). There are various physical and biological feedbacks (independent as well as overlapping) that exist in the ecosystem, which regulates the characteristics of the regional climate. The hydrological parameters that primarily control the precipitation processes are evapotranspiration, soil moisture, snow cover (and permafrost), and runoff, etc. These hydrological parameters independently as well as in conjunction with each other control the regional climate through various feedbacks.

In addition, the land use–land cover (LULC) of the region can also influence the nature of the precipitation; for example, an urban region has more runoff and less evapotranspiration than the vegetated grassland or forests. Table 1 provides a brief review of the research that shows the influence of land-surface feedback on ISMR.

Here, it is important to note that although there exist various other land-surface feedback mechanisms such as snow–albedo–precipitation feedback, soil moisture—precipitation, land–sea thermal contrast, and other topographical features that influence the ISMR, in this chapter, we restrict ourselves to the discussion only the role of evapotranspiration–precipitation feedback in Indian monsoon. In the next section, we shall discuss the impact of precipitation recycling on Indian summer monsoon rainfall.

Table 1 Impact of land–atmosphere interactions on summer monsoon rainfall in India: the literature review

Feedback	Author	Key finding	Study area
Snow–precipitation	Blanford (1884)	Discovered the relationship between Himalayan winter snow cover and monsoon rainfall in India	Indian subcontinent
Snow–precipitation	Hahn and Shukla (1976)	The study observed that a significant negative correlation exists between the snow cover anomaly over Eurasia during the winter and the subsequent monsoon rainfall in India	Eurasia and India
Snow–albedo–precipitation	Yasunari et al. (1991)	A moderately weak monsoon signal is observed in the simulations with excess snow conducted using the five MRI GCM runs. This weakening in monsoon activity is accompanied by the development of conditions similar to the ENSO. During summer, the decrease in surface temperature over the mid-latitudes in the excess runs results in reduction of the total diabatic heating near surface which is counterbalanced by the increased evaporation	Eurasia
Soil moisture–evapotranspi-ration–precipitation	Meehl (1994)	The role of land-surface processes in Asian monsoon is studied by comparing the relative contribution of internal feedbacks (soil wetness) and external conditions (albedo), using different GCM simulation experiments and sensitivity experiments. A positive feedback was observed between soil wetness and precipitation	Global tropics
Precipitation recycling	Trenberth (1999)	The recycling ratio over India is observed to be higher during September, October, and November (high evaporation months). Whereas during the summer season (i.e., June, July, and August), due to moisture transport from the southern hemisphere, the average recycling ratio is less	Global
Snow–albedo–precipitation	Kripalani and Kulkarni (1999)	Snow cover all-India monsoon rainfall relationship	Indian Monsoon region

(continued)

Table 1 (continued)

Feedback	Author	Key finding	Study area
Snow cover–precipitation	Bamzai and Shukla (1999)	The inverse relationship between the monsoon and the snow cover is observed only in those years when the snow cover is anomalously different (either high or low) than the normal conditions. However, it was also concluded that the seasonal snow cover over the Himalaya has no significant relation with the upcoming monsoon rainfall	Indian summer monsoon rainfall region
Soil moisture–precipitation	Douville et al. (2002)	Assessed the influence of soil moisture on seasonal and interannual variability of Asian and African monsoon. The impact of soil moisture anomaly is significant in dry regions such as the Indian subcontinent and northeast Asia. In the regions where the role of soil moisture is significant, the soil moisture feedback plays an important role during onset rather than the boundary condition	Asian and African monsoon region
Topography	Abe et al. (2003)	Studied the importance of Tibetan Plateau in monsoon penetration and intensity. Maximum enhancement is observed in 40% of the present global orography	Asian monsoon region
Soil moisture–precipitation	Koster et al. (2004)	Strong soil moisture–precipitation feedback over the relatively drier regions such as Indian subcontinent, Yellow River	Global study
Aerosol–precipitation	Niyogi et al. (2007)	Proposed a framework and discussed the role of aerosols in the Indian monsoon	Indo-Ganges Basin
Snow cover–ENSO–precipitation	Peings and Douville (2009)	The snow–monsoon relationship: Concluded that instead of direct snow–monsoon relationship, the monsoon rainfall during winter and summer season also has a strong teleconnection with the ENSO	Indian summer monsoon rainfall region
Land-use change–precipitation	Lee et al. (2009)	The discussed the relation between the variability in early summer monsoon rainfall over India and its association with irrigation or intensification in agriculture	Indian subcontinent

(continued)

Table 1 (continued)

Feedback	Author	Key finding	Study area
Land-use change (agricultural intensification)–precipitation	Douglas et al. (2009)	The agricultural changes including irrigation modify the mesoscale convection, convective available potential energy, and the rainfall in the Indian region	Indian subcontinent
Land-use change (agricultural intensification)–precipitation	Niyogi et al. (2010)	The study concluded the reduction in the rainfall over some regions in India are associated with changes in land-use change due to agricultural intensification	Indian subcontinent
Soil moisture–precipitation	Asharaf et al. (2011)	Both positive as well as negative feedback were detected during the summer monsoon season; also, the study highlighted the evidence of dominant negative feedback mechanism over the western and northern part of India	India
Local recycling	Gimeno et al. (2012)	The neighboring oceanic sources are the main sources of water vapor for monsoon precipitation over Indian subcontinent. Furthermore, the contributions from local recycling to the precipitation are less as compared to the oceanic sources	Global
Local recycling	Ordóñez et al. (2012)	It was also observed that recycling within western Indian subcontinent contributes toward the end of summer monsoon season	Western and southern India
Vegetation (land-use change)–precipitation	Yamashima et al. (2015)	The study concluded that the historical land-use changes resulted into reduction in rainfall by more than 2 mm day^{-1} during the onset phase. In addition, the land-use change has a significant impact on monsoon onset and the delay of approximately four pentads was observed over the Indian subcontinent	Indian summer monsoon region
Vegetation (land-use change)–precipitation	Devaraju et al. (2015)	Investigated the effects of large-scale deforestation on ITCZ and its implications for the northern hemisphere and southern hemisphere monsoon regions. The reduction in Indian summer monsoon due to global deforestation case is maximum (18%)	Global monsoon regions

(continued)

Table 1 (continued)

Feedback	Author	Key finding	Study area
Urbanization–precipitation	Shastri et al. (2014)	The study concluded that on an overall scale, the impact of urbanization on extreme precipitation is spatially non-uniform. At zonal level, especially over the central India and western India, there exists a significant impact of urbanization on rainfall extremes	India

4 Precipitation Recycling: Evapotranspiration–Precipitation Feedback

The evapotranspiration over a region (land surface) supplies additional moisture to the adjoining atmospheric column of water vapor, which contributes to the precipitable water to generate precipitation over the same region, and this is known as '*precipitation recycling*.' The extent, to which local source of moisture through evapotranspiration dominates the regional precipitation, decides the potential role of land-surface processes in regional climate. A significant variation in the land-surface feedback mechanisms through evapotranspiration can affect regional precipitation characteristics. Therefore, an understanding of the land-surface feedbacks and its association with the rainfall is of major importance in any regional climate system study. A brief introduction to precipitation recycling, its definition, and method used to quantify its impact is described in the next subsections.

4.1 *Precipitation Recycling: A Definition*

The atmospheric water vapor 'w' (also known as precipitable water) can be defined in terms of atmospheric water column as the vertically integrated water vapor depth above the region of interest. As the evaporation rate is inversely proportional to the specific humidity '$\bar{q}(p)$,' the water vapor content can be represented as vertical integral of specific humidity '$\bar{q}(p)$' from the earth surface to a height where pressure 'p' approaches zero:

$$w = \frac{1}{g} \int_{p_0}^{0} \bar{q}(p)\mathrm{d}p \tag{1}$$

where 'g' is acceleration due to gravity. The surface evaporation 'E' from land and ocean provides water vapor to the atmosphere. The fraction of water vapor present in the atmosphere can be classified into two categories based upon the source of

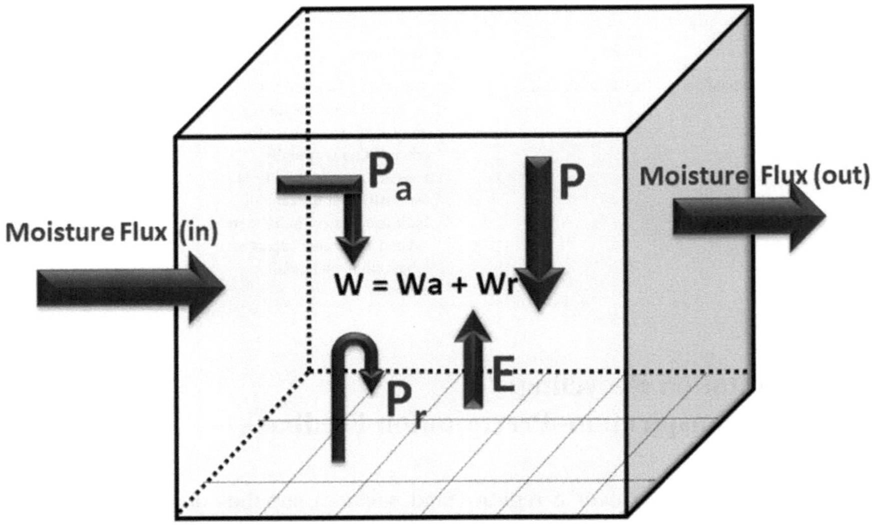

Fig. 1 Schematic representation of various fluxes in a region

evaporation: a local or internal component and an external (also called as advective) component, which is evaporated outside the region's boundary. Thus, the precipitable water present in the atmospheric moisture column is written in terms of addition of advective 'w_a' and local component 'w_r' of water vapor:

$$w = w_a + w_r \tag{2}$$

The precipitation generated from local component of water vapor 'w_r' to the same region is known as recycled precipitation 'P_r.' Therefore, the total precipitation originating from these two sources (i.e., due to advection 'P_a' and recycling 'P_r') can be written as,

$$P = P_a + P_r \tag{3}$$

Figure 1 illustrates the schematic diagram, showing various fluxes inside region's atmospheric volume. The water vapor is being added to the atmospheric water column by the incoming moisture carrying winds, and evaporation 'E' within the region, whereas the precipitation 'P' and outgoing moisture weighted winds remove the moisture from the region. The precipitation 'P' over the region is composed of recycled precipitation 'P_r' and advective precipitation 'P_a,' respectively. Here, the advective moisture fluxes are represented by '*Moisture Flux (in)*' and '*Moisture Flux (out),*' respectively.

The precipitation recycling is characterized by local recycling ratio denoted by 'R_l,' which is given by the ratio of recycled precipitation to the total precipitation at a particular grid. For example, the local recycling ratio 'R_{l_m}' at a particular grid point 'm' of a uniformly spaced gridded region is given by:

$$R_{l_m} = \frac{P_{r_m}}{P_m} = \frac{W_{r_m}}{w_m} \tag{4}$$

Globally, the recycling ratio ranges from its limiting value ***0*** (local evaporation contributed zero) by considering a same grid point as source to ***1*** (when all water evaporated from earth surface and precipitated back on the earth) by considering entire earth as one region. The recycling ratio can be used as a diagnostic tool to quantify the land–atmosphere interactions through evapotranspiration–precipitation feedback. A large value of recycling ratio corresponds to the large potential of land–atmosphere interactions.

4.2 Common Assumptions

Most precipitation recycling studies make following basic assumptions:

- In the planetary boundary layer (*PBL*) of earth's atmosphere (Eltahir and Bras 1996), water vapor is well mixed. Therefore, we can write:

$$\frac{P_a}{P_r} = \frac{w_a}{w_r} \quad \text{or} \quad \frac{P_a}{P} = \frac{w_a}{w} \quad \text{or} \quad \frac{P_r}{P} = \frac{w_r}{w} = R_l \tag{5}$$

- The mixing in the PBL occurs at very short timescales. In general, it takes about 15 min for water vapor molecule evaporating at the surface to mix in the atmosphere up to a height of 1 km (Paluch 1979).
- At sufficiently long-time scale (monthly or longer), the change in atmospheric moisture storage is very small compared with other moisture fluxes (Eltahir and Bras 1994). However, at daily scale, change in atmospheric moisture storage is non-negligible.
- Time-averaged (sub-daily, daily, monthly) data is used in the study.

4.3 Model Description: Dynamic Recycling Model (DRM)

The concept of precipitation recycling is a potential measure of understanding of land–atmosphere interaction. In the past, studies involving precipitation recycling and source-sink analysis have been dealt in a variety of ways, such as numerical tracer experiments, physical analysis using the isotopes, and analytical models. A detailed description of the precipitation recycling and model development can be found in Budyko (1974), Lettau et al. (1979), Brubaker et al. (1993), Eltahir and Bras (1994), Eltahir and Bras (1996), Burde and Zangvil (2001), and Dominguez et al. (2006). The readers are advised to go through these articles for an in-depth understanding of the precipitation recycling. Although there are various techniques used in the literature to quantify the impact of local evaporation to the precipitation, here, we restrict ourselves to a physics-based bulk recycling model such as DRM (Dominguez et al. 2006). Another reason to implement DRM in recycling studies is its applicability at daily timescale. As the ISMR has significant variability at intraseasonal timescale (such as active and break phase of duration ~3 to 8 Days), hence, the timescale to be used for recycling studies of ISMR should be shorter than month or week. Therefore, we restrict ourselves to dynamic recycling model (Dominguez et al. 2006).

The dynamic recycling model like any other bulk recycling models is based on conservation of atmospheric water vapor mass with the assumption of a well-mixed atmosphere. The equation for conservation of atmospheric water vapor mass is given by:

$$\text{(Change in atmospheric moisture storage)} + \text{(zonal moisture fraction)}$$
$$+ \text{(meridional moisture fraction)} = \text{(evaporation)} - \text{(precipitation)}$$

$$\frac{\partial w}{\partial t} + \frac{\partial (w u_m)}{\partial x} + \frac{\partial (w v_m)}{\partial y} = E - P \tag{6}$$

Here, the amount of total water vapor, advective water vapor, and locally evaporated water vapor present in the column of atmosphere is represented by 'w,' 'w_a,' and 'w_r,' respectively. The moisture weighted zonal and meridional winds are represented by 'u_m' and 'v_m,' respectively. Decomposed form of Eq. 6 for advective and recycled components is written as,

$$\frac{\partial w_a}{\partial t} + \frac{\partial (w_a u_m)}{\partial x} + \frac{\partial (w_a v_m)}{\partial y} = -P_a \tag{7}$$

$$\frac{\partial w_r}{\partial t} + \frac{\partial (w_r u_m)}{\partial x} + \frac{\partial (w_r v_m)}{\partial y} = E - P_r \tag{8}$$

Here, the evaporation, precipitation, advective and recycled component of precipitation are represented by 'E,' 'P,' 'P_a,' and 'P_r,' respectively.

The dynamic recycling model (Dominguez et al. 2006) is based on a Lagrangian trajectory approach which runs backward in time along the trajectory of the water vapor for a duration equal to the moisture residence time. The moisture residence time corresponds to the time starting from the day of precipitation at location (χ, ξ, τ) to ending at the day on which moisture entered the boundary of the region. The ratio, evaporation by precipitable water, 'ε/w' is calculated along a trajectory for a duration of moisture residence time. Then, the integration (Eq. 9) is performed back in time from the current time step ('τ') to the time step at which water vapor crossed the boundary of the zone (i.e., starting point). The relative contribution of recycled precipitation (i.e., derived from local evaporation) to total area precipitation is characterized by the local recycling ratio denoted by 'R' or 'R_l.' The local recycling ratio at each grid point is given by (9) as,

$$R(\chi, \xi, \tau) = 1 - \exp\left[- \int_{0}^{\tau} \frac{\varepsilon(\chi, \xi, \tau)}{\omega(\chi, \xi, \tau)} \partial\tau' \right] \quad (9)$$

The recycled precipitation value for a particular grid point is calculated by multiplying the local recycling ratio with total precipitation at that grid point as shown in Eq. 5. The regional recycling ratio, a ratio of total recycled precipitation to the total precipitation in the region for each region (zone), is given by:

$$R_z = \frac{P_r}{P} = \frac{\sum_{i=1}^{n} \left(R_{l_i} * P_i * \Delta A \right)}{\sum_{i=1}^{n} \left(P_i * \Delta A \right)} \quad (10)$$

where 'P_r' is recycled precipitation in a region (mm), 'P' is total precipitation in a region (mm), 'R_{li}' is local recycling ratio at grid point 'i,' 'ΔA' is the area of the individual rectangular grid (km^2), and 'n' is a total number of grids in a region.

4.4 *Precipitation Recycling and Subcontinental Evapotranspiration*

In the context of the Indian subcontinent, it is observed that the evapotranspiration increases as monsoon progresses with least amount prior to the monsoon onset reaches its maximum during September and October. The increased evapotranspiration during the monsoon can significantly contribute to the recycled precipitation. Pathak et al. (2014) studied the role of entire subcontinental evapotranspiration to the monsoon precipitation by applying dynamic recycling model (Dominguez et al. 2006).

Figure 2 shows the recycled precipitation (top row), local recycling ratio (middle row), and total precipitation (bottom) at each grid point during June, July, August, and September (JJAS). It can be seen here that the advective moisture from the oceans (non-local) has a main role in monsoon initiation, but the moisture supplied from the local evapotranspiration enhances the subsequent precipitation through recycling. Generally, the peak of monsoon rainfall is observed in the month of July and August, which is followed by a drastic reduction in rainfall during September. However, the pattern of intraseasonal variations in recycling (recycling ratio) is slightly different. The recycling ratio is observed to be increasing as monsoon progresses over the Indian subcontinent with the highest recycling in September due to increased evapotranspiration. Additionally, the moisture carrying winds during the September traverses a larger extent of the land area prior to the precipitation event which results in increased moisture residence time (τ-0 in Eq. 9) within a region. Therefore, the

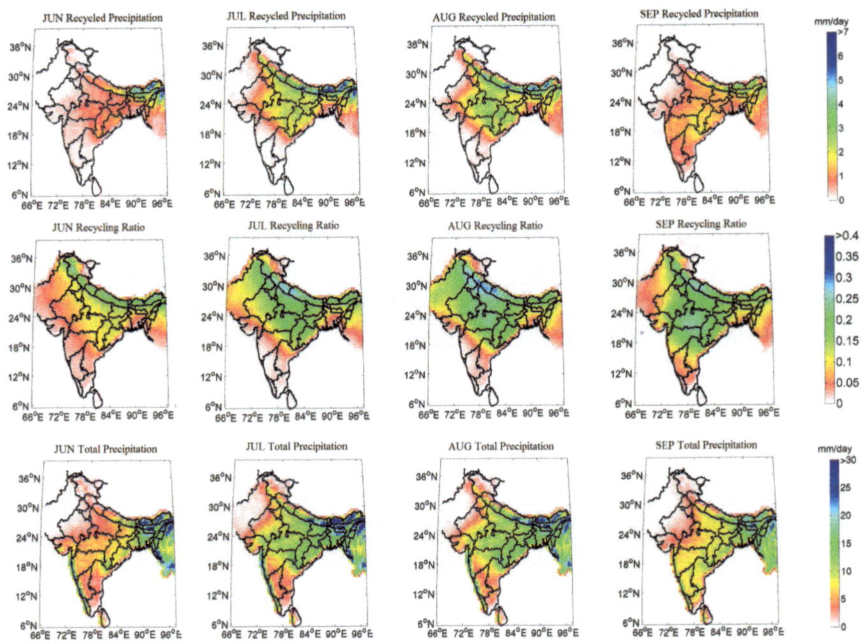

Fig. 2 Climatological mean of recycled precipitation in mm/day (top row), recycling ratio (middle row), and total precipitation in mm/day (bottom row) over Indian subcontinent during the summer monsoon months (JJAS) (Reproduced from: Pathak et al. 2014, under the copyright of © American Meteorological Society)

enhancement of recycling in September month is accompanied by the weakening of oceanic moisture flux and increased moisture residence time.

A prominent impact of recycling is observed specifically in central India, Ganges Plain, and northeast India during August and September. Contrary to that, the precipitation over the regions such as Western Ghats and peninsular India has relatively lower influence of recycling due to dominant moisture flux from the Indian Ocean. We therefore can say that the precipitation recycling is an integral part of Indian summer monsoon rainfall and significantly contributes to the seasonal total and variations within.

4.5 Precipitation Recycling and Monsoon Withdrawal

Here, it is important to note that although the precipitable water is relatively less during the end of ISMR due to the weak moisture flux from oceans, the contribution from terrestrial evapotranspiration to the precipitable water is quite significant.

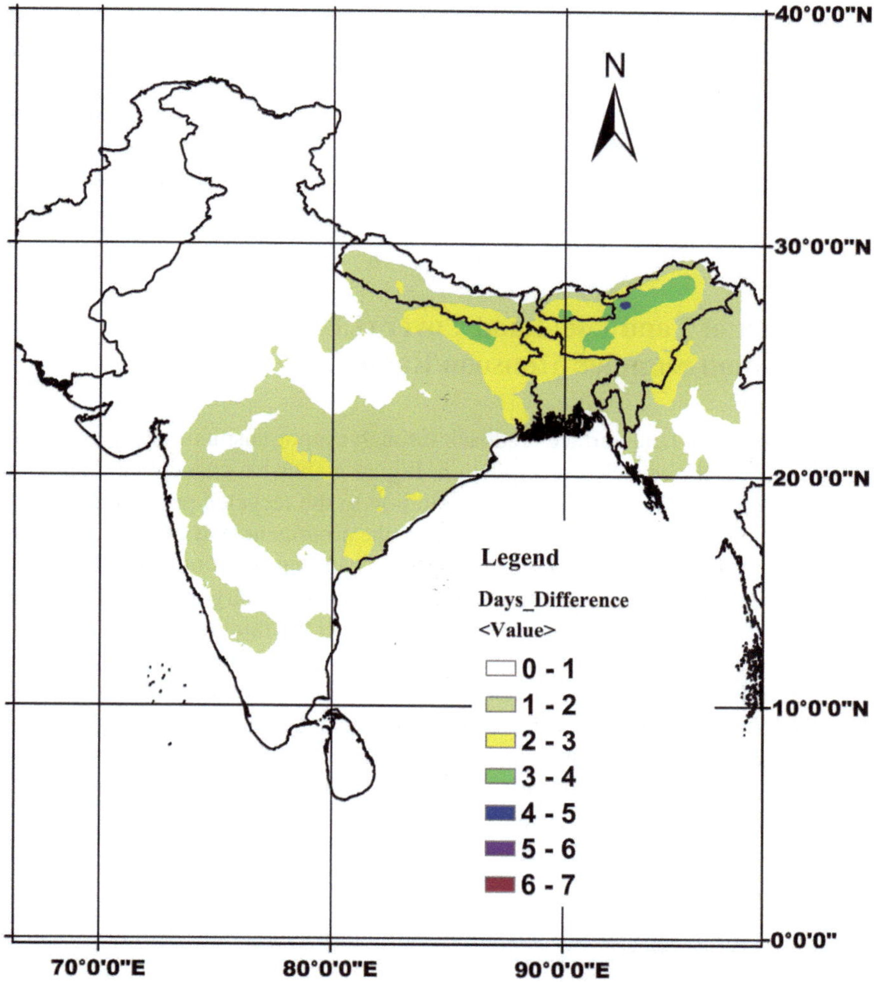

Fig. 3 Precipitation recycling and its impact on summer monsoon withdrawal (Reproduced from: Pathak et al. 2014, under the copyright of © American Meteorological Society)

Therefore, a higher amount of evapotranspiration over the entire subcontinent during September provides strength to the weakened monsoon.

Pathak et al. (2014) investigated the impact of precipitation recycling on monsoon withdrawal. A set of summer monsoon withdrawal dates were calculated separately by considering total precipitation (actual) and advective precipitation independently. The monsoon withdrawal over the area was defined as cessation of rainfall activity for continuous 5 days after September 1. Figure 3 shows the mean difference in days (i.e., prolongation) calculated by subtracting the two time series (advective precipitation and total precipitation) at each grid point. Here, it is important to note that the method used for computing withdrawal date has the limitation that it does not consider indicators such as vertically integrated moisture transport, wind shear, and outgoing longwave radiation.

A delay in monsoon withdrawal due to precipitation recycling by approximately ~3–5 days over the northeastern region and ~2–3 days over the eastern region is observed by Pathak et al. (2014). In other words, we can say that precipitation recycling prolongs the summer monsoon over these specific regions for few more days.

5 Impact of Land Use–Land Cover (LULC) Change on Indian Summer Monsoon Rainfall

We have seen that land-surface feedback through evapotranspiration can contribute significantly to the recycled precipitation. Figure 4 shows a conceptual diagram to represent the contribution of evapotranspiration to the recycled precipitation under most common LULC type. The regions where the prominent land cover is forest land or the agriculture have a relatively higher rate of evapotranspiration than other LULC types. Therefore, a significant change in the LULC due to large-scale deforestation can result in low evapotranspiration and hence affect the rainfall.

It is observed that the Indian subcontinent has experienced a remarkable and widespread LULC changes between two recent decades of the 1980s to the 2000s (Fig. 5). Woody savanna (mostly forest land) which was a prominent LULC type over central India, peninsular India, and northeast India during the 1980s has been converted into cropland in the recent decade of 2000s. The drastic changes in the LULC type are mostly associated with the human activities such as agricultural intensification and large-scale deforestation.

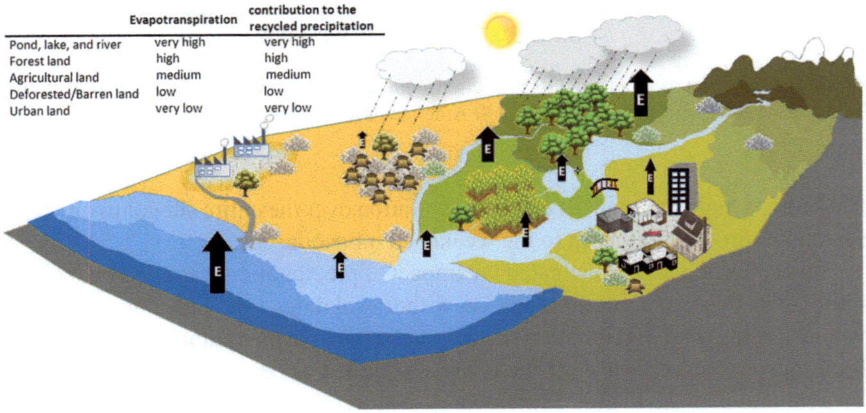

Fig. 4 Contribution of evapotranspiration to the recycled under different land use–land cover types

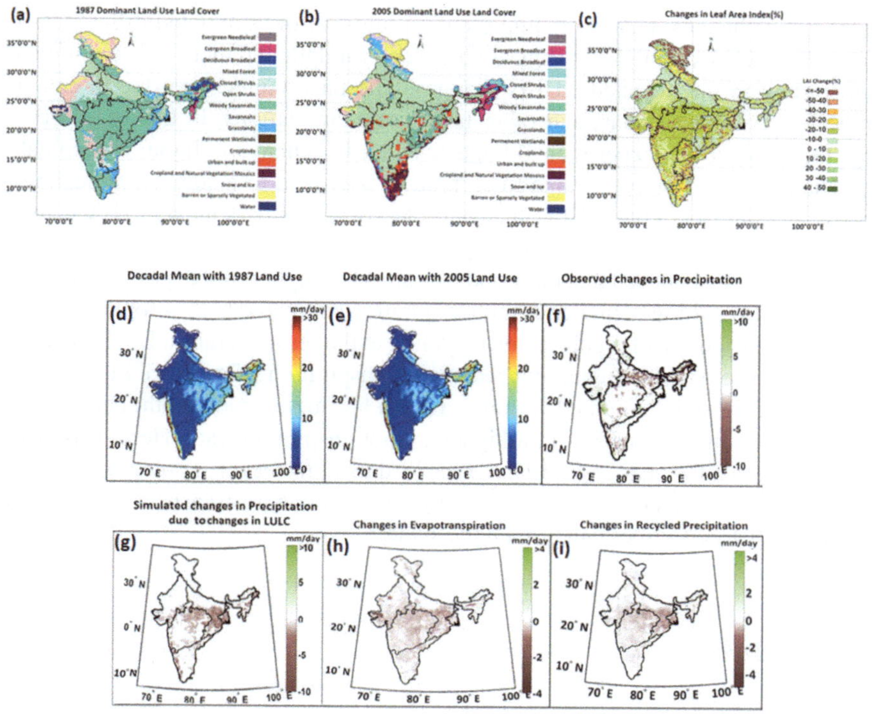

Fig. 5 **a** LULC classification for 1987, **b** and 2005, **c** percentage changes in leaf area index in 2005 from 1987. The LULC classification of 1987 is obtained from AVHRR, and 2005 is obtained from MODIS. Furthermore, the simulated mean JJAS precipitation using **d** 1987 and **e** 2005 LULC and **g** their differences, which are to some extent consistent with observed changes in ISMR, **f** Similarly, the reduction in (**h**) evapotranspiration and **i** recycled precipitation is observed in the simulated runs of the period 2000–10 using 1980s and 2000s LULCs (Adapted from Paul et al. 2016, under Creative Commons Attribution 4.0 License)

Paul et al. (2016) performed the sensitivity analysis experiment using weather research and forecasting (WRF) model to study the changes in the monsoon rainfall as a result of LULC changes. In their experiment, the WRF model coupled with Community Land Model (CLM4.0) was forced with two LULC type representative experiment (the 1980s: LULC map of 1985, and 2000s: LULC map of 2005) to study the changes in rainfall during summer the monsoon in India.

The simulated mean summer monsoon rainfall obtained with both the LULCs for a time period of 2000–2010 is presented in Fig. 5d, e. The differences in simulated precipitation estimate resulting from the two LULCs represent the changes in precipitation due to large-scale deforestation in India. We can see that differences in the observed rainfall are mostly negative over Ganga Basin, central India, and northeast India (Fig. 5f), which is mostly reflected in the difference of simulated precipitation estimates (Fig. 5g). It means that the significant decline in ISMR in the recent decade is associated with the changes in LULCs in addition to the other large-scale changes. Here, it is important to note that the simulated differences do not completely explain the changes in monsoon rainfall, and there exists significant uncertainty in the detected LULC changes. The uncertainties are due to changes in SST, aerosol emissions, and other large-scale circulations. In this case, we see that the effects of an experimental large-scale LULC change in India—specifically, the change from woody savanna to cropland—may lead to a significant decline in ISMR. The impact of this LULC change is clearly visible over the regions where the recycled precipitation is maximum (Fig. 2).

It is generally believed that the woody savanna has higher root depth, hence higher water intake resulting in higher evapotranspiration compared to the cropland. Therefore, the contribution of precipitation recycling to the seasonal rainfall is less over the converted cropland due to the reduced evapotranspiration (Fig. 5h, i). Furthermore, in comparison to the cropland, woody savanna has higher root depth; hence, it has higher water intake resulting in a higher evapotranspiration rate. Here, it is worth mentioning that the changes in LULC type lead to a low recycled precipitation, even though the large-scale circulation in the form of moisture transport from remote oceanic moisture sources can remain the same.

6 Conclusions

The present chapter highlights the importance of land-surface feedbacks in the Indian summer monsoon rainfall. We believe that by improving the representation of land surface processes in the climate model, the biases in the current generation climate model outputs can be rectified to some extent. This chapter provides the brief description of the precipitation recycling process in the Indian subcontinent and its association with the Indian summer monsoon rainfall. We have seen that the precipitation recycling is an integral part of monsoon rainfall and significantly contributes to the precipitation during the peak and end phase of the summer monsoon in India. The regions such as central India, Ganga Basin, and northeast India receive a considerable amount of precipitation through the recycling. However, the southern peninsula, western and southeastern parts receive relatively less amount of recycled precipitation owing to the dominance of oceanic moisture and high advection of the evaporated moisture. The weakened monsoon during September gets its additional strength from the recycling. The precipitation recycling is also responsible for delaying the monsoon withdrawal over eastern and northeastern part of the Indian subcontinent.

We have also seen that alteration in the existing LULCs type can have profound consequences in terms of changes in the precipitation amount. In the context of India, it observed that the changes in the LULC type from woody savanna (mostly forest land) to the cropland can result in a weakening of simulated precipitation over central India. The reduction in the rainfall is considerably linked with the weakening of recycling mechanism due to a reduction in the evapotranspiration under the changed

LULC type. The other reasons that may account for this weakening in summer monsoon rainfall are aerosols and warming of Indian Ocean SST, etc.

Furthermore, in addition to the deforestation, change in major LULC types such as increased urbanization and large water resources structures may also contribute to the precipitation anomaly. Hence, it is advised that sufficient sensitivity experiments by considering different climate scenarios should be conducted before planning of any policy-level decisions which considers such large scale of changes in the LULC types.

References

Abe M, Hitoh A, Yasunari T (2003) An evolution of the Asian summer monsoon associated with mountain uplift, simulation with the MRI atmosphere—ocean coupled GCM. J Weteovol Soc Japan 81:909–933

Asharaf S, Doble A, Ahrens B (2011) Soil moisture initialization effects in the Indian monsoon system. Adv Sci Res 6:161–165

Bamzai AS, Shukla J (1999) Relation between Eurasian snow cover, snow depth, and the Indian summer monsoon: an observational study. J Climate 12:3117–3132. https://doi.org/10.1175/1520-0442(1999)012<3117:RBESCS>2.0.CO;2

Blanford HF (1884) On the connection of the Himalayan snowfall with dry winds and seasons of drought in India. Proc Roy Soc 37:3–22

Brubaker KL, Entekhabi D, Eagleson PS (1993) Estimation of continental precipitation recycling. J Clim 6:1077–1089

Budyko MI (1974) Climate and life. Academic Press, p 508

Burde GI, Zangvil A (2001) The estimation of regional precipitation recycling. Part I: review of recycling models. J Clim 14:497–2508

Devaraju N, Bala G, Modak A (2015) Effects of large-scale deforestation on precipitation in the monsoon regions: remote versus local effects. Proc Natl Acad Sci U S A 112:3257–3262. https://doi.org/10.1073/pnas.1423439112

Dominguez F, Kumar P (2008) Precipitation recycling variability and ecoclimatological stability—A study using NARR data. Part I: Central U.S. plains ecoregion. J Clim 21:5165–5186

Dominguez F, Kumar P, Liang X, Ting M (2006) Impact of atmospheric moisture storage on precipitation recycling. J Clim 19:1513–1530

Douglas EM, Beltrán-Przekurat A, Niyogi D, Pielke RA Sr, Vörösmarty CJ (2009) The impact of agricultural intensification and irrigation on land-atmosphere interactions and Indian monsoon precipitation: a mesoscale modeling perspective. Glob Planet Change 67:117–128. https://doi.org/10.1016/j.gloplacha.2008.12.007

Douville H, Chauvin F, Planton S, Royer JF, Salas-Melia D, Tyteca S (2002) Sensitivity of the hydrological cycle to increasing amounts of greenhouse gases and aerosols. Clim Dynamicz 2:45–68

Eltahir EAB, Bras RL (1994) Precipitation recycling in the Amazon basin. QJR Meteorol Soc 120:861–880

Eltahir EAB, Bras RL (1996) Precipitation recycling. Rev Geophys 34:367–378

Gimeno L, Stohl A, Trigo RM, Dominguez F, Yoshimura K, Yu L, Drumond A, Durán-Quesada AM, Nieto R (2012) Oceanic and terrestrial sources of continental precipitation. Rev Geophys 50:RG4003. https://doi.org/10.1029/2012rg000389

Hahn DG, Shukla J (1976) An apparent relationship between Eurasian snow cover and Indian monsoon rainfall. J Atmoz Sci 33:2461–2462

Koster RD, Dirmeyer PA, Guo ZC, Bonan G, Chan E, Cox P, Gordon CT, Kanae S, Kowalczyk E, Lawrence D, Liu P, Lu CH, Malyshev S, McAvaney B, Mitchell K, Mocko D, Oki T, Oleson K, Pitman A, Sud YC, Taylor CM, Verseghy D, Vasic R, Xue YK, Yamada T (2004) Regions of strong coupling between soil moisture and precipitation. Sci 305:1138–1140

Kripalani RH, Kulkarni A (1999) Climatology and variability of historical Soviet snow depth data, some new perspectives in snow—Indian monsoon teleconnections. Clim Dynamicz 15:475–489

Lee E, Chase TN, Rajagopalan B, Barry RG, Biggs TW, Lawrence PJ (2009) Effects of irrigation and vegetation activity on early Indian summer monsoon variability. Int J Climatol 29:573–581. https://doi.org/10.1002/joc.1721

Lettau H, Lettau K, Molion LCB (1979) Amazonia's hydrologic cycle and the role of atmospheric recycling in assessing deforestation effects. Mon Wea Rev 107:227–238

Meehl GA (1994) Influence of land surface in the Asian summer monsoon: external conditions versus internal feedbacks. J Clim 7:1033–1049

Niyogi D, Chang H, Chen F, Gu L, Kumar A, Menon S, Pielke RA Sr (2007) Potential impacts of aerosol-land-atmosphere interaction on the indian monsoonal rainfall characteristics. Nat Hazards 42:345–359. https://doi.org/10.1007/s11069-006-9085-y

Niyogi D, Kishtawal C, Tripathi S, Govindaraju RS (2010) Observational evidence that agricultural intensification and land use change may be reducing the Indian summer monsoon rainfall. Wat Resour Res 46:W03533

Ordóñez P, Ribera P, Gallego D, Peña-Ortiz C (2012) Major moisture sources for Western and Southern India and their role on synoptic-scale rainfall events. Hydrol Process 26:3886–3895. https://doi.org/10.1002/hyp.8455

Paluch IR (1979) The entrainment mechanism in Colorado cumuli. J Atmos Sci 36:2467–2478

Pathak A, Ghosh S, Kumar P (2014) Precipitation recycling in the Indian subcontinent during summer monsoon. J Hydrometeor 15:2050–2066. https://doi.org/10.1175/JHM-D-13-0172.1

Paul S, Ghosh S, Oglesby R, Pathak A, Chandrasekharan A, Ramshankaran R (2016) Role of changes in land use land cover to the recent decadal weakening of Indian summer monsoon rainfall. Sci Rep 6. https://doi.org/10.1038/srep32177

Peings Y, Douville H (2009) Influence of the Eurasian snow cover on the Indian summer monsoon variability in observed climatologies and CMIP3 simulations. Clim Dyn. https://doi.org/10.1007/s00382-009-0565-0

Shastri H, Paul S, Ghosh S, Karmakar S (2014) Impacts of urbanization on Indian summer monsoon rainfall extremes. J Geophys Res Atmos 119. https://doi.org/10.1002/2014jd022061

Trenberth KE (1999) Atmospheric moisture recycling: role of advection and local evaporation. J Clim 12:1368–1381. http://dx.doi.org/10.1175/1520-0442(1999)012<1368:AMRROA>2.0.CO;2

Yamashima R, Matsumoto J, Takata K, Takahashi HG (2015) Impact of historical land-use changes on the Indian summer monsoon onset. Int J Climatol 35:2419–2430

Yasunari T, Hitoh A, Tokioka T (1991) Local and remote responses to excessive snow mass over Eurasia appearing in the northern spring and summer climate a study with the MRI GCM. J Weteovol Soc Japan 69:473–487

Aerosol Influences on Cloud Modification and Rainfall Suppression in the South Asian Monsoon Region

Prashant Dave, Nitin Patil, Mani Bhushan and Chandra Venkataraman

Abstract The South Asian monsoon (SAM) is critical to the agricultural economy of the region. Large-scale processes affecting the monsoon have been traditionally been well studied; however, much less is known about the role of local processes, especially one governed by levels of atmospheric aerosols, or the mix of pollutant particles and dust. This chapter delineates the aerosol influence on cloud modification and short-term rainfall inhibition over India, employing observational data for the period of 2000–2009. Aerosols can change cloud-related properties in contrasting ways in years of deficient and abundant monsoon. In deficient years, increased levels of atmospheric aerosols correlate with smaller cloud drops, shallower cloud heights, and less cloud-ice formation. In contrast, in abundant rainfall years, higher levels of aerosols correlate with larger cloud droplet size, taller clouds, and greater ice-cloud formation. Further, causality was established in high aerosol and low rainfall regions spanning across the Indian subcontinent, wherein enhancement in aerosol levels caused suppression of daily precipitation anomaly multiple times in a monsoon season with lags of a few days. The studies also suggest aerosol effects on increases in the frequency and length of monsoon breaks, with implications for rainfall deficit and food grain production. This work makes important linkages between enhanced air pollution and perturbations in the timing and short-term suppression of regional monsoon rainfall, implying the need for greater synergy in policies addressing air pollution and climate change.

P. Dave · N. Patil · M. Bhushan · C. Venkataraman (✉)
Interdisciplinary Program (IDP) in Climate Studies, Indian Institute of Technology Bombay, Mumbai 400076, Maharashtra, India
e-mail: chandra@iitb.ac.in

P. Dave
e-mail: prashantdave25@gmail.com

N. Patil
e-mail: nitinpatil85@gmail.com

M. Bhushan
e-mail: mbhushan@iitb.ac.in

M. Bhushan · C. Venkataraman
Department of Chemical Engineering, Indian Institute of Technology Bombay, Mumbai 400076, Maharashtra, India

C. Venkataraman et al. (eds.), *Climate Change Signals and Response*,
https://doi.org/10.1007/978-981-13-0280-0_2

1 The Role of Local and Regional Processes in South Asian Monsoon Change

The South Asian monsoon (SAM) is the main factor influencing availability of water and affects water management concerning rainfed agricultural practices across India (Gadgil and Gadgil 2006; Prasanna 2014). Rainfall is having an important role in lives of people in the region, with a majority of the population directly affected by it, from their association with agriculture-related activities. About 80% of the annual mean precipitation over India is provided by SAM, spanning across four months of summer, i.e., from June to September, with large-scale consequences to agriculture, health, water resources, economy, and ecosystems throughout South Asia (Webster et al. 1998). Many complex factors can influence SAM ranging from global-scale atmospheric circulation to regional and local physical processes. While large-scale processes have been traditionally well studied, the role of local processes is yet to be understood. Recently, local and regional physical processes affecting the SAM have received attention, including those governed by changing land-use patterns or levels of atmospheric aerosols.

Atmospheric aerosols, which include a mix of pollution particles and dust, consist of trace gases and a variety of liquid and solid constituents that exist as dispersed phases in air. Aerosols result from both natural and anthropogenic sources. Main sources of natural aerosols are wind-blown dust, volcanic ash, and sea salt, while those of anthropogenic, or human made aerosols include wide-ranging industrial and energy-use activities. Thus, pollution particles, along with dust, constitute atmospheric aerosols. The presence of aerosols controls the cooling or heating effect on the earth surface and in turn warming or cooling of the atmosphere. An important aspect of climate perturbation is the interaction between atmospheric aerosols and the SAM. Aerosols mainly affect precipitation patterns through radiative effects on large-scale atmosphere and ocean dynamics or by modulating cloud formation processes and cloud structure. As delineated by the Intergovernmental Panel on Climate Change (Boucher et al. 2013), aerosol impacts on climate change occur through two main pathways (i) ERFari or effective radiative forcing from aerosol-radiation interactions and (ii) ERFaci or effective radiative forcing from aerosol-cloud interactions (Boucher et al. 2013). ERFari includes direct aerosol perturbation of Earth's radiation budget by absorbing and/or scattering of solar and terrestrial radiation, emanating from the sun or from Earth's surface. Aerosols such as sulfate particles in the atmosphere can scatter the sunlight and cool the surface, while absorbing aerosols such as black carbon (BC) and dust can scatter as well as absorb radiation. ERFari may also act through aerosols present inside clouds and influence cloud properties, by increasing cloud temperatures, thus reducing cloud drop sizes. ERFaci includes aerosol changes in cloud microphysics and dynamics, which can lead to smaller, but more numerous drops, increasing cloud reflectance and to increased lifetime of clouds, by suppressing large drop growth and consequent rainout. Overall, the role of aerosols in climate change is associated with large uncertainties and constitutes "a hole in climate-science research (Schiermeier 2010)."

Atmospheric aerosols have been linked, in recent modeling studies, to suppression of rainfall on long timescales and to the early onset or increase in monsoon on short timescales. Different studies have focused on changes in monsoon that are mediated through changes in sea surface temperatures, reduction of the north-south temperature gradient in the northern Indian Ocean that lead to circulation changes causing reduction in the monsoon (Ramanathan et al. 2005; Bollasina et al. 2011). Such changes in Indian monsoon rainfall, linking slow adjustments to aerosol radiative forcing, explained above, are known as slow responses. These changes are induced by thermodynamical adjustments of mean sea surface temperature and atmospheric moisture content (Ramanathan et al. 2005; Meehl et al. 2008), diminishing the land–sea temperature contrast along with zonal winds, and altering regional energy imbalances (Bollasina et al. 2011; Ganguly et al. 2012; Krishnan et al. 2016), and are consequently linked with precipitation reduction at multidecadal timescales.

In contrast, short-term changes in SAM precipitation that are linked to aerosols, known as fast responses, mediate through strengthening of meridional pressure gradient or surface temperature (Ganguly et al. 2012; Vinoj et al. 2014) and mid-tropospheric diabatic heating (Lau and Kim 2006), causing increased northward transport of moisture, which in turn lead to the onset and enhanced precipitation over a timescales ranging from days to month. More recently, due to both anthropogenic and natural aerosols (Lau et al. 2006; Ganguly et al. 2012; Vinoj et al. 2014), rapid changes in radiative forcing have been associated with increased northward moisture transport and as a consequence of which, enhancement of rainfall on daily to monthly timescales is observed. Over northern India, changes in aerosols were also linked with changes in precipitation asymmetrically, with increased precipitation in the west and decreased precipitation in the east of 80°E (Ganguly et al. 2012). Studies have also identified the relationship of spatially distant aerosols (in the middle to upper troposphere) with observed increases of net diabatic heating rates and with an intensified northward pressure gradient over the Arabian Sea and a subsequent increase in moisture convergence over India at synoptic scale.

In studies not related to the SAM, mechanisms of aerosol-induced modulation of cloud and precipitation development have been proposed. Absorbing aerosols, including dust and pollution particles containing soot or black carbon, could lead to near-surface atmospheric stabilization that leads to positive feedback causing reduction in cloudiness. The theory that black carbon aerosols inhibit development of warm clouds finds support from observational evidence collected from the Amazon region during biomass burning season. Absorptive dust aerosol outbreaks over the Taklamakan Desert and East Asia's arid regions have been associated with atmospheric warming effects and with a reduction in the liquid water path and ice water path in clouds contaminated with dust. On contrary, increase in the cloud condensation nuclei at base of the cloud could enhance cloud "invigoration" and increase rainfall intensity (Rosenfeld et al. 2008). Aerosol-mediated enhanced transition to convective regime from stratocumulus regime of the cloud and intensity of rainfall has received support from observational evidences.

During SAM season, significant aerosol concentrations over the Indian subcontinent have been found. While atmospheric aerosols have been linked in previous modeling studies to monsoon rainfall changes on varying timescales, such signals are not well understood in observations. Observation based studies are needed to identify possible existing effects, which have manifested in recent decades. In this chapter, we present results related to the modulation of monsoon clouds and rainfall by both spatially and temporally coincident aerosols, developed from a set of recent studies (Dave et al. 2017; Patil et al. 2017). The questions asked included what effects aerosols coincident in the SAM region induce on cloud and rainfall, the time and spatial scales of these effects, whether aerosols affect clouds in similar or dissimilar ways, in deficient and abundant monsoon years, and whether we can go beyond correlation and establish causation in a system as complex as the atmosphere to understand linkages between aerosols and monsoon rainfall changes using mainly observational data. We also discuss strategic knowledge gained and possible policy implications emanating from the new findings.

2 Observational Datasets and Analysis Methodology

For the analysis, satellite- and ground-based observations for aerosol, precipitation, cloud properties, over India (6.5°–40°N and 66.5°–100°E) for June –July–August–September (JJAS) at $1^\circ \times 1^\circ$ resolution for the period of 2000–2009, were used. Aerosol optical depth (AOD), which is a measure of atmospheric aerosol, Level-3 (L3) data were retrieved from the moderate resolution imaging spectroradiometer (MODIS) on board the Earth Observing System's (EOS) Terra and Aqua satellites. Data are made available through the National Aeronautics and Space Administration (NASA) Deep blue (Collection-6) algorithm (Remer et al. 2005). This data also include AOD retrievals over the regions with highly reflective surface. Absorbing aerosol index (AAI) data were retrieved using the Earth Probe TOMS (McPeters et al. 1998) and OMAERUV (Torres et al. 2007) algorithms for the total ozone mapping spectrometers (TOMS) on board the Earth Probe satellites and the ozone monitoring instrument (OMI) sensor on board the EOS Aura satellite. AAI values are within the range of 0.5 and 3.5 km (Herman et al. 1997) and neglect the residues pertaining to completely scattering aerosols (Torres et al. 2007). For cloud droplet effective radius (CDER), MODIS L3 data collected from EOS Terra and Aqua satellites were used. For lapse rate calculation, temperature of nine layers of atmosphere between 1000 and 750 hPa, from the European Centre for Medium-Range Weather Forecasts (ECMWF) reanalysis (ERA)-interim dataset (Dee et al. 2011), was used. Precipitation data were obtained from India Meteorological Department (IMD). The data were released as gridded product, which were interpolated using 1803 irregularly located meteorological stations spread across India (Rajeevan et al. 2006). Further cloud properties such as liquid water path (LWP), ice water path (IWP), cloud fraction (CF), column water vapor (CWV), cloud top pressure (CTP), and cloud top temperature (CTT) were also obtained using MODIS L3 satellite.

In each region, deficient and abundant rainfall years were identified using a seasonal normalized precipitation anomaly for JJAS, for years 2000–2009. Daily pixel-level data were transformed to normalize anomalies, and using the normalized anomalies, hierarchical clustering were performed to formulate high AOD-low precipitation (HL) and low AOD-low precipitation (LL) clusters for each of the 10 years. Spatially inhomogeneous clusters, which extended over the subcontinent from 75°–90°E to 10°–35°N, typically contained 50–150 pixels in individual years. Averaged time series data were used to analyze the aerosol causal influence on precipitation, individually for the HL and LL clusters in each year, using Granger causality (Granger 1969). The stationarity of the time series, an essential condition for causality detection (Granger 1969), was tested using the Kwiatkowski–Phillips–Schmidt–Shin (KPSS) test (Kwiatkowski et al. 1992). In case of non-stationary original time series, first-order differencing was used. Significance tests were performed at $\alpha = 0.1$ during the whole study.

3 Opposing Effects of Aerosols on Cloud Properties in Contrasting Monsoon Years

In order to understand the role of aerosol in modulating cloud properties, out of six homogenous monsoon rainfall regions demarcated by IMD, three regions were identified. These regions were identified on the basis of similarity in regional rainfall characteristics and association of sub-divisional monsoonal rainfall with regional/global circulation parameters (Patil et al. 2017). Using IMD rainfall observations from 2000 to 2009 for JJAS, the aerosol modulation of clouds and precipitation during both abundant and deficient years was investigated. Deficient and abundant rainfall years were identified from the 10-year time period and differed in the three regions studied (R1-59 pixels; R2-87 pixels; and R3-63pixels). At each pixel, the combined rainfall anomalies were calculated for the "deficient" and "abundant" years for each region. For deficient years, rainfall anomaly varied from −3 to 0, while for abundant regions, rainfall anomaly varied from 0 to +3 mm day^{-1}, for over 95% of the pixels (Fig. 1a, b). These deficient and abundant years were found to exhibit largely positive anomalies in AOD (83% of pixels) and largely negative (69% of pixels), respectively (Fig. 1c, d). Along with this, prominent cloud properties coincident with aerosols showed marked difference between both the contrasting deficient and abundant monsoon years. The cloud drop radius, measured as CDER, had negative anomalies in all regions during the deficient monsoon years (Fig. 1e; 99% pixels), while CDER had positive anomalies during the abundant monsoon years (Fig. 1f; 84% pixels). It has been reported that in order to initiate drop formation, followed by drop growth and finally precipitation to set in, cloud droplet radius has to reach a critical radius. Once the drop radius reaches critical radius, autoconversion processes get initiated. These processes ultimately lead to growth of droplets and onset of the rainfall. These findings were found to be consistent as increased frequency of observations with smaller CDER in July month was made in deficient rainfall years (Ramachandran et al. 2013). Along with microphysical effects, large presence of absorbing aerosols, discussed previously, can potentially exert radiative effect by stabilizing the atmosphere near the surface (Koren et al. 2010) and can restrict vertical moisture transport. Analysis of cloud property with aerosol variation showed that deficient rainfall years were also associated with large negative anomalies of the IWP (91% of pixels) and positive anomalies (98%) in CTP (Fig. 1g, i). On the contrary, abundant monsoon years exhibited positive anomalies in IWP (93%) and negative anomalies in CTP (86%) (Fig. 1h, j). Along with this, it was observed that the mean cloud top extent during abundant monsoon years (with a CTP of 410–580 mb), temperatures reached below freezing point (257–318 K), while indeficient rainfall years (CTP of 460–650 mb), the temperatures were largely above freezing (312–317 K). Other studies (Rajeevan et al. 2006) have also shown similar results, i.e., larger cloud vertical extent and greater liquid and ice water contents during active monsoon spells.

In order to understand the temporal variability of aerosols with cloud properties, temporal correlation analysis at pixel level of daily mean values was carried out for the period of 2000–2009 (Fig. 2). For correlation analysis, absolute value of AOD and cloud properties (LWP, CDER, CTP and IWP) were used. $\alpha = 0.10$ was selected as threshold to test the statistical significance of correlation coefficient. Further for a given pixel, cumulative frequency of occurrence was identified. In this, a positive correlation coefficient was assigned as +1 and negative correlation coefficient was assigned as -1. Sum of these values was calculated for abundant and deficient years separately to get the cumulative frequency, larger positive value of which indicated presence of greater positive correlations and vice versa.

Between AOD and CDER, negative correlations occurred during deficient monsoon years, while predominantly positive correlations were observed during abundant monsoon years (Fig. 2a, b). In R2 and R3, this contrast was most prominent. Thus, greater aerosol abundance was found to be coincident with decreased (increased), CDER in deficient (abundant) monsoon years, at daily timescale. Thus, an opposing influence of aerosol abundance on CDER is evident; as in deficient monsoon years, enhanced aerosol levels lead to smaller cloud drop sizes while in abundant monsoon year lead to larger size. In all the three regions, the correlations between AOD and CDER were found to be statistically significant (Fig. 3a–c), based on the composited analysis.

Similar to AOD-CDER correlation analysis, pixel-wise analysis of AOD-IWP was also examined. For this purpose, pixels with CTT less than 0 °C were selected in all the regions (Fig. 2c, d). The behavior differed among these regions. In regions R1 and R2, negative correlations between AOD-IWP occurred with increased frequency during deficient monsoon years, and in abundant monsoon years, positive correlations were found to be prominent (Fig. 3d–f). In region R3, during deficient years, due to insufficient data point, relationship between AOD and IWP could not be identified; however, no significant correlations were found during the abundant monsoon years. Thus, aerosols were found to be related to a decreased ice processes in deficient and with increased ice processes in abundant monsoon years.

In AOD-LWP relationship, correlations were found to be positive in both abundant and deficient monsoon years (Fig. 2e, f), with increased frequency of positive correlations during the abundant monsoon years. This suggests that with increased

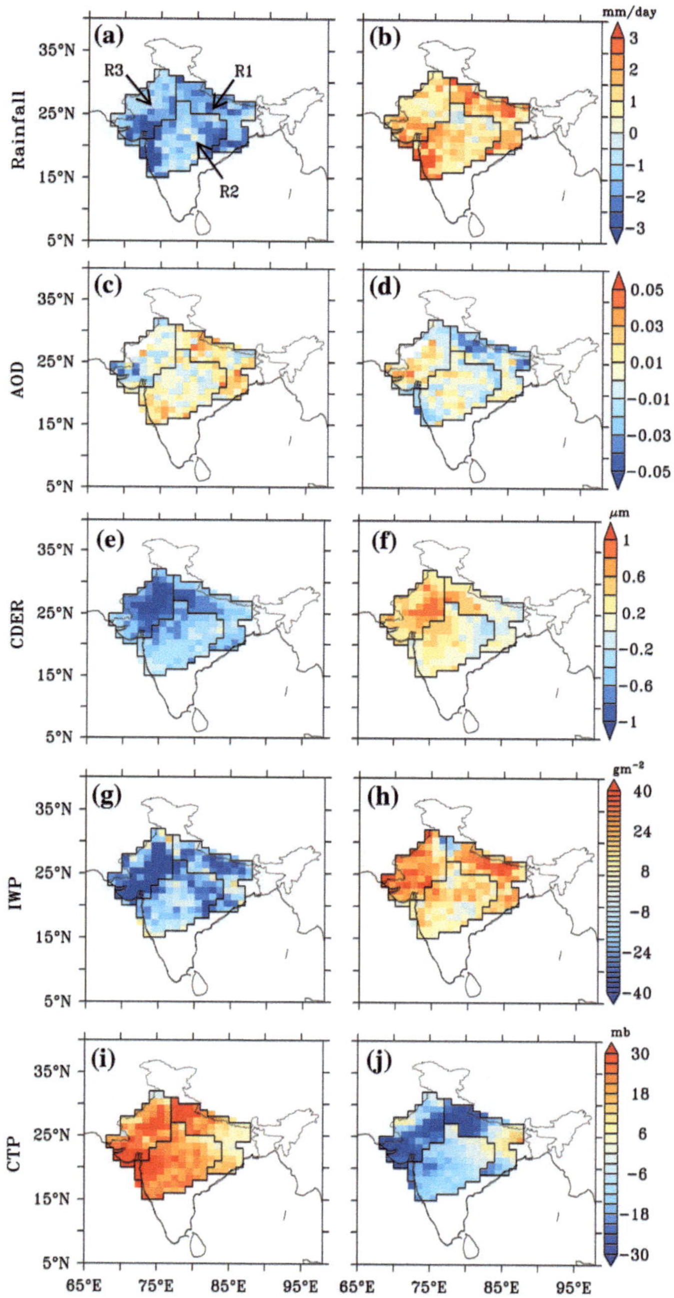

Fig. 1 Region-specific spatial distribution of anomalies in June–September (JJAS) during deficient and abundant monsoon years. **a**, **b** Rainfall (mm day^{-1}), **c**, **d** AOD, **e**, **f** CDER (μm), **g**, **h** IWP (g m^{-2}) and **i**, **j** CTP (mb) (Reproduced from Patil et al. 2017)

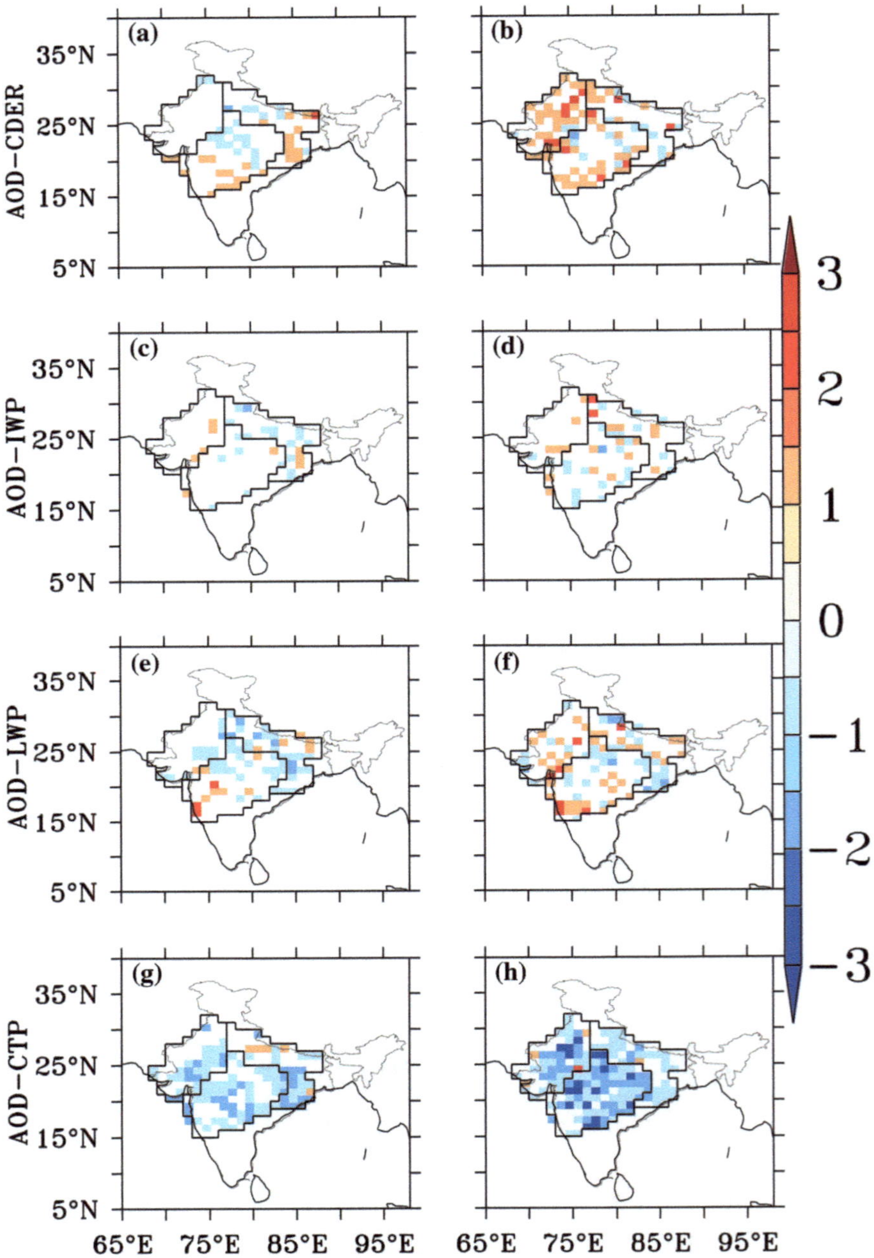

Fig. 2 Analysis of cumulative frequency of temporal correlation during deficient and abundant monsoon years. **a**, **b** AOD-CDER; **c**, **d** AOD-IWP, (CTT < 0 °C); **e**, **f** AOD-LWP; and **g**, **h** AOD-CTP (Significance was tested at $p < 0.1$) (Reproduced from Patil et al. 2017)

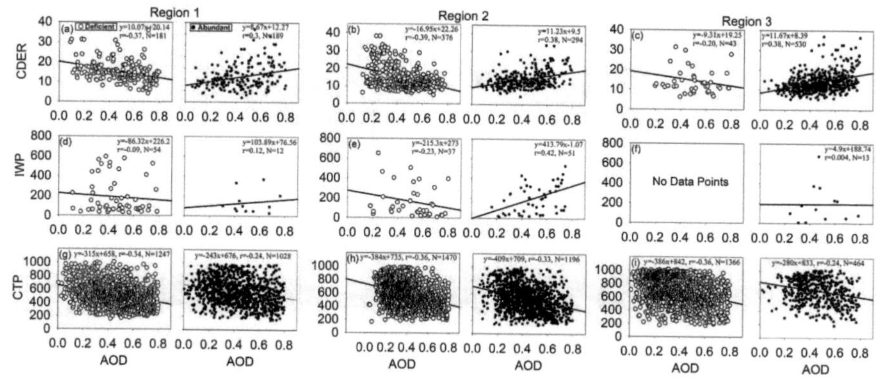

Fig. 3 Region-specific pixel-level scatter plot of correlations between AOD–cloud property during deficient and abundant monsoon years (data points with only statistically significant correlation are displayed). **a–c** AOD-CDER; **d–f** AOD-IWP; and **g–i** AOD-CTP. Reproduced from Patil et al. (2017)

aerosol abundance (due to increased availability of CCN, particularly of sea salt aerosols entrained in the strong westerly monsoon flows (Vinoj and Satheesh 2004)) were correlated with increased amounts of liquid water in the clouds.

The negative anomalies shown by AOD in the abundant monsoon years are consistent with the washout of aerosols by rain. With this diminished aerosol levels, correlations between AOD-CTP pair (Fig. 2g, h) were frequently negative in abundant years as compared to deficient years. This suggests that during abundant monsoon years, with increased aerosol abundance, lower CTP or higher cloud heights occur. However, the correlations between CTP-AOD (Fig. 3g–i) did not differ significantly between the deficient and abundant monsoon years, indicating that the magnitude of this effect does not change much between abundant and deficient years.

4 Short-Term Precipitation Suppression by Aerosols

To investigate the aerosol-induced causal effects on changes in rainfall anomaly, observational data were first subjected to hierarchical clustering to form high aerosol (AOD)-low precipitation (HL) and low aerosol (AOD)-low precipitation (LL) clusters for each of the 10 years. The effects of aerosols were examined and evaluated in the different clusters to understand if high levels of aerosol in the HL cluster had any bearing on low levels of rainfall. The other two clusters, with high rainfall, were not further evaluated. It was found that HL clusters spanned across most part of India for years 2004, 2005 and 2009 (Fig. 4a, b). A leading positive AOD anomaly was found to cause a negative anomaly of precipitation, where AOD anomaly led precipitation anomaly by 1–5 days, in the JJAS months in 2004, 2005, and 2009 (Fig. 4a, b) in the HL clusters. However, contrary to HL cluster, causal influence

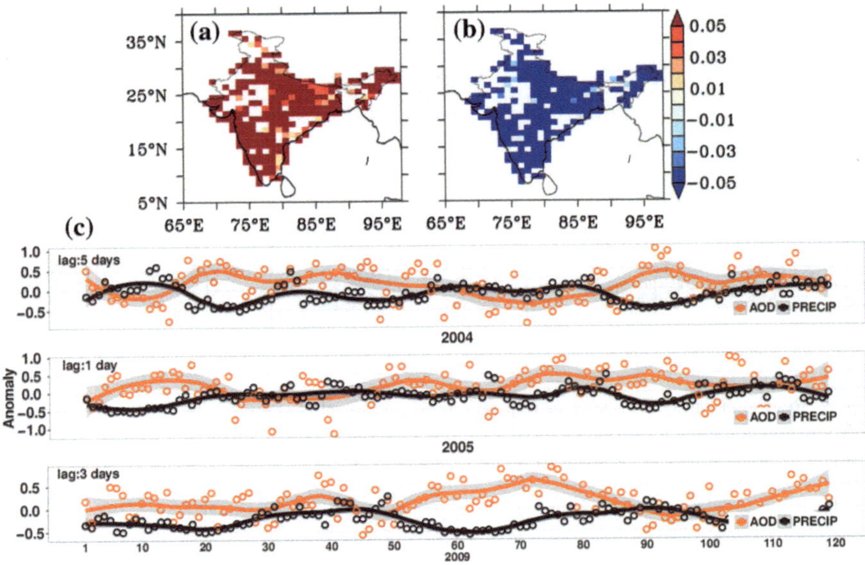

Fig. 4 Pixel distribution for AOD anomaly and precipitation anomaly across India in HL cluster (**a**, **b**); temporal series averaged for the cluster for years 2004, 2005 and 2009 (**c**). Reproduced from Dave et al. (2017)

was absent in the LL clusters (formulated with low level of aerosol). Estimated lag times between enhanced aerosols and suppressed precipitation varied among the years (2004:days 2–5; 2005:days 1–2; 2009:days 2–5), with the strongest causal influence, quantified as correlation coefficient, captured on separate days in different years (2004:day 5; 2005:day 1; 2009:day 3). The AOD and precipitation anomaly time series (Fig. 4c) showed multiple intra-seasonal periods of high AOD leading periods of low precipitation.

In addition to AOD, effects of absorbing aerosols (AAI) on precipitation were also examined in both HL and LL clusters. A leading positive AAI anomaly exerted a strong negative causal influence on precipitation anomaly, with varying lag times (2004 and 2005:2–5 days; 2009:1 day). The maximum causal influence was noted on separate days in different years (2004:day 5; 2005:day 5; 2009:day 1). As the clustering was done based on AOD, the LL clusters contained approximately 30–40% positive AAI anomaly values and thus included multiple days of high absorbing aerosol levels. The data showed causality between AAI and precipitation, with a lag of 1–5 days in 2005 and 2009. Overall, enhanced levels of both total and aerosols had a lagged causal effect on diminished precipitation, with a lag of up to 4–5 days occurring 3–5 times during monsoon months.

4.1 Development and Validation of Cause–Effect Model

To understand the mechanism of AOD–and/or AAI–modulated suppression of precipitation, two main pathways, i.e., cloud microphysics and radiative pathways, were studied. In microphysical pathway, AOD was linked with precipitation via the CDER. In the radiative pathway, AOD and AAI were linked to precipitation mediated through lapse rate (as a measure of atmospheric stability). Pairwise causality tests were performed between AOD–CDER, AOD–lapse rate, and AAI–lapse rate, using Granger causality. Further path analysis was used to segregate and quantify the aerosols effect on precipitation into these two mediating pathways. These pathways were compared with direct pathways that linked column water vapor (CWV) to both precipitation (PRECIP) and CDER, in order to compare the strength of the aerosol-induced suppression and the enhancing effects of CWV on CDER and PRECIP (Lebsock et al. 2011).

Causality was established between AOD enhancement, and both reduction in lapse rate for 2004, and reduction in CDER for 2005, with a lag of 5 days (panels 1 and 3, Fig. 4c), while effect of AOD on CDER was found to be absent in 2004 and 2009. A compelling finding was the shorter lag of 1 day, in the suppression of lapse rate by the enhancement of AAI (panel 2, Fig. 4c) was found. However absence of causal effect of AOD on the suppression suggests that the radiative effect on lapse rate change was primarily influenced by absorbing aerosols. Similarly, no causal influences were found for the LL clusters, which implies that such precipitation suppression was absent in regions of lower aerosol abundance.

4.1.1 Cloud Microphysical Pathway

In path analysis, using the cause–effect model along with lagged correlation coefficients, overall effect was segregated into individual pathways. The magnitude and sign of path coefficients indicated the strength and direction of the causal influence (Fig. 5a, b). A cloud microphysical pathway (defined as AOD–CDER–PRECIP) acting on the HL clusters (Fig. 5a) showed that enhanced AOD caused reductions (blue color) in CDER (arrow direction). A positive causal influence of CDER (red color) on PRECIP (arrow direction), found in 2004, 2005 and 2009, indicated the influence of a larger CDER on precipitation. A statistically significant positive causal influence of CWV was found on both PRECIP and on CDER, as expected. The strength of direct positive pathway between CWV–PRECIP was significantly larger than the negative AOD–CDER–PRECIP pathway, as found from the magnitude of the overall path effects. In contrast, the cloud microphysical pathway showed no such causal influence of AOD on CDER for the LL clusters (can be seen by missing causal lines from AOD to CDER in Fig. 5b). The positive causal effects of CWV on PRECIP on the LL clusters were retained, both directly (CWV–PRECIP, Fig. 5b) and indirectly through CDER (CWV–CDER–PRECIP); this indicated that aerosol mediation of

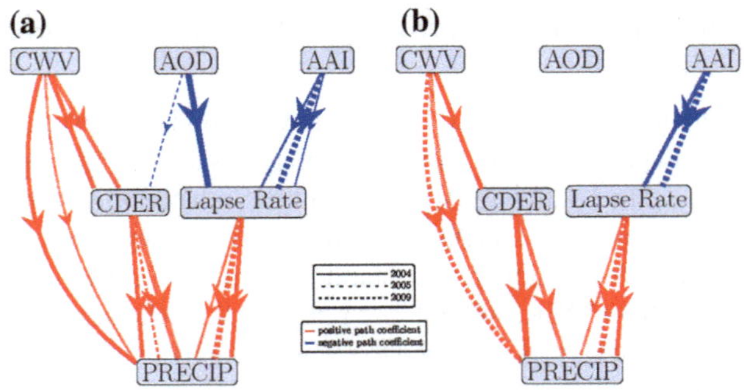

Fig. 5 Cause–effect model for years 2004, 2005 and 2009: **a** HL, **b** LL (here, absolute magnitude is represented with line-width, width is proportional to absolute magnitude of coefficient). Causal influence is represented by arrow direction, sign of causal influence in indicated by color (red representing positive and blue representing negative sign); line styles correspond to different years. Reproduced from Dave et al. (2017)

precipitation was not significant in regions with low aerosol levels. Thus, the cloud microphysics pathway acts to reduce CDER and precipitation, with a lag time.

Cloud microphysical processes that influence precipitation suppression due to increase in aerosol at cloud levels lead to moisture redistribution, resulting into increased number of smaller drops. Increased number of smaller drops cause coalescence efficiency to decrease causing slow down of the conversion process of cloud drops to graupel or raindrops. Due to vapor condensation and collision/coalescence processes, raindrop formation is initiated on timescale of minutes. In the next 15–20 min, intensification of scavenging of small cloud drops by larger drops due to gravitational settling (termed autoconversion) can lead to onset of precipitation. Such fast scale microphysical effects, in pristine marine clouds, have been previously linked to rainfall shut-off along ship tracks. The two pathways, i.e., microphysical and radiative pathways, are mostly independent and occur at different values of AOD. The time response of the microphysical processes is much shorter as compared to radiative processes. Hence, the presence of 1–5 day lag times between AOD enhancement and suppression of CDER and precipitation suppression indicates that these observed causal relationships might not be an outcome of a microphysical effect directly and, accordingly the involvement of a radiative pathway was investigated.

4.1.2 Radiative Pathway

Cloud radiative pathways (defined as AOD–lapse rate–PRECIP and AAI–lapse rate–PRECIP) acting on the HL clusters (Fig. 5a) showed that enhanced AOD caused reductions (blue color) in the lapse rate (arrow direction). Typically, a smaller magnitude of lapse rate, calculated as the slope of potential temperature, indicates higher atmospheric stability and vice versa. Enhanced AOD and AAI both caused reductions (blue color) in lapse rate (arrow direction), or a stabilization effect, at lag times of

1–3 days, as discussed above. The causal influence of AAI on lapse rate was strong in all years and can be seen by larger magnitudes of negative path coefficients. A statistically significant causal influence between AOD and lapse rate was found in 2004 (Fig. 5a). Aerosols, represented by AAI, were expected to have a stronger atmospheric stabilization effect than total aerosols, represented by AOD. A positive causal influence of lapse rate (red color) on PRECIP (arrow direction), found in 2004, 2005, and 2009, indicated the influence of atmospheric stabilization (lower lapse rate) on decreases in PRECIP. A positive causal influence of CWV on PRECIP was comparable to the negative AAI–lapse rate–PRECIP pathway, with the strength of the AAI–lapse rate–PRECIP pathway being of 15–100% of that of the CWV—PRECIP pathway. Larger negative values of the path effects between AAI and lapse rate than between AOD and CDER indicated a stronger influence of the radiative pathway as compared to cloud microphysical pathway on short-term suppression of monsoon precipitation. The radiative pathway also showed significant effects in the LL clusters as well; path coefficients were negative in 2005 and 2009 and statistically significant (as the clusters were based on AOD), but they included several regions with positive AAI anomalies. By inclusion of all the pathways, i.e., radiative, cloud microphysical, and CWV pathways, approximately half of the precipitation variability (PRECIP $R^2 > 0.50$) could be explained. When the radiative pathway was excluded from the model, a significant drop in PRECIP R^2 for the HL and LL clusters was observed. This highlights the strong effects of aerosol-induced atmospheric stabilization on suppression of precipitation and establishes radiative pathway as significant mechanism by which aerosol causes repeated short-term suppression of SAM precipitation.

Short-term radiative effects, such as those seen here, after an increase in aerosol levels, can manifest within a day and can last for two or more days. Local atmospheric heating along with surface cooling has been linked with perturbation of incoming solar radiation by aerosols. This has been also linked with reduction in atmospheric lapse rate and a consequent increase in atmospheric stability. The reported phenomena are consistent with the radiative pathway seen in the current study. Further, the effect is linked with diminished moisture and heat fluxes from the surface, reduced convection, and suppressed vertical mixing of moisture.

The overall mechanism behind aerosol-induced precipitation suppression can be understood under the presence of high and low aerosol fields. With increased aerosol loading, the atmosphere was more stable, causing an increased divergence of moisture, along with reduced convection leading to decreased uplift of moisture and subsequent suppression of precipitation. The two effects, i.e., reduced moisture availability and inhibited convection, conflate and add to precipitation suppression.

4.2 Implications On Monsoon Break Spells

Break spells are one of the main characteristics of the Indian monsoon. Duration and intensity of break spells play a significant role in as certaining deficient rainfall and

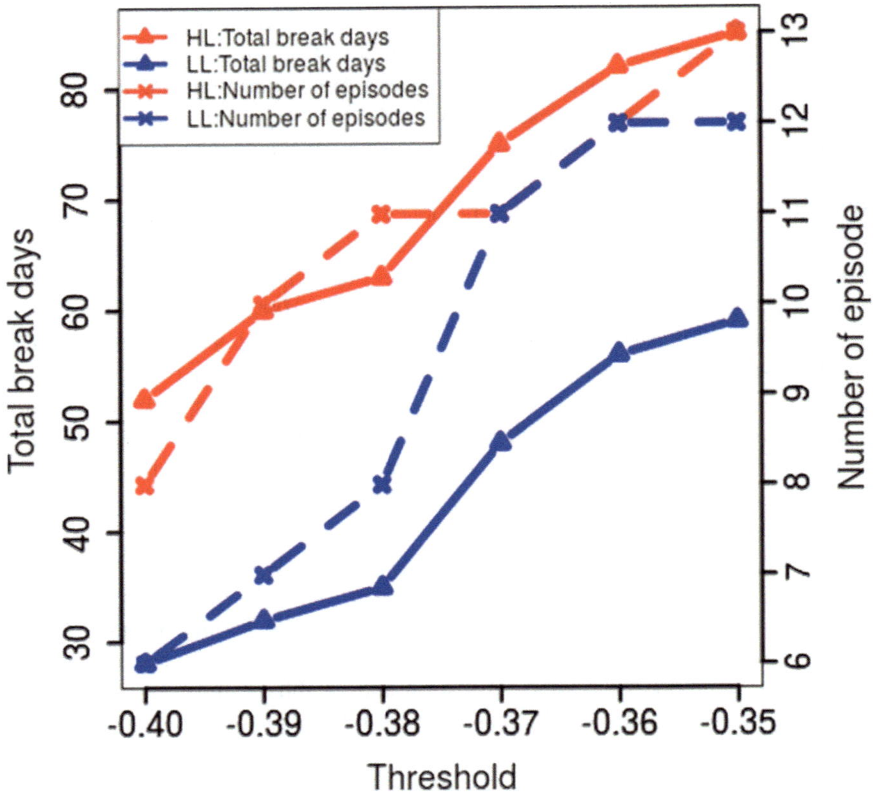

Fig. 6 In HL and LL clusters, comparison of duration and frequency of break spells with varying AOD threshold. Reproduced from Dave et al. (2017)

drought conditions (Gadgil and Joseph 2003; Prasanna 2014). Several definitions have been used to define a break spells, out of those one definition is based on a normalized anomaly threshold of one standard deviation below the mean, sustaining for at least three consecutive days has been widely used.

The potential influence of aerosols on break spells was investigated by comparing characteristics of breaks in HL regions with that of the LL regions (Fig. 6). The precipitation anomaly chosen has a threshold one negative standard deviation and more stringent thresholds. An increased duration along with increased frequency of break spells was found in HL clusters, at all the threshold values examined. Further, it has been accepted that extended breaks (>7 days) usually result in droughts (Gadgil and Joseph 2003; Prasanna 2014). Break spells lasting for at least seven days or more in HL regions occurred more frequently (i.e., 3–4 times in a given season, averaged across thresholds) as compared to LL regions (1–2 times in a given season). During the last 50 years, studies examining monsoon variability over India have also found an increased duration and frequency of break spells, while the studies undertaken here suggest increases in aerosol levels, as one among several factors influencing these changes.

5 Strategic Knowledge, Future Perspective, and Policy Implications

In this chapter, we discuss influence of pollution particles, along with dust, namely atmospheric aerosols, on recent changes in SAM rainfall. Using a 10-year dataset, from 2000 to 2009, of satellite- and ground-based observations of aerosol abundance and cloud properties, we confirm that local physical processes, mediated by aerosols, have exerted important effects on cloud modifications and short-term rainfall suppression over India in recent decades. Over the core monsoon region of India, opposing relations between aerosol levels and cloud properties were found in years of contrasting monsoon rainfall. Overall, during deficient monsoon years, increased aerosol levels correlated with decreased cloud drop size, lower development of ice processes and reduced cloud heights, indicating aerosol-mediated inhibition of cloud development. In contrast, during abundant rainfall years, increased aerosol levels correlated with increased cloud drop size, tall clouds and increased ice processes implying cloud invigoration mediated by the coincidence of aerosols and enhanced convective fields.

In high aerosol-low rainfall regions covering over the whole India, increased aerosols levels acted to suppress daily precipitation anomaly, multiple times in a given season, where aerosol enhancement leads precipitation suppression by a few days. In addition, a higher frequency of prolonged rainfall breaks (>7 days) occurred in these regions. The suppression mechanism was acted primarily through an aerosol radiative pathway mediated by atmospheric stability, reduced convection, and increased divergence of moisture horizontally. Thus, a fast response caused by coincident aerosol forcing, through repeated stabilization of atmosphere from aerosols, resulted in short-term monsoon precipitation suppression and aggravation of monsoon break conditions. Interestingly, causal influences of aerosols on suppression of precipitation were not limited to years characterized by a large-scale weak monsoon, suggesting the likelihood of a widespread presence of identified phenomenon.

The causal influence of aerosol-induced suppression of precipitation is relevant to the monsoonal inter-annual variability and monsoon break spells. Elongated and intense breaks spells in the monsoon season have been associated with precipitation deficits, which have been linked to diminished food grain production during 1951–2003. Thus, aerosol-induced precipitation suppression and intensification of break spells could affect precipitation deficits and agricultural vulnerability in India, with implications for policies related to water resources management.

The work described here emphasizes the important influences of aerosols in the Indian region, on alteration of cloud properties and suppression of monsoon rainfall through a set of mechanisms, acting on mesoscale spatial extents and on timescales of days to months. Observations from 1950 onwards suggest an increase in frequency of occurrence of high-intensity rainfall, but a decrease in frequency of drizzle and moderate rainfall. The studies also suggest aerosol effects on short-term rainfall suppression, increasing the length of monsoon breaks. Thus, the work makes important linkages between enhanced air pollution and climate change, in regard to short-term suppression of monsoon rainfall. It is important to understand how such disruptions would evolve, with expected future increases in aerosol emissions driven by human activities. Air pollution-induced rainfall changes, revealed in this work, indicate that air pollution and climate regulation need to be better linked. Since anthropogenic aerosols arise from human activities related largely to energy-use and land-use changes, which are drivers of both regional air quality and climate change, these studies point to the need to seek synergy in strategies which can simultaneously mitigate both air quality and regional climate change.

References

Bollasina M, Ming Y, Ramaswamy V (2011) Anthropogenic aerosols and the weakening of the South Asian summer monsoon. Science (80-)334:502–505. https://doi.org/10.1126/science.1204994

Boucher O, Randall D, Artaxo P, Bretherton C, Feingold G, Forster P, Kerminen V.-M. V.-M, Kondo Y, Liao H, Lohmann U, Rasch P, Satheesh S. K, Sherwood S, Stevens B, Zhang X. Y, and Zhan, X. Y (2013) "Clouds and Aerosols," Climate Change 2013: The Physical Science Basis. Contribution of Working Group I to the Fifth Assessment Report of the Intergovernmental Panel on Climate Change, 571–657.

Dave P, Bhushan M, Venkataraman C (2017) Aerosols cause intraseasonal short-term suppression of Indian monsoon rainfall. Sci Rep 7:17347

Dee DP, Uppala SM, Simmons AJ et al (2011) The ERA-Interim reanalysis: configuration and performance of the data assimilation system. Q J R Meteorol Soc 137:553–597. https://doi.org/10.1002/qj.828

Gadgil S, Gadgil S (2006) The Indian monsoon GDP and agriculture. Econ Polit Wkly 4887–4895

Gadgil S, Joseph PV (2003) On breaks of the Indian monsoon. Proc Indian Acad Sci Earth Planet Sci 112:529–558. https://doi.org/10.1007/BF02709778

Ganguly D, Rasch PJ, Wang H, Yoon JH (2012) Fast and slow responses of the South Asian monsoon system to anthropogenic aerosols. Geophys Res Lett 39:1–5. https://doi.org/10.1029/2012GL053043

Granger CWJ (1969) Investigating causal relations by econometric models and cross-spectral methods. Econometrica 37:424–438. https://doi.org/10.2307/1912791

Herman JR, Bhartia PK, Torres O et al (1997) Global distribution of UV-absorbing aerosols from Nimbus 7/TOMS data. J Geophys Res 102:16911. https://doi.org/10.1029/96JD03680

Koren I, Remer LA, Altaratz O et al (2010) Physics aerosol-induced changes of convective cloud anvils produce strong climate warming, pp 5001–5010. https://doi.org/10.5194/acp-10-5001-2010

Krishnan R, Sabin TP, Vellore R et al (2016) Deciphering the desiccation trend of the South Asian monsoon hydroclimate in a warming world. Clim Dyn 47:1007–1027. https://doi.org/10.1007/s00382-015-2886-5

Kwiatkowski D, Phillips PCB, Schmidt P, Shin Y (1992) Testing the null hypothesis of stationarity against the alternative of a unit root. J Econom 54:159–178. https://doi.org/10.1016/0304-4076(92)90104-Y

Lau KM, Kim KM (2006) Observational relationships between aerosol and Asian monsoon rainfall, and circulation. Geophys Res Lett 33:1–5. https://doi.org/10.1029/2006GL027546

Lau KM, Kim MK, Kim KM (2006) Asian summer monsoon anomalies induced by aerosol direct forcing: the role of the Tibetan Plateau. Clim Dyn 855–864. https://doi.org/10.1007/s00382-006-0114-z

Lebsock MD, L'Ecuyer TS, Stephens GL (2011) Detecting the ratio of rain and cloud water in low-latitude shallow marine clouds. J Appl Meteorol Climatol 50:419–432. https://doi.org/10.1175/2010JAMC2494.1

McPeters R, Bhartia PK, Krueger A et al (1998) Earth probe total ozone mapping spectrometer (TOMS) data product user's guide

Meehl GA, Arblaster JM, Collins WD (2008) Effects of black carbon aerosols on the Indian monsoon. J Clim 21:2869–2882. https://doi.org/10.1175/2007JCLI1777.1

Patil N, Dave P, Venkataraman C (2017) Contrasting influences of aerosols on cloud properties during deficient and abundant monsoon years. Sci Rep 7:44996

Prasanna V (2014) Impact of monsoon rainfall on the total food grain yield over India. J Earth Syst Sci 123:1129–1145. https://doi.org/10.1007/s12040-014-0444-x

Rajeevan M, Bhate J, Kale JD, Lal B (2006) Development of a high resolution daily gridded rainfall data set for the Indian region. Curr Sci 91:296–306. https://doi.org/10.1007/s12040-007-0019-1

Ramachandran S, Ghosh S, Verma A, Panigrahi PK (2013) Multiscale periodicities in aerosol optical depth over India. Environ Res Lett 8:14034

Ramanathan V, Chung C, Kim D et al (2005) Atmospheric brown clouds: impacts on South Asian climate and hydrological cycle. Proc Natl Acad Sci U S A 102:5326–5333. https://doi.org/10.1073/pnas.0500656102

Remer LA, Kaufman YJ, Tanré D et al (2005) The MODIS aerosol algorithm, products, and validation. J Atmos Sci 62:947–973. https://doi.org/10.1175/JAS3385.1

Rosenfeld D, Lohmann U, Raga GB et al (2008) Flood or drought: how do aerosols affect precipitation? Science 321:1309–1313. https://doi.org/10.1126/science.1160606

Schiermeier Q (2010) The real holes in climate science. Nat News 463:284–287

Torres O, Tanskanen A, Veihelmann B et al (2007) Aerosols and surface UV products form ozone monitoring instrument observations: an overview. J Geophys Res Atmos 112:1–14. https://doi.org/10.1029/2007JD008809

Vinoj V, Rasch PJ, Wang H et al (2014) Short-term modulation of Indian summer monsoon rainfall by West Asian dust. Nat Geosci 7:308–313. https://doi.org/10.1038/NGEO2107

Vinoj V, Satheesh SK (2004) Direct and indirect radiative effects of sea-salt aerosols over Arabian Sea. Curr Sci 86:1381–1390

Webster PJ, Magaña VO, Palmer TN et al (1998) Monsoons: Processes, predictability, and the prospects for prediction. J Geophys Res Ocean 103:14451–14510. https://doi.org/10.1029/97JC02719

Future Hydrologic Scenarios in India Under Climate Change

Tarul Sharma, Aditya Gusain and Subhankar Karmakar

Abstract It is evident that global warming and overexploitation of natural resources by the human interventions such as urbanization and irrigation have accelerated changes in the regional hydrological cycle. These changes result in more frequent and severe occurrence of hydro-climatic extremes such as floods and droughts, with devastating impacts on livelihood of the region. In order to understand the future hydrological response at a regional or watershed scale, it is important to understand the changes in several hydro-meteorological variables such as run-off, stream flow and precipitation under various climate change scenarios. However, quantification of the changes in hydro-meteorological variables and their associated uncertainties through an understanding of the global- and regional-scale factors under extreme climatic conditions makes it a rather challenging task. The key focus of this chapter is to understand the changes in regional hydrological response of different meteorological variables under the effect of changing climate and quantify the uncertainties in climatic projections under data-scarce conditions, over the Indian subcontinent. The chapter also reviews the impact of climate change on hydrological cycle and its effect on water availability and security, community livelihood under hydro-climatic extremes and agriculture.

Keywords Agriculture · Climate change · Hydrology · Uncertainty Vulnerability

---

T. Sharma · S. Karmakar (✉)
Interdisciplinary Programme in Climate Studies, Indian Institute of Technology Bombay, Mumbai 400076, India
e-mail: skarmakar@iitb.ac.in

T. Sharma
e-mail: tarul.0110@gmail.com

A. Gusain · S. Karmakar
Centre for Environmental Science and Engineering, Indian Institute of Technology Bombay, Mumbai 400076, India
e-mail: aditya140492@gmail.com

S. Karmakar
Centre for Urban Science and Engineering, Indian Institute of Technology Bombay, Mumbai 400076, India

C. Venkataraman et al. (eds.), *Climate Change Signals and Response*,
https://doi.org/10.1007/978-981-13-0280-0_3

1 Background: Hydrological Cycle and Climate Change

The interplay of optimum conditions for gases, temperature and water are responsible for the existence of life on earth. Among all factors, water forms a crucial component in supporting life in all the life forms, from simple single-celled organisms to complex organisms as human. Although our earth is covered with 70% of water, roughly only 3% is available as freshwater where most of it is trapped under ice sheets and glaciers. Therefore, a very limited freshwater supply is available for our use in the form of rivers, lakes, groundwater storage, etc., and its judicious use is vital for sustenance of life on earth. However, the overexploitation of water resources for a multitude of purposes has increased the demand for clean water supply, especially for the densely populated South Asian region. This has already feud water-related disputes as a consequence of environmental degradation of water quality, which is further worsened by the changing climate (Lal 2005). Eventually, the questions that arise in the minds of the scientific community are as follows: (1) How are the changes in climate affecting the hydrological cycle? (2) Can the extreme events be attributed to anthropogenic climate change or are they just the mere consequences of natural climate variability? (3) What adaptive measures need to be considered to deal with water stress conditions caused by climate change? Several similar questions are yet to be answered in a scientifically satisfactory manner. In the past few decades, declining water quality and scarcity of water have drawn the attention of water managers, regional planners and policymakers at different levels of government to devise strategic management practices.

To understand the complex mechanisms that are influencing the hydrological cycle, it is important to get familiar with terms such as "climate system", "hydrological cycle" and "climate change". A climate system can be described as a complex system where abiotic (land, water and atmosphere) and biotic (plants, animals and micro-organisms) factors interact through different physical, chemical and/or geological phenomena. This climate system is subjected to several natural and anthropogenic factors that influence the changes in the climate. Sun being the prime source of energy on earth, it drives the climate and hydrological system through solar radiations. In a hydrological cycle, the surface of water body gets warmed up by the incoming solar radiations causing evaporation of water. Warm air being lighter is pushed upwards, and as it is being lifted up, the moist air cools down causing the condensation of water vapour to form clouds. The moisture-carrying clouds are carried across the globe through atmospheric circulation and precipitate water back to the surface. The precipitated water either evaporates back to the atmosphere or else joins the water bodies as run-off or base flow which seeps into rivers, lakes and ultimately to the oceans (Fig. 1).

However, the changing climate and increased anthropogenic activities have altered the hydrological cycle, thereby changing the rainfall and temperature patterns, melting of glaciers, sea level rise, increased extreme precipitation events and changes in run-off or stream flow across the globe (Stocker et al. 2013). The indirect effects of climate change are the matter of major concern in hydrological cycle. Higher the

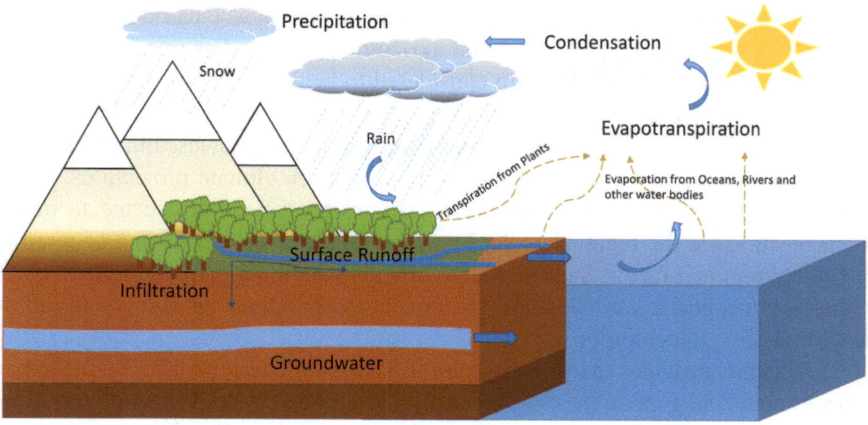

Fig. 1 Pictorial representation of hydrological cycle

temperature above the regional mean, higher is the rate of evaporation, thus increasing the rate of accumulation of water vapour in the atmosphere. Since water vapour acts as an aerosol having absorbing effect, it traps the incoming/outgoing solar radiation by further warming the atmosphere. This warm moisture-filled atmosphere results in the intensification of the precipitation event, which is distributed unevenly throughout the globe. These changes have profound impacts on surface water run-off or stream flow, which results in increased incidences of floods and droughts (Gupta 2012).

Unrestricted increase in the greenhouse gas (GHG) concentrations is considered to be the main factor responsible for climate change (Mitchell et al. 1995), in addition to the other factors such as land use/land cover change including urbanization and change in agricultural practices (Pielke 2005). Seneviratne et al. (2006) found that increase in GHG concentration can lead to inter-annual variability of the climate of any region, with more frequent extreme events resulting from land–atmosphere interactions. Post-1850, the succeeding decades have been warmer than any of the preceding decades, with ocean warming dominating in increasing the energy stored in climate system (Trenberth 1999). Loarie et al. (2009) cautioned that as the climate changes, the current distribution of climatic conditions will get reshuffled on the globe, with some climates disappearing entirely and new (no-analogue) climates originating in wide regions. Apart from these signals, it has been proved to be affecting the hydrological cycle by spatial and temporal variations in the water availability (Huntington 2006) and hence by directly influencing the dependent agricultural resources of the region.

For assessing the impacts at a computational level, World Meteorological Organization (WMO) defines climate change as the change in statistical description of mean and variability of relevant quantities of surface weather parameters such as temperature and precipitation, over a period of time (generally 30 years). The impact assessment is usually done by understanding the effect of changes in statistical properties of climate on the land–atmosphere system, using physics-based mathematical models such as hydrological, flood and agricultural model. For this purpose, general circulation models (GCMs) are considered to be the base which provides with a time series of multiple spatially distributed land–atmospheric variables (Stocker et al. 2013). The GCM variables are then given as input to downscaling techniques like statistical and dynamical, to obtain fine-resolution climate projections. Given the importance of the study, multiple researches have been conducted to derive the projections at the regional scale, by considering the effects of one or more of the global/regional factors such as stationarity, non-stationarity, land use/land cover changes, atmospheric variables, sub-grid spatial and temporal variability and carbon dioxide (CO_2) concentration.

This chapter provides an insight into the projected climate conditions for the Indian subcontinent under possible climate change scenarios by reviewing some of the breakthrough studies. Essentially, the focus has been made on the climate projections and possible uncertainties that arise with the use of multiple GCMs. The uncertainty gets transmitted into impact assessments; thus, the need is to incorporate and optimize climate uncertainty before addressing the needs of the end-users such as farmers. The organization of this chapter is as follows: the first section provides a general introduction to the concept of climate change and its implication on the hydrological cycle that ultimately could bring drastic effects to the availability of water in future. In Sect. 2, the future projections for different hydro-climatic variables using well-established downscaling approaches have been described. Also, the uncertainties associated with projections from different GCMs and downscaling approaches are considered. In Sect. 3, a brief survey about historical and possible future impacts of climate change on hydrological system, human health, socio-economic aspects, livelihood and agriculture has been described. Section 4 presents some concluding remarks and recommendations for future research.

2 Projected Regional Changes in Hydrological Cycle Over the Indian Subcontinent

Appreciable research efforts have been made in the past, which specifically aimed at understanding of climate change impacts globally. However, at regional scale such as for the Indian subcontinent, the land–atmosphere interactions are only partially understood. Factors accountable are sub-regional land–atmosphere and ocean–land interactions. It has been widely believed that one of the major consequences of changing climate will be the alterations in regional hydrological cycles (Gleick 1989). GCMs are considered to be reliable outcomes of the climate system simulations that can significantly demonstrate the nature and extent of the climate changes. The outcomes have been widely used to assess the change impacts. GCMs lack in capturing the non-smooth variables such as precipitation (Kannan and Ghosh 2011), which is the driving component of the hydrological cycle. India being an agrarian country, its agricultural output is highly dependent on surface and groundwater storages, which in turn depend on the seasonality and variability of the Indian monsoon. Thus, this makes the Indian agrarian sector extremely vulnerable to the time- and space-dependent monsoonal behaviour. Spatially downscaling the GCMs through statistical or dynamical methods enhances its skill by resolving the sub-regional features such as convention and topography (Pielke and Wilby 2012; Shashikanth et al. 2014). The following subsections depict the climatic and hydrological projections over the nation and the uncertainties associated with it, which can arise from different sources such as inter-model spread among GCMs, downscaling methods, impact assessment models or due to the climate system itself.

2.1 Climatic and Hydrological Projections

GCMs available under the framework of coupled model inter-comparison project (CMIP) phase 5 are the most credible tool for assessing the changes in climatological variables due to changing global circulation patterns at coarse resolution, but limit direct applicability when associated with local-scale rainfall projections. Reason can be due to the poor understanding of the physics behind local climatology, especially the precipitation extremes (Ghosh and Mujumdar 2006). The limitation can be overcome by bridging the gap between the course resolution of GCMs and the resolution of impact assessment, through means of the downscaling techniques. Broadly, statistical and dynamical are the two classes of downscaling techniques. Dynamical downscaling usually considers the methodology of nesting a regional climate model within a GCM boundary condition (Leung and Wigmosta 1999). Statistical downscaling, on the other hand, is a transfer function-based technique that relates local climate to GCM output (Wood et al. 2004).

Numerous efforts have been made in the past, to understand the climate change impacts on variables such as precipitation and temperature and consequently, on hydrology. The assessment study is important specifically for India, as any changes in spatio-temporal variability of hydro-climatic factors affect the agricultural production. The observed meteorological data sets have revealed a decreasing trend in rainfall, whereas the temperature trend has increased over India (Kothyari and Singh 1996). Dash et al. (2007) used the same observed data sets to show that the temperature enhancement is double for the months of winter (January and February) and post-monsoon (October to December). This inter-seasonal increase in the variability of climate has a huge impact on the Indian economy, which is majorly driven by agriculture.

For the future projections, several studies have demonstrated the use of different downscaling techniques to simulate the climate variables over India. For example, Anandhi et al. (2009) showed an increase in the projected rainfall, by using support vector machine for statistical downscaling. Salvi and Ghosh (2013) established a robust statistical downscaling technique, which establishes a statistical relationship between coarse resolution climate variables and observed fine-resolution rainfall data and then applies the relationship on the GCM variables. The methodology projected spatially non-uniform changes in rainfall, which demanded a detailed hydrological impact assessment study for strategic management. This methodology was adapted from Kannan and Ghosh (2011) which applied kernel regression technique on GCM data, to statistically downscale rainfall over Mahanadi River Basin of India. In the study, a significant increase in the number of dry and wet rainfall events as compared to medium rainfall events was attributed. In the case of dynamical downscaling, multiple regional climate models have been used to understand the physics of Indian climatic conditions. Ashfaq et al. (2009) used the dynamical approach to downscale precipitation and observed that there will be an overall suppression of summer monsoon along with the increase in the occurrence of monsoon breaks. Furthermore, Ashfaq et al. (2011) showed that the monsoonal precipitation is significantly influenced by the pre-monsoonal soil moisture content of the region.

Apart from the mean precipitation, it has been cited globally about the increase in the occurrence of hydro-climatic extremes because of the direct implication on global atmospheric circulation pattern by the global temperature rise (Stocker et al. 2013). Several studies have demonstrated the use of different downscaling approaches to simulate mean rainfall conditions (Bardossy and Plate 1991; Kannan and Ghosh 2011; Wilby and Dawson 2013; Sachindra and Perera 2016; and references therein) though less efforts have been made on simulating extremes (Abraham et al. 2011; Fowler et al. 2005; Friederichs and Hense 2007). To illustrate a few, Revadekar et al. (2011) evaluated an increase in precipitation in future projections of precipitation extremes over India, by using a dynamical downscaling approach. Similarly, Mohan and Rajeevan (2017) used a statistical approach to understand the precipitation pattern over the core monsoon region of India and suggested a significant increase in the precipitation intensity and hydro-climatic intensity. Box 1 depicts a case study of comparison of the conventional statistical downscaling approach with the modified approach where the concept of extreme precipitation has been incorporated by the inclusion of specific indicators that capture the extremity of the precipitation.

With the projected increase in climatic variables such as precipitation and temperature, its direct impact will be on the hydrological cycle of the region. Global land evapotranspiration has been observed to have decreased due to limited moisture supply (Jung et al. 2010). Along with that, the stream flow has shown an increasing trend globally. Prominently, hydrological cycle of any region directs socio-economic whereabouts, by directly affecting the agriculture, land use and land cover and other geophysical characteristics of the region. Mishra and Lilhare (2016) have warned about the intensification of the hydrological cycle of the Indian basins under the impact of future climate change, with an increase in run-off and evapotranspiration. Moreover, Chattopadhyay and Hulme (1997) have shown increased projection of evaporation and potential evapotranspiration, with uneven spread between regions and seasons. The Himalayan rivers are expected to have harder impact on climate change with increase in the snowmelt and stream flow, as compared to glacier melt (Singh and Kumar 1997; Singh and Bengtsson 2004). Albeit it should be noted that with increase in the projected water availability, it is the spatio-temporal variability/availability of precipitable water that will be more difficult to understand.

Box 1 Improvement in Extreme Rainfall Projections Using a Modified Statistical Downscaling Approach for Mahanadi River Basin in India

This section of the chapter is based on a smaller case study area. In the past several decades, the Mahanadi River Basin has experienced an increasing number of extreme precipitation events that have resulted in the increased incidences of floods. The highly dense river network and the heavy influence of the Bay of Bengal on the meteorology and climatology of the basin make it more vulnerable to several extreme events that are associated with enormous socio-economic or ecological losses. Therefore, in the context of trend and behaviour of hydro-climatic variables like rainfall, temperature and stream flow, many researchers have undertaken studies on Mahanadi River Basin (Rao 1993; Gosain et al. 2006; Asokan and Dutta 2008; Raje and Mujumdar 2010; Jena et al. 2014).

In this study, a comparison of conventional and modified statistical downscaling has been performed. In terms of accuracy, the conventional statistical downscaling approach captures the spatial variability of rainfall over the basin well but fails to simulate extremes for each grid, in terms of 30-year return level (RL) values, reliably. In terms of absolute percentage error, the deviation of 30-year RL extreme precipitation values for National Centres for Environmental Prediction (NCEP)-simulated precipitation from observed precipitation was obtained as high as 80–100%. Thus, it necessitates the usage of improved version of statistical downscaling approaches that can easily emulate the behaviour of atmospheric processes that describes the mechanism for heavy rainfall during extreme events. As shown in Fig. 2, a remarkable reduction in error was observed while projecting precipitation with conventional and modified downscaling approach.

Further, to address the issues of inefficacy of conventional statistical downscaling approach to simulate extreme precipitation days, a modified algorithm developed by Shashikanth et al. (2017) was demonstrated over the study area for historical (1951–2005) and near future (2021–2055) period under RCP 4.5 and RCP 8.5 climate change scenarios using 6 GCMs. Also, a small experiment was performed to examine the role of dynamic factors in improving simulation of extreme events in terms of return period values. The study concluded that use of simple downscaling might be useful in simulating expected conditions over a period, though for analysis of extremes, the methodology fails to capture the effect of extreme values.

Hence, implementing a modified statistical downscaling algorithm will surely improve the results and the fine-resolution products which may serve as better inputs to drive a mesoscale hydrological model to assess the impacts of climate on hydrology and riverine ecosystem.

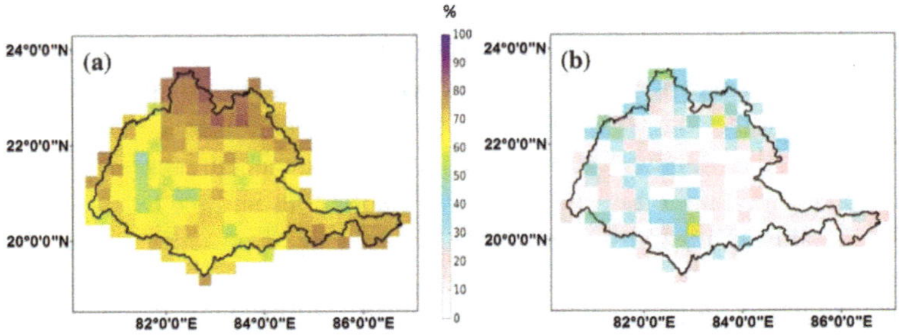

Fig. 2 Absolute percentage error of the simulated extreme rainfall with respect to observed data corresponding to 30-year RL estimated by peak over threshold approach in the Mahanadi Basin at 0.25° resolution by using **a** conventional statistical and **b** modified statistical downscaling

2.2 Sources of Uncertainty in Climate Projections

Hydro-climatic projections play an important role in strategizing the adaptive measures that can be implemented to cope with the changing climate. Although the climate projections can be derived from multiple sources, such as GCM and downscaling techniques, there lies a disagreement herein, which can be due to the incomplete understanding of the geophysical processes of the climate system. Given the multiple efforts made by researchers to project the climate, a debate about the reliability of the projections always exists. For example, dynamical downscaling has the advantage of consideration of physics to link between large-scale phenomena and sub-grid features. However, Racherla et al. (2012) showed that RCMs do not add significant value to the skill of GCMs in capturing the climatological averages of precipitation and temperature. Apart from this, PaiMazumdar and Done (2015) showed improvement in RCM skill, by modifying the outputs through bias correction. In case of statistical downscaling, the skill is mainly dependent on factors such as selection of predictors (Timbal et al. 2008) and calibration (Teutschbein et al. 2011). Although statistical downscaling has an advantage over dynamical downscaling of deriving statistical relationship between large- and local-scale variables, the main limitation is its assumption of stationarity (Salvi et al. 2016). Considering the above advantages and disadvantages of both the downscaling techniques, the outcome from a set of GCM and downscaling technique may differ from the other, in terms of uncertainties arising from inter-GCM, inter-downscaling or the selection of inappropriate predictors or nested region. Thus, incomplete understanding about the global- and local-scale features may generate some spurious results that are not competent enough to explain the possible realistic scenario(s).

Sharma et al. (2018) showed a detailed analysis of the climatic projection uncertainties and their propagation in hydrological variables, over India. Three GCMs and two downscaling (statistical and dynamical) techniques were used to derive the climate projection. It was observed that both the downscaling techniques revealed different climatic projections. Rainfall mainly showed increased changes. However, it was varying for both the downscaling techniques. For example, core monsoon region of India showed significant rainfall changes through dynamical downscaling, but statistical showed no change. Similarly, temperature projections were similar for different downscaling techniques but varied significantly in magnitude. Overall, the results were inconsistent for multiple downscaling techniques applied on multiple GCMs. Further analysis was performed to understand and quantify these uncertainties. Among the different sources of uncertainties, two were considered in the study: GCM and downscaling uncertainty. GCM uncertainty was represented as the uncertainty arising from the climate projections of multiple GCMs through the same downscaling technique, whereas downscaling uncertainty was represented as the uncertainty arising from multiple downscaling techniques applied on the same GCM. The mathematical expressions that were used in the study for deriving the two types of uncertainties were as follows:

$$DU_j = \frac{1}{n}\sum_{i=1}^{n}\left(\max\left(C_{G_i D_{1:2}}\right) - \min\left(C_{G_i D_{1:2}}\right)\right). \tag{1}$$

$$GU_j = \frac{1}{2}\left[\left(\max\left(C_{D_1 G_{1:n}}\right) - \min\left(C_{D_1 G_{1:n}}\right)\right) + \left(\max\left(C_{D_2 G_{1:n}}\right) - \min\left(C_{D_2 G_{1:n}}\right)\right)\right]. \tag{2}$$

Here, *DU* and *GU* were symbolized to represent the downscaling and GCM uncertainty, respectively, for the *j*th grid. *C* was denoted as the projected changes derived from *n* number of GCMs *G* and D_1 and D_2 as the types of downscaling techniques. To understand the total impact of the two uncertainties, total uncertainty was derived by considering the equation as follows:

$$TU_j = \max\left(C_{D_{1:2} G_{1:n}}\right) - \min\left(C_{D_{1:2} G_{1:n}}\right). \tag{3}$$

These equations were applied over all the climatic variables. Additionally, an impact assessment hydrological model (Variable infiltration capacity) was applied over India, by considering multiple climatic projections. The hydrological projections showed similar disagreement in deriving the changes. The same uncertainty expressions were then applied to understand the cascade of uncertainties from climatic to hydrological variables. The results showed that inter-GCM uncertainty had a significant influence on the total uncertainty of the hydro-climatic projections. Figure 3 shows the comparison of the multiple sources of uncertainties for all the hydro-climatic variables. The study also revealed that the uncertainty was spatially non-uniform and was region-specific, which directly imparts to the conclusion that neglecting the uncertainty analysis will not only increase the horizon of multiple policy options to cope with the changing climate, but will also impart significant regional vulnerability to the hydro-climatic resources of India.

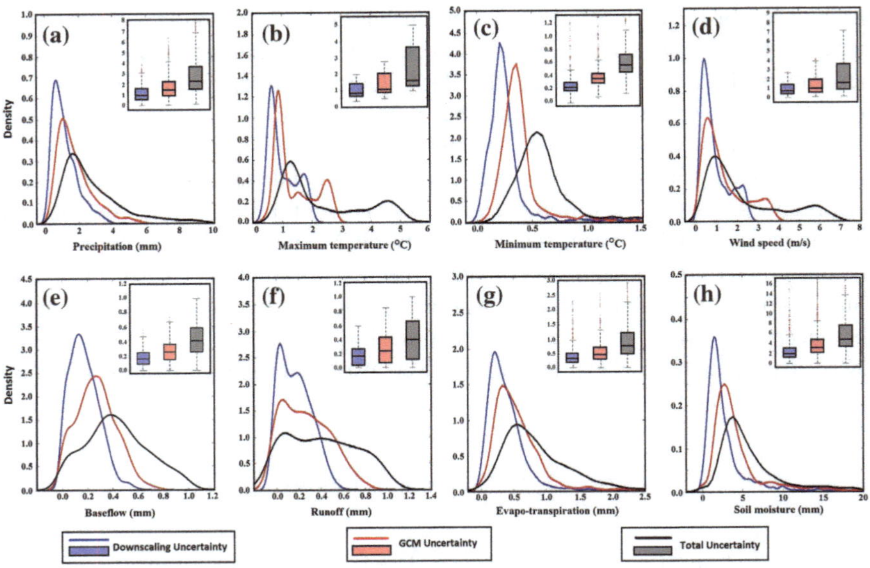

Fig. 3 Comparison of multiple sources of uncertainties in hydro-climatic projections over India

3 Impacts of Changes in Hydrological Cycle

Different components of the hydrological cycle are highly sensitive towards changing climate and land use. Such changes are therefore responsible for making the atmospheric interactions among hydro-climatic variables more complex. Increased concentration of GHGs and aerosols in the atmosphere due to natural or anthropogenic sources has brought about significant changes to the climate. The impacts of these changes are unequivocal across the globe, and the effects vary from region to region, especially causing major alterations in different processes of the hydrological cycle (Fig. 4). This has led to increased precipitation in some areas making them prone to high inundation risk due to frequent floods and rising sea levels.

The warming effect of aerosol has resulted in rise in global temperature, thus warming the atmosphere to accommodate more moisture which further increases the temperature. Over the past few decades, increased incidences of hydro-climatic extremes such as floods, droughts and cyclones have been observed which are directly or indirectly hampering the growth in sectors like agriculture, industries, housing and water supply infrastructure. Impacts of climate change on hydrological cycle have drastic implication on water availability (demand–supply), socio-economic damage due to hydro-climatic extremes, agriculture sector (irrigation) and food security, human health and damage to environment. Therefore, it is important to understand the impacts of these changes not only on the characteristics of hydrological cycle but also on the availability of water for use, especially in a developing country like India.

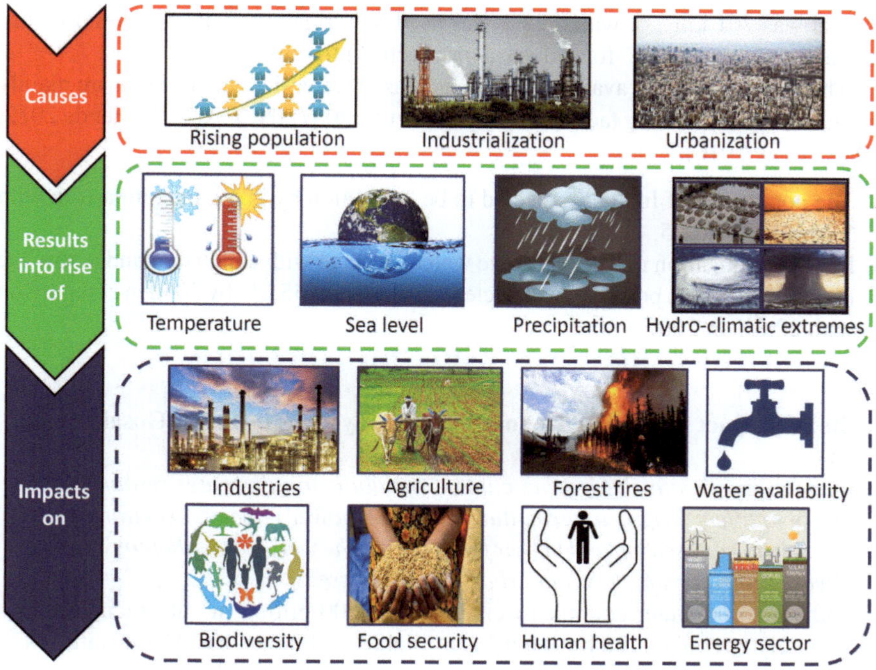

Fig. 4 Pictorial representation of the causes, results and impacts of climate change on different sectors

3.1 *Impacts on Water Availability and Security*

Water availability in the Indian subcontinent is dependent upon the monsoon rainfall, which is driven by south-west winds in summers and north-east winds during winters, and glacial snowmelts. Therefore, the abundance of water throughout the year in different sources of water (rivers, lakes, ponds and/or underground water supply) in a catchment/basin is indeed dependent on the precipitation (rainfall and glacial snow accumulation), which is approximately 4000 billion m^3 per annum of which only 1123 billion m^3 is usable (Government of India 2009). The climate variability or long-term changes in the climate has varying impacts on water availability in different regions. Some regions will experience an abundance of water, while others may experience shortage or water stress conditions. The most plausible causes for such conditions are the changes in precipitation pattern, in terms of frequency and intensity, spatially due to the changing climate.

The major sectors that pose a serious demand for adequate water supplies are agriculture, industrial manufacturing units, commercial and domestic purposes. However, water demand and its sector-wise usage differ among developed and developing countries. For example, a developing nation like India uses 80–90% of the total water withdrawal for agriculture and the rest for other purposes. (In 2010, total water withdrawal was 761 km^3, of which 91% was used for irrigation, 7.4% for domestic use and approximately 1.6% for industrial production) (Syaukat 2012).

The decreased water availability and increased water demand in India are mainly because of the following facts (Amarasinghe et al. 2007; INCCA 2010; Hegde 2012; Ghosh et al. 2016):

- The population of India is expected to be 1.6 billion by 2050, in comparison with 1.1 billion in 2005.
- Rapid urbanization may cause acute water shortage with a high demand across the nation, with urban population expected to increase to 55.2% by 2050 in comparison with 22.8% in 2007.

Box 2 Impact of Climate Change on River Systems in India (Gosain et al. 2006)

Impacts of climate change and climate variability on the water resources are likely to affect irrigated agriculture, installed power capacity, environmental flows in the dry season and higher flows during the wet season, thereby causing severe droughts and floods in urban and rural areas.

India being a water country receives about 4000 billion m^3 of precipitation annually but only 1123 billion m^3 is available for human use. These statistics tend to change with underlying uncertainties of climate change predictions. The study was performed to quantify the potential impacts of the changing climate on different water resources available in the Indian River System to determine the present water availability. The PRECIS daily weather data were used to determine the spatio-temporal water availability in the river systems. A distributed hydrological model, Soil and Water Assessment Tool (SWAT), has been used to simulate the river basins across the country, for evaluating vulnerable hot spots of droughts and floods. It was reported an increase in precipitation for most of the river basins but a decreasing trend in precipitation and associated water yield in the basin was observed for Brahmaputra, Cauvery and Pennar. Similarly, varying trends for other hydrological variables like evapotranspiration, run-off, water storage across different basins were observed. The study was conducted on behalf of the Ministry of Environment and Forests, Government of India, as a part of the NATCOM II study (Second National Communication of India to UNFCCC), which would be useful for making any national- or state-level plans for management of water resources.

- Industrialization may increase its contribution in the gross domestic product (GDP) from 29.1% in 2000 to 40% in 2050, which will cause a higher water demand of 161 billion m^3 as compared to 30 billion m^3 in 2000.
- A significant decrease in monsoon rainfall over major water surplus river basins in India and therefore reduction in the availability of water for human use. The spatio-

temporal variations in mean and extreme rainfall have resulted in the imbalance of water supply in the natural and/or man-made reservoirs.
- Increase in overall rainfall but decrease in number of rainy days has been predicted, with temperature projected to rise to 1.7–2 °C by 2030s.
- Increase in effective area under irrigation to over 104 million ha by 2025 would require more freshwater supplies.
- The agricultural water demand will increase by 80% by 2050 with more focus on the development of water-intensive cash crops.
- Excessive use of chemical fertilizers and pesticides and over-irrigation have already turned over 9 million ha fertile land into contaminated wastelands unfit for human use or agricultural purpose, and it has also contaminated the groundwater sources due to percolation of chemicals through the soil.

The rising water demand poses a serious threat to the population and may fuel into various interstate disputes for sharing of water resources in near future. Therefore, it is important to assess the impacts of climate change and human interventions on water availability in future to meet the challenge of fulfilling the demands of growing population (Grover 2016).

3.2 Impacts on Socio-economic Factors and Livelihood Under Hydro-climatic Extremes

The changing climate tends to increase the occurrences of hydro-climatic extreme events such as rainfall, cyclones, floods and droughts which are associated with ecological and socio-economic losses at the cost of human life, property, agriculture and infrastructure in the affected areas (Stocker et al. 2013). India experiences flash floods, heavy rainfall and glacial melting, droughts, heat waves, and tropical cyclones and depression in the Bay of Bengal almost every year causing a death toll of over hundred thousand, eventually accounting for one-fifth of the total global deaths (WRIS 2015). According to a United Nations Global Assessment Report (GAR) on Disaster Risk (2015), India faces an average annual economic loss of USD 9.8 billion due to natural disasters out of which flood accounts for more than USD 7 billion. As global temperatures are rising, the associated impacts have serious implications on the livelihood of major population in India (Halgamuge and Nirmalathas 2017). A majority of the population in India is still living in close proximity to hazard-prone regions making them highly vulnerable to these hydro-climatic extremes. Overpopulation and space constraint for living have forced a mass of population to encroach to vulnerable regions that lack in structural protection measures. An increasing trend in the frequency of floods due to precipitation extremes and glacier lake outburst floods has been observed in South Asia. Bihar floods in 2004 and 2007 are considered to be one of the worst floods that killed 885 people and affected the lives of over 21 million with crops damaged worth nearly USD 80 million. Maharashtra and Gujarat faced one of the greatest flood miseries in 2005. Mumbai was flooded by an extremely heavy rainfall that led to death of nearly 5000 people. In Gujarat floods, over 60% of the state was inundated by floodwaters that caused a net loss of property of worth USD 1.5 billion. In 2007, extreme rainfall events have resulted in severe flood events in Bangladesh, India and Nepal killing more than 2000 persons and rendered more than 20 million persons homeless, thus disturbing the normal lives of the people (Dhara et al. 2013). Other incidences of floods in Ladakh (2010), Assam (2012) and Jammu and Kashmir (2014) submerged several villages under water, thus displacing thousands. In 2013, Uttarakhand witnessed one of the worst natural calamities of Indian history after 2004 Tsunami which claimed more than 5700 lives. Rising temperature and scorching heat have resulted to erratic weather conditions leading to frequent instances of heatwaves in the past decade with over 98 incidences of severe heatwave days recorded. Severe heatwaves hit the parts of Telangana and Andhra Pradesh in 2013, 2014, 2015 and 2016, and over 5000 deaths were recorded (Mazdiyasni et al. 2017). In 2010, over 1344 deaths were caused by severe heatwaves in Ahmedabad City of India (Azhar et al. 2014). Apart from floods and heatwaves, India also experiences several cyclonic and depression activities in the Bay of Bengal due to its geographical location and coastline of approximately 7516 km. The damage to property, livestock, agriculture and livelihood cost more than million USD in previous cyclonic events and displaced over thousand people along the coastline. Agriculture was hit hard by these events and the associated floods causing greater damage to both, farmers and fishermen. Previous cyclonic incidences including Laila (in 2010), Jal (in 2010), Phailin (in 2013), Hudhud (in 2014) and Nada (in 2016) are among the major ones that have hit India over the last few years.

3.3 Impacts on Agriculture and Food Security

In general, the greenhouse effect is an essential natural phenomenon in order to sustain life on earth by maintaining the apt living conditions such as optimum temperature, balanced composition of gases in the air and protection of life forms from harmful radiations. However, the rise in global temperature that continues to warm the atmosphere will have a profound impact on the lives of agricultural producers and their associated sectors. Importantly, the effects of climate change may have disparate impacts; that is, it may have positive or negative effects on crop yield of different crops. Therefore, the increasing trend in the concentration of GHGs is a matter of concern, and it is important to assess the direct (change in temperature, humidity and concentration of CO_2) and indirect (occurrence of hydro-climatic extremes such as floods and droughts) effects of future climate changes on agriculture. Crop growth and yield are highly influenced by the climatic conditions which are suitable for the crop, but the slight deviations from normal conditions in the growing season may hamper the food production drastically, especially the rainfed or unirrigated crops that comprise nearly 60% of the Indian cropland (Shankar 2011). Also, the agricultural and irrigation practices or strategies adapted fail to work efficiently and will lead to higher food prices against lower food production, causing serious impairment to food security. The Indian economy where agriculture sector contributes 15% to India's GDP is strongly dependent on the impacts of climate change on Indian monsoon. Climate change has an impact of 4–9% per year on agricultural production causing a damage of about 1.5% loss in India's GDP each year. The Indian crop-growing season is classified into kharif (July–October) and rabi (October–March) which are mainly irrigated through south-west and north-east monsoon, respectively. Climate variability can have severe consequences on these cropping seasons causing huge crop losses threatening the food security of the nation. The Kharif crops are mainly affected by the variability in Indian Summer Monsoon Rainfall (ISMR), while the minimum temperature forms the governing factor in crop yield generated during Rabi cropping season. Among all the crops, rice and wheat are the major dominant crops that are grown during Kharif and Rabi seasons across the entire nation, respectively, and have an important share in total food grain production in India. Although overall rice yields have increased in India, a significant loss in production has been attributed due to unfavourable rainfall and temperature conditions during growing season. Once being a major global wheat contributor, India now faces a stagnant yield despite improved management practices such as fertilization. Schellnhuber et al. (2013) observed extremely high temperatures in northern India as a responsible factor for a substantial reduction in wheat yields, and rising temperatures can only aggravate the situation.

Even the IPCC, which is scarcely alarmist, depicts a 0.5° rise in winter temperature would reduce wheat yield by 0.45 tons per hectare in India. Naturally, any change in rice and wheat yields will have a significant impact on food security of the country. Figure 5 shows the trends in rainfall, temperature and crop yield in the present and the futuristic scenarios, derived from Sinha and Swaminathan (1991), Gangadhar Rao and Sinha (1994), Saseendran et al. (2000), Aggarwal (2000), Rathore et al. (2001), Aggarwal and Mall (2002), Attri and Rathore (2003).

In future, agriculture will likely remain an important sector of the economy in terms of food production as well as employment generation (Mirza and Ahmad 2005). Though it is inevitable to adopt top-down approaches for regions already affected by adverse impacts, there is also a need to understand the effect of various strategies related to bottom-up approach that can reduce agricultural risk in the long run and help in better management and planning under projected climate change scenarios. This will not only help the scientists to keep the farmers informed with long-term adaptation practices such as change of crop, crop variety, shift sowing date and increasing fertilizer, but also help them to identify the regions that are more vulnerable and hence should be given prior importance.

3.4 Impacts on Health

The climate affects the water availability and quality, food and air quality in any region. The variability in climate deteriorates the physical comforts and health of all living organism which are directly or indirectly getting affected by such changes.

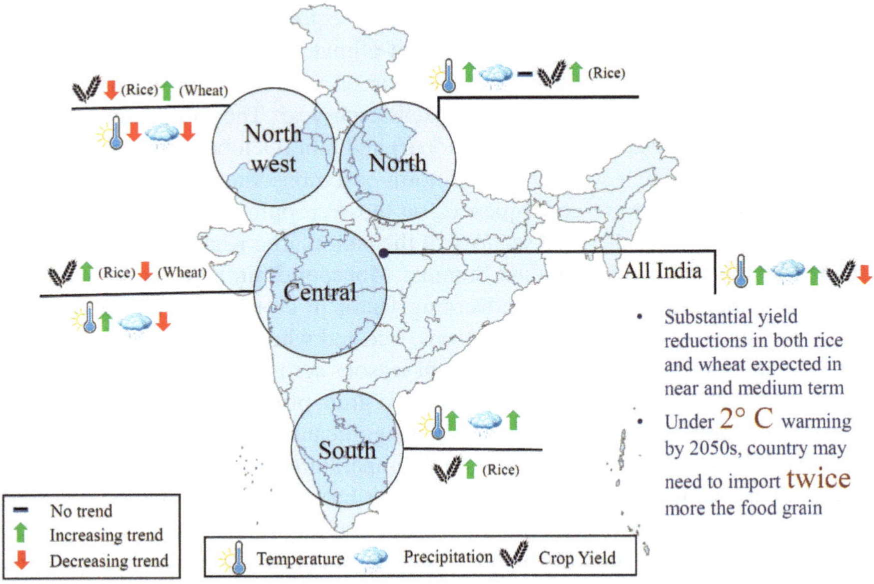

Fig. 5 Changes in crop yield under changing climate conditions for different regions in the Indian subcontinent

However, the major question that arises is how the climate changes are affecting human health? With rising global temperatures and increased occurrence of hydro-climatic extremes, a significant rise in associated health-related losses, in terms of exposure to diseases or deaths, has been observed in India (Singh and Dhiman 2012). Human health due to such changes is greatly hampered both, physically and mentally. The major health effects that render after the occurrence of hydro-climatic extreme (such as floods, cyclones and droughts) include the occurrence of numerous diseases and illness that could be traumatic as well as infectious. Many prevalent diseases that occur after extreme events are linked to loss of drinkable water, hygiene and sanitation, loss of shelter and belongings, migration of population, exposure to toxic chemicals and biological agents, and agricultural losses that lead to hunger and malnutrition risk. For example, flooding in West Bengal in 1988 led to outbreak of diarrhoea due to cholera and claimed lives of more than 270 persons (Dhara et al. 2013). Agricultural impacts due to climate change are expected to cause several health disorders such as malnutrition and child stunting which is projected to increase by 35% by 2050. Several cardiovascular diseases associated with mortality and respiratory illnesses due to heatwaves are related to the altered climate and pose profound effects on these diseases. Transmission of infectious vector-borne diseases such as malaria and dengue has greatly increased due to changing climate which is favourable for the growth of disease-causing organisms and is strongly affected by fluctuations in temperature (Patz et al. 2005). The mortality rate has risen substantially with the tendency of occurrence of extreme weather events. Each year, the mortality rate for children and elderly has seen a fluctuating, but increasing trends and this situation have been exacerbated by climate change. As numerous studies have reported an increasing trend of extremes in India, as projected under different climate change scenarios, there are going to be serious implications on human health.

4 Concluding Remarks

Hydrological resources are important dynamic natural resources that play a crucial role in supporting India's economic growth. The impacts due to climate change pose serious threats to this central arena which must be managed and attenuated by research institutions and implementing agencies, be it non-government or government. For a country like India, understanding the socio-economic damage in terms of agricultural production losses/gains, and loss of property, lives and livelihood due to regional vulnerability to climate change becomes crucial. In addition to this, hydrological extremes such as recurrent droughts and floods threaten the livelihood of millions, especially the small and marginal farmers, landless poor, women, children and elderly. Furthermore, the climate sensitivity of hydrological regime is uncertain, as there are regional variations in rainfall, temperature, soil moisture, evapotranspiration and water resource management practices. Given this scenario, it becomes essential to understand the role of climate projections in leading the water resources of a region along with the level of uncertainty in these projections.

GCMs pragmatically simulate large-scale circulation patterns and are hence considered to be the base for any climate change impact assessment study pertaining to the future. However, its use is limited because of coarser resolution, due to which it needs to be downscaled before applying it at regional scale. Albeit the uniqueness of both the downscaling techniques (statistical—less computational requirement, assumption of stationarity; dynamical—physics-based, region-specific) gets reflected in the climatic variables, uncertainties exists among these projections. Apart from the downscaling techniques, the selection of global-scale variables plays a pivotal role in capturing the climatology of a region like the Mahanadi River Basin of India which is exceptionally influenced by extreme hydrological events. This consequently emphasizes the importance of follow-on research to reduce the various uncertainties in the agro-hydro-climatic projections, such as selections of downscaling techniques, GCMs, impact assessment models, variables, thus recommending a Code of Practice to provide formal guidance, particularly for climate change impact assessment studies that can help stakeholders/policymakers to understand certain adverse impacts and tactically define the adaptation strategies for end-users.

References

Abraham Z, Tan PN, Xin F (2011) Smoothed quantile regression for statistical downscaling of extreme events in climate modeling. In: CIDU, pp 92–106

Aggarwal PK (2000) Application of systems simulation for understanding and increasing yield potential of wheat and rice, Ph.D. thesis, Wageningen University, The Netherlands, p 176

Aggarwal PK, Mall RK (2002) Climate change and rice yields in diverse agro-environments of India. II. Effect of uncertainties in scenarios and crop models on impact assessment. Clim Change 52(3):331–343

Amarasinghe UA, Shah T, Turral H, Anand BK (2007) India's water future to 2025-2050: business as usual scenario and deviations. Research Report 123, IWMI, 52 pp

Anandhi A, Srinivas VV, Kumar DN, Nanjundiah RS (2009) Role of predictors in downscaling surface temperature to river basin in India for IPCC SRES scenarios using support vector machine. Int J Climatol 29(4):583–603

Ashfaq M, Shi Y, Tung WW, Trapp RJ, Gao X, Pal JS, Diffenbaugh NS (2009) Suppression of south Asian summer monsoon precipitation in the 21st century. Geophys Res Lett 36(1)

Ashfaq M, Skinner CB, Diffenbaugh NS (2011) Influence of SST biases on future climate change projections. Clim Dyn 36(7–8):1303–1319

Asokan SM, Dutta D (2008) Analysis of water resources in the Mahanadi River Basin, India under projected climate conditions. Hydrol Process 22:3589–3603

Attri SD, Rathore LS (2003) Simulation of impact of projected climate change on wheat in India. Int J Climatol 23:693–705

Azhar GS, Mavalankar D, Nori-Sarma A, Rajiva A, Dutta P, Jaiswal A, Sheffield P, Knowlton K, Hess JJ (2014) Heat-related mortality in India: excess all-cause mortality associated with the 2010 Ahmedabad heat wave. PLoS ONE 9(3):e91831

Bardossy A, Plate EJ (1991) Modeling daily rainfall using a semi-Markov representation of circulation pattern occurrence. J Hydrol 122:33–47

Chattopadhyay N, Hulme M (1997) Evaporation and potential evapotranspiration in India under conditions of recent and future climate change. Agric For Meteorol 87(1):55–73

Dash SK, Jenamani RK, Kalsi SR, Panda SK (2007) Some evidence of climate change in twentieth-century India. Clim Change 85(3–4):299–321

Dhara VR, Schramm PJ, Luber G (2013) Climate change & infectious diseases in India: implications for health care providers. Ind J Med Res 138(6):847

Fowler H, Ekström M, Kilsby C, Jones P (2005) New estimates of future changes in extreme rainfall across the UK using regional climate model integrations. 1. Assessment of control climate. J Hydrol 300:212–233

Friederichs P, Hense A (2007) Statistical downscaling of extreme precipitation events using censored quantile regression. Mon Weather Rev 135:2365–2378

Gangadhar Rao D, Sinha SK (1994) Impact of climate change on simulated wheat production in India. In: Rosenzweig C, Iglesias I (ed) Implications of climate change for international agriculture: crop modelling study. USEPA230-B-94-003. USEPA, Washington, D.C., pp 1–17

Ghosh S, Mujumdar P (2006) Future rainfall scenario over Orissa with GCM projections by statistical downscaling. Curr Sci 90:396–404

Ghosh S, Vittal H, Sharma T, Karmakar S, Kasiviswanathan KS, Dhanesh Y, Sudheer KP, Gunthe SS (2016) Indian summer monsoon rainfall: implications of contrasting trends in the spatial variability of means and extremes. PLoS ONE 11(7):e0158670

Gleick PH (1989) Climate change, hydrology, and water resources. Rev Geophys 27(3):329–344

Gosain AK, Rao S, Basuray D (2006) Climate change impact assessment on hydrology of Indian River basins. Curr Sci 90:346–353

Government of India (2009) Background note for consultation meeting with policy makers on review of National Water Policy. Ministry of Water Resources, 50 pp

Grover VI (2016) Impact of climate change on the water cycle. In: Shrestha S, Anal AK, Salam PA, Van der Valk M (eds) Managing water resources under climate uncertainty. Springer International Publishing, Switzerland, pp 3–30
Gupta J (2012) Climate change and water law: the regimes compared. In: Grover (ed) Impact of climate change on water and health. CRC Press, Boca Raton
Halgamuge MN, Nirmalathas T (2017) Analysis of large flood events: based on flood data during 1985–2016 in Australia and India. Int J Disas Risk Reduct
Hegde NG (2012) Water scarcity and security in India. In: BAIF-Indian science congress
Huntington KW, Blythe AE, Hodges KV (2006) Climate change and Late Pliocene acceleration of erosion in the Himalaya. Earth Plan Sci Lett 252(1–2):107–118
INCCA (2010) Climate change and India: a 4 × 4 assessment. A sectoral and regional analysis for 2030s. Ministry of Environment and Forests, INCCA report 2010, New Delhi. Available from: http://moef.nic.in/downloads/others
Jena PP, Chatterjee C, Pradhan G, Mishra A (2014) Are recent frequent high floods in Mahanadi basin in eastern India due to increase in extreme rainfalls? J Hydrol 517:847–862
Jung M, Reichstein M, Ciais P, Seneviratne SI, Sheffield J, Goulden ML, Bonan G, Cescatti A, Chen J, De Jeu R, Dolman AJ (2010) Recent decline in the global land evapotranspiration trend due to limited moisture supply. Nature 467(7318):951
Kannan S, Ghosh S (2011) Prediction of daily rainfall state in a river basin using statistical downscaling from GCM output. Stoch Env Res Risk Assess 25:457–474
Kothyari UC, Singh VP (1996) Rainfall and temperature trends in India. Hydrol Process 10(3):357–372
Lal M (2005) Climatic change—implications for India's water resources. In: Mirza MMQ, Ahmad QK (eds) Climate change and water resources in South Asia. A.A. Balkema Publishers, Leiden, The Netherlands, pp 155–196
Leung LR, Wigmosta MS (1999) Potential climate change impacts on mountain watersheds in the Pacific Northwest. JAWRA J Am Water Resour Assoc 35(6):1463–1471
Loarie SR, Duffy PB, Hamilton H, Asner GP, Field CB, Ackerly DD (2009) The velocity of climate change. Nature 462(7276):1052
Mazdiyasni O, AghaKouchak A, Davis SJ, Madadgar S, Mehran A, Ragno E, Sadegh M, Sengupta A, Ghosh S, Dhanya CT, Niknejad M (2017) Increasing probability of mortality during Indian heat waves. Sci Adv 3(6):e1700066
Mirza MMQ, Ahmad QK (2005) Climate change and water resources in South Asia: an introduction. In: Mirza MMQ, Ahmad QK (eds) Climate change and water resources in South Asia. A.A. Balkema Publishers, Leiden, The Netherlands, pp 1–22
Mishra V, Lilhare R (2016) Hydrological sensitivity of Indian sub-continental river basins to climate change. Glob Planet Change 139:78–96
Mitchell JF, Johns TC, Gregory JM, Tett SFB (1995) Climate response to increasing levels of greenhouse gases and sulphate aerosols. Nature 376(6540):501
Mohan TS, Rajeevan M (2017) Past and future trends of hydroclimatic intensity over the Indian monsoon region. J Geophys Res Atmos 122(2):896–909
PaiMazumder D, Done JM (2015) The roles of bias-correction and resolution in regional climate simulations of summer extremes. Clim Dyn 45(5–6):1565–1581
Patz JA, Campbell-Lendrum D, Holloway T, Foley JA (2005) Impact of regional climate change on human health. Nature 438(7066):310
Pielke RA (2005) Land use and climate change. Science 310(5754):1625–1626
Pielke RA, Wilby RL (2012) Regional climate downscaling: what's the point? Eos, Transactions American Geophysical Union 93(5):52–53
Racherla PN, Shindell DT, Faluvegi GS (2012) The added value to global model projections of climate change by dynamical downscaling: a case study over the continental US using the GISS-ModelE2 and WRF models. J Geophys Res Atmos 117:D20
Raje D, Mujumdar P (2010) Reservoir performance under uncertainty in hydrological impacts of climate change. Adv Water Resour 33:312–326

Rao PG (1993) Climatic changes and trends over a major river basin in India. Clim Res 2:215–223
Rathore LS, Singh KK, Saseendran SA, Baxla AK (2001) Modelling the impact of climate change on rice production in India. Mausam 52(1)
Revadekar JV, Patwardhan SK, Rupa Kumar K (2011) Characteristic features of precipitation extremes over India in the warming scenarios. Adva Meteorol
Sachindra DA, Perera BJC (2016) Statistical downscaling of general circulation model outputs to precipitation accounting for non-stationarities in predictor-predict and relationships. PLoS ONE 11(12):e0168701
Salvi K, Ghosh S (2013) High-resolution multisite daily rainfall projections in India with statistical downscaling for climate change impacts assessment. J Geophys Res Atmos 118(9):3557–3578
Salvi K, Ghosh S, Ganguly AR (2016) Credibility of statistical downscaling under nonstationary climate. Clim Dynam 46(5):1991–2023
Saseendran AS, Singh KK, Rathore LS, Singh SV, Sinha SK (2000) Effects of climate change on rice production in the tropical humid climate of Kerala, India. Clim Change 44:495–514
Schellnhuber HJ, Hare B, Serdeczny O, Schaeffer M, Adams S, Baarsch F, Schwan S, Coumou D, Robinson A, Vieweg M, Piontek F (2013) Turn down the heat: climate extremes, regional impacts, and the case for resilience. Turn down the heat: climate extremes, regional impacts, and the case for resilience. International Bank for Reconstruction and Development, World Bank. ISBN: 9781464800566
Seneviratne SI, Lüthi D, Litschi M, Schär C (2006) Land-atmosphere coupling and climate change in Europe. Nature 443(7108):205
Shankar PV (2011) Towards a paradigm shift in India's rainfed agriculture. Innov Dev 1(2):321–322
Sharma T, Vittal H, Chhabra S, Salvi K, Ghosh S, Karmakar S (2018) Understanding the cascade of GCM and downscaling uncertainties in hydro-climatic projections over India. Int J Climatol 38:e178–e190
Shashikanth K, Salvi K, Ghosh S, Rajendran K (2014) Do CMIP5 simulations of Indian summer monsoon rainfall differ from those of CMIP3? Atmos Sci Lett 15(2):79–85
Shashikanth K, Ghosh S, Vittal H, Karmakar S (2017) Future projections of Indian summer monsoon rainfall extremes over India with statistical downscaling and its consistency with observed characteristics. Clim Dyn 1–15
Singh P, Bengtsson L (2004) Hydrological sensitivity of a large Himalayan basin to climate change. Hydrol Process 18(13):2363–2385
Singh PK, Dhiman RC (2012) Climate change and human health: Indian context. J Vector Borne Dis 49(2):55
Singh P, Kumar N (1997) Impact assessment of climate change on the hydrological response of a snow and glacier melt runoff dominated Himalayan river. J Hydrol 193(1–4):316–350
Sinha SK, Swaminathan MS (1991) Deforestation, climate change and sustainable nutrients security. Clim Change 16:33–45
Stocker T, Qin D, Plattner G, Tignor M, Allen S, Boschung J, Nauels A, Xia Y, Bex B, Midgley B (2013) IPCC, 2013: climate change 2013: the physical science basis. Contribution of working group I to the fifth assessment report of the intergovernmental panel on climate change, p 1535
Syaukat Y (2012) Irrigation in Southern and Eastern Asia in figures. FAO the United Nation
Teutschbein C, Wetterhall F, Seibert J (2011) Evaluation of different downscaling techniques for hydrological climate-change impact studies at the catchment scale. Clim Dyn 37(9–10):2087–2105
Timbal B, Hope P, Charles S (2008) Evaluating the consistency between statistically downscaled and global dynamical model climate change projections. J Clim 21(22):6052–6059
Trenberth KE (1999) Conceptual framework for changes of extremes of the hydrological cycle with climate change. In: Weather and climate extremes. Springer, Dordrecht, pp 327–339
Wilby RL, Dawson CW (2013) The statistical downscaling model: insights from one decade of application. Int J Clim 33:1707–1719
WRIS-India (2015) Water resource information system of india—a joint venture of the Central Water Commission (CWC), Ministry of Water Resources, Government of India and Indian Space

Research Organization (ISRO), Department of Space, Government of India. http://india-wris.nrsc.gov.in/wrpinfo/index.php?title=Mahanadi. Accessed 25 Aug 2017
Wood AW, Leung LR, Sridhar V, Lettenmaier DP (2004) Hydrological implications of dynamical and statistical approaches to downscaling climate model outputs. Clim Change 62(1–3):189–216

Links Between Global Climate Teleconnections and Indian Monsoon Rainfall

S. Adarsh and M. Janga Reddy

Abstract The knowledge of teleconnections elucidates the mechanisms behind the local meteorological processes that are influenced by the ocean/atmospheric circulations, which operate at global scale. As the local meteorological variables such as regional temperature/rainfall and global climate signals possess multiscaling properties, it is highly desired to examine the teleconnections in multiple process scales rather than following conventional statistical correlation analysis and/or based on periodicity estimation. In this context, this chapter presents an investigation of hydroclimatic teleconnections of All India Summer Monsoon Rainfall (AISMR) with large-scale climate oscillations in a multiscaling framework by employing Hilbert–Huang Transform (HHT)-based Time-Dependent Intrinsic Correlation (TDIC) analysis. The study investigated the teleconnections of AISMR for the period 1950–2012, with the four global climate oscillations such as Quasi-Biennial Oscillation (QBO), El Niño Southern Oscillation (ENSO), Equatorial Indian Ocean Oscillation (EQUINOO), and Atlantic Multidecadal Oscillation (AMO) in different process scales. The study found that both the strength and nature of association between global climate oscillations and ISMR vary with process scale and there could be multiple switchovers in the character of such associations over the time domain.

S. Adarsh · M. Janga Reddy (✉)
Department of Civil Engineering, Indian Institute of Technology Bombay, Mumbai 400076, Maharashtra, India
e-mail: mjreddy@civil.iitb.ac.in

S. Adarsh
e-mail: adarsh_lce@yahoo.co.in

S. Adarsh
Department of Civil Engineering, TKM College of Engineering, Kollam, Kerala, India

M. Janga Reddy
Interdisciplinary Program (IDP) in Climate Studies, Indian Institute of Technology Bombay, Mumbai 400076, Maharashtra, India

C. Venkataraman et al. (eds.), *Climate Change Signals and Response*,
https://doi.org/10.1007/978-981-13-0280-0_4

1 Background

Teleconnection of hydrometeorological variables refers to a connection or association between local meteorological variable in a region with global atmospheric circulations which occur a long distance apart. Investigations on teleconnections may help in elucidating the true mechanisms behind those variables and exposing role of atmospheric circulations operating at global scales and the dynamics of the earth system. For example, the increase in sea surface temperature (SST) over the central and eastern equatorial Pacific (so-called the phenomenon of El Niño), and the corresponding decrease (known as La Niña), contribute to seasonal climate fluctuations in many parts of the world. During the summer seasons of northern hemisphere, El Niño results in severe/hot temperatures over the Indian subcontinent and substantial rains over the western USA, whereas La Niña may result in reverse situations over these regions, with a few exceptions.

Indian monsoon is influenced by many complex factors including many large-scale atmospheric/oceanic circulations. The knowledge of teleconnections between Indian Summer Monsoon Rainfall (ISMR) and large-scale climatic signals is of prime importance for the hydroclimatologists, as it may help for performing reliable predictions of ISMR. The investigations on such linkages have a longer history from the times of Blanford (1884). The link between ISMR and El Niño Southern Oscillation (ENSO) is the most debated one by the researchers (Walker 1923; Shukla and Paolino 1983; Kripalani and Kulkarni 1997a, b; Krishna Kumar et al. 1999; Gadgil et al. 2004; Kumar et al. 2006). The links of ISMR with other climate oscillations such as Quasi-Biennial Oscillation (QBO) (Rao and Lakhole 1978; Chattopadhyay and Bhatla 2002; Claud and Pascal 2007), Indian Ocean Dipole (IOD) (Saji et al. 1999), Equatorial Indian Ocean Oscillation (EQUINOO) (Gadgil et al. 2004; Maity and Kumar 2006a, b), Atlantic Multidecadal Oscillation (AMO) (Goswami et al. 2006; Feng and Hu 2008) were also reported in some of the research works. Many times such teleconnections were investigated merely based on statistical correlation between ISMR and the climatic signals (Maity and Kumar 2006a, b, 2007). This approach is not always reliable as both of the signals possess multiscaling properties, and it is more appropriate to investigate such teleconnections in a multiscaling framework.

Understanding the association of ISMR with global climate oscillations in multiple timescales may eventually help in improving the predictability efforts of ISMR. To analyze the teleconnections in multiscale perspective, any time–frequency analysis method can be used and the continuous wavelet transform (CWT) and allied techniques such as "wavelet coherence" were widely adopted in the past (Massei et al. 2007, 2011; Rossi et al. 2009; Nalley et al. 2016) and some of them investigated the association between ISMR and global climatic/atmospheric oscillations (Torrence and Webster 1999; Azad et al. 2008; Narasimha and Bhattacharya 2010). However, the application of CWT for hydroclimatic teleconnection studies is rather challenging because it involves the complex step of selecting the most appropriate wavelet function. Moreover, when such analysis is extended to the wavelet coherence, smoothing operation becomes inevitable which in turn may influence the localization in time domain (Liu 1994; Grinsted et al. 2004). In this context, the spectral analysis method propounded by Huang et al. (1998), called as HHT method becomes a viable alternative technique for hydroclimatic teleconnection studies. The HHT integrates the procedures of data adaptive decomposition process called Empirical Mode Decomposition (EMD) with the classical Hilbert Transform (HT). The former facilitates the separation of time series signal into modes of specific periodicity while latter facilitates for the time–frequency characterization at different timescales.

In recent past, few studies used HHT to investigate the link between hydrologic variables with that of global climate variables by the comparison of periodicities of the modes obtained by decomposition (Iyengar and Raghukanth 2005; Huang et al. 2009a; Antico et al. 2014). But, a quantitative assessment to establish the teleconnections remained as an open problem. The multiscale dynamic (running) correlation analysis method called Time-Dependent Intrinsic Correlation (TDIC) can be explored to handle such issues (Chen et al. 2010). Even though some of the studies attempted to investigate the teleconnections of ISMR using the HHT (Iyengar and Raghukanth 2005), the possible associations of ISMR with ENSO, QBO, sunspot cycles, and tidal forcing were reported based on periodicity of resulting modes. Performing an in-depth running correlation analysis between different oscillatory modes (of specific periodicity) of monsoon rainfall and climatic oscillation may give more insights to such links and TDIC analysis is believed to be a viable tool to perform such analysis. Therefore, this chapter presents investigation of the hydroclimatic teleconnections of ISMR with global atmospheric/climate oscillations such as ENSO, AMO, QBO, and EQUINOO in different process scales by invoking the HHT-based TDIC analysis.

2 Methodology

HHT is an advanced spectral analysis method, which combines a data adaptive multiscale decomposition process and Hilbert transform used for the characterization of the geophysical time series. The decomposition stage, so-called EMD, results in the evolution of a set of orthogonal time series known as Intrinsic Mode Functions (IMFs) and a final residue. The noise assisted and ensemble averaged variants makes the traditional EMD more effective by separating the time series to distinct timescales. One such variant namely the complete ensemble EMD with adaptive noise (CEEMDAN) proposed by Torres et al. (2011) is used in the present study, details of which can be found elsewhere (Rao and Hsu 2008; Torres et al. 2011).

In the procedure of HHT, the IMFs of the signal $X(t)$ obtained from decomposition process are further subjected to HT, to get instantaneous frequencies and amplitudes. For investigating the hydroclimatic teleconnections in different process scales, a scale-specific correlation analysis is more appropriate. The results of HHT can be effectively utilized to perform a running correlation analysis between associated time series in different timescales. Chen et al. (2010) proposed a method namely

TDIC for determining scale-specific associations between two time series having comparable periodicity. In this method, the window size is selected adaptively (from the instantaneous frequencies obtained by HHT), ensuring stationarity of the series within the sliding window. Subsequently, Student's t-test is performed to examine statistical significance of correlation coefficients obtained from the analysis. The theoretical details of the method can be found in some of the past works (Chen et al. 2010; Huang and Schmitt 2014; Adarsh and Janga Reddy 2017). The output of the TDIC analysis can be represented as a triangular-shaped TDIC plot, in which the entities of horizontal axis (i.e., time axis) are the midpoints of the sliding window, while that of vertical axis are the size of the sliding window. Appropriate color scheme can be used to represent the significant correlation between the two correlated series. The value at the apex of the triangle will be the correlation coefficient between the time series pairs chosen considering the complete data length, if maximum window size is equal to the data length (Chen et al. 2010).

3 Database

In this study, the AISMR data for the period 1950–2012 is collected from Indian Institute of Tropical Meteorology (IITM) Pune (http://www.tropmet.res.in). The monthly records/data of QBO was acquired from the Web site of Earth System Research Laboratory of National Oceanic and Atmospheric Administration (NOAA) organization for the same period. The El Niño events are generally quantified based on the average sea surface temperature (or SST) anomalies over different Niño regions (defined by specific latitudes and longitudes) in the Pacific Ocean. In the past studies, it was also found that ISMR is best correlated with temperature anomaly from Niño3.4 region, which overlaps between Niño 3 and Niño 4 (Maity and Kumar 2006a; Kumar et al. 2007). The data related to Niño3.4 region (120°W–170°W, 5°S–5°N) called as Oceanic Niño Index (ONI) for the period 1950–2012 was obtained from National Weather Service Climate Prediction Center of NOAA and engaged as the ENSO index. The data of monthly AMO for the period of study (1950–2012) was also obtained from the same centre. The negative of the anomaly of the zonal component of surface wind in the equatorial Indian Ocean region (60–90°E, 2.5°S–2.5°N) was used as EQUINOO index (Maity and Kumar 2006a) and this data was extracted from the Web site of National Centre for Environmental Prediction (NCEP).

4 Results and Discussion

4.1 Multiscale Decomposition and Correlation Analysis

First, the monthly AISMR time series and four of climate oscillation indices series for the period 1950–2012 are subjected to multiscale decomposition operation using the CEEMDAN method, after setting the noise parameter as 0.2 for rainfall, 0.02 for climate indices, and number of realizations as 500 for all cases, following the past studies (Antico et al. 2014). The decomposition process has resulted in six IMFs and a final residue for different series and the mean period of the resulted oscillatory modes are presented in Table 1.

Table 1 shows that the period of evolution of AISMR and climate oscillations are similar. For higher order modes, there exist some differences in periodicities, which may be due to the limited number of cycles in these modes. The correlation between the modes of AISMR time series and that of climate oscillations is computed and the values are presented in Table 2. From Table 2, it is noted that large value of correlation exists only between the trend components or higher order modes of the time series pairs chosen. In short, it can be inferred that the linear association between climate oscillations and monsoon rainfall gets displayed in the low-frequency part of their spectra.

From the above correlation analysis, it can be postulated that since the nature of association between monsoon rainfall and climate oscillations is very complex, certain processes involved may be correlated with each other only at some of the process scales. Also, the correlation coefficients between rainfall and climate oscillations are not alike at different scales. This is evident in the link between ENSO and ISMR. From Table 2 it can be noticed that the residue components of the two series are negatively correlated (correlation coefficient of −0.992), while a strong positive correlation (correlation coefficient of 0.945) is noticed between IMF6 of the two time series. This is a remarkable observation, as the positive correlation at one scale would have been counteracted by the negative correlation at another scale,

Table 1 Mean periodicity (in months) of modes of monthly AISMR and different climate oscillations for the period 1950–2012

Mode number	Rainfall	Climate oscillations			
	AI	QBO	ENSO	AMO	EQUINOO
1	3.231	3.150	3.600	3.360	3.150
2	6.300	9.333	8.400	7.636	6.811
3	13.263	18.000	15.750	14.000	14.000
4	28.000	31.500	28.000	28.000	25.200
5	42.000	63.000	50.400	84.000	50.400
6	84.000	126.00	126.00	126.00	126.000
7	252.000	252.00	252.00	252.00	252.000

Table 2 Correlation coefficients between orthogonal modes of monthly AISMR (1950–2012) and that of climate oscillations at the corresponding scale (italic figures indicate significant correlation, p value < 0.05)

Mode of rainfall	Correlation coefficient with the mode of			
	QBO	ENSO	AMO	EQUINOO
IMF1	−0.085	−0.031	0.085	*−0.349*
IMF2	0.019	−0.347	−0.044	−0.154
IMF3	*−0.407*	−0.399	0.074	−0.151
IMF4	−0.115	−0.100	*−0.206*	*−0.256*
IMF5	*−0.395*	−0.154	*−0.401*	−0.041
IMF6	***0.879***	***0.945***	−0.026	*−0.496*
Residue	***0.999***	***−0.992***	***−0.992***	***−0.985***

which eventually leads to overall low correlation between ISMR and a specific climate oscillation. In this case, the overall correlation between ISMR and ENSO is only −0.21, while the correlation values are 0.01, 0.013, and −0.33, respectively, in the relationships with QBO, AMO, and EQUINOO. The above-stated values are quite low numerically as the correlation coefficient depicts only the linear association while the true associations might be of nonlinear in characteristics. It may also be noted that the climate forcing of varying periodicities may be in action for many shorter time spells only and it may have significant impact on the rainfall processes. To capture such associations, a correlation analysis considering shorter time spells are very useful; therefore, this study adopted TDIC analysis for an in-depth running correlation analysis of IMFs, which may give more insights into the existing linkages between ISMR and climate oscillations.

4.2 TDIC Analysis of AISMR and Climate Oscillations

TDIC analysis is performed between the different IMF pairs (of monsoon rainfall and climate oscillation) and developed the TDIC plots. It is to be noted that the TDIC plots may not show visible color if the Student's t-test for the correlation fails or if there is considerable difference between the periodicities of the pair of modes under consideration. To show the usefulness of TDIC analysis, first the ENSO-rainfall link is considered and the TDIC plots for different IMFs are provided in different panels of Fig. 1.

The bottom contour of the triangular plots portrays the instantaneous frequencies and an upward shift is noticed in different plots when we consider the plots of low-frequency IMFs. The blank spaces within the triangular plots convey the information that such correlations fail to satisfy the Student's t-test and hence are not statistically significant. From the results of study, it is noticed that ENSO shows a long-range negative association with AISMR at process scales of IMF2

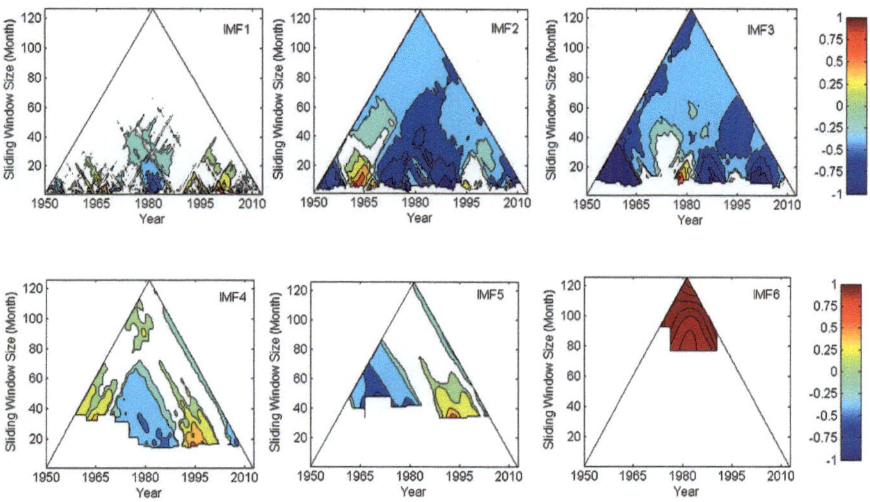

Fig. 1 TDIC plots of IMFs of ENSO and that of monsoon rainfall. Only significant correlations (at 5% significance level) are shown in different TDIC plots

(periodicity of $8.4 \times 4/12 = 3$ year) and IMF3 (periodicity of $15.7 \times 4/12 = 5$ year). Thus, it can be concluded that both oscillatory modes may be contributed by similar physical processes like westerly wind bursts and/or same oscillatory patterns such as Madden Julian Oscillations (MJO) which vary in intra-seasonal scales (Wang and Picaut 2004). On examining the correlations along the time domain, it is noted that at many of the shorter time spells the correlations are positive (e.g., in IMF2 during ~1964–66; in IMF3 during ~1979–82; in IMF4 during 1995–1998). It is well known that the strongest El Niño of the past century (1997–98) elicited an Indian Ocean Dipole (IOD), which resulted in above-average rainfall during the period 1997–98 (Kumar et al. 2006). From Fig. 1, it is noted that for IMF4 (of ~7-year periodicity), the correlation between the AISMR and ENSO is strongly positive during 1995–98. This result agrees with the results presented by Chen et al. (2010) who investigated the link between IOD and ENSO using the TDIC method. Also, it is well understood that in 2001–2002, Indian monsoon weakened due to the effect of ENSO, and it is observed that IMF2 and IMF3 show very strong negative correlation with ISMR during this period. Moreover, it is observed that there is rich dynamics in the nature of correlations (switching from positive to negative and vice versa) between ENSO and AISMR in particular at the high-frequency mode IMF1 and such associations are more apparent for smaller window size. The TDIC plots of QBO, EQUINOO, and AMO with ISMR are presented in Figs. 2, 3, and 4, respectively.

Figure 2 shows that the correlations between QBO and AISMR are negative at almost all scales except for IMF6. However, a direct correlation between QBO and ISMR is observed in the period ~1967–1994 for the IMF2. For IMF3, the relation between QBO and AISMR is primarily long-range negative association. But, some

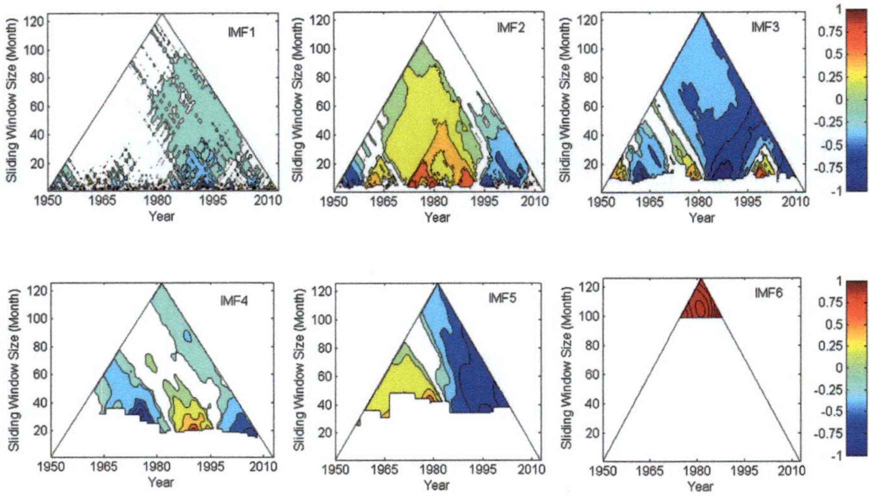

Fig. 2 TDIC plots of IMFs of QBO and that of monsoon rainfall. Only significant correlations (at 5% significance level) are shown in different TDIC plots

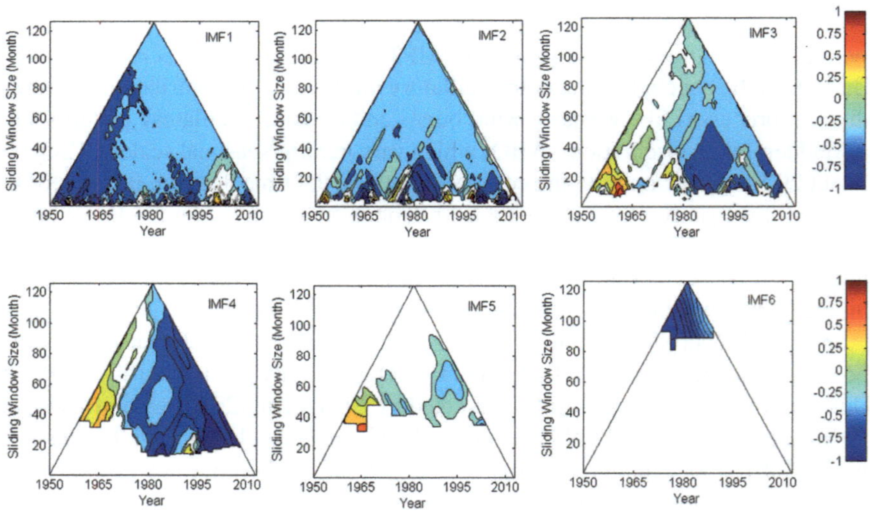

Fig. 3 TDIC plots of IMFs of EQUINOO and that of monsoon rainfall. Only significant correlations (at 5% significance level) are shown in different TDIC plots

significant direct relation between the two is noticed during short-term spells of 1953–55, 1978–80, and 1997–98.

From Fig. 3, it is noticed that for most of timescales the EQUINOO–AISMR relation is strongly opposing nature, and also very localized direct correlations exist

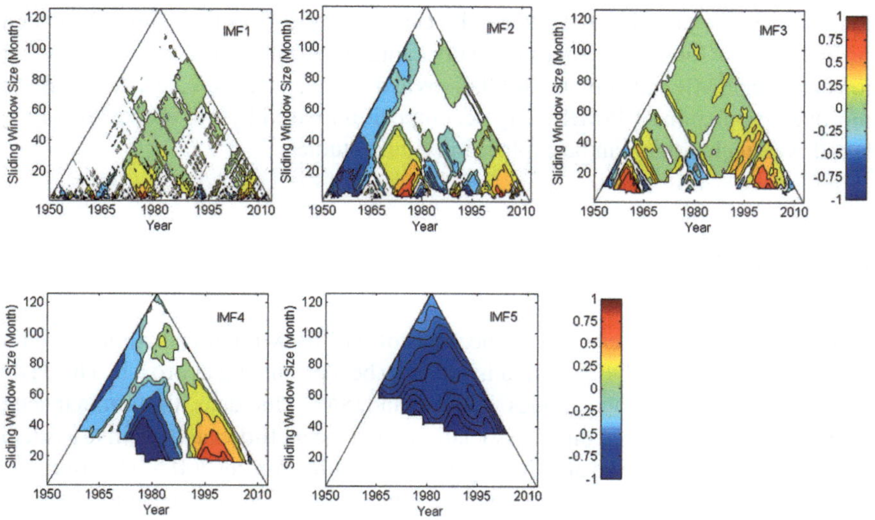

Fig. 4 TDIC plots of IMFs of AMO and that of monsoon rainfall. Only significant correlations (at 5% significance level) are shown in different TDIC plots

between the two for some shorter timescales (e.g., prior to 1960s in IMF3, IMF4, and IMF5, IMF1 in 1995–1998).

From Fig. 4, it is noticed that a weak positive correlation exists between AMO and AISMR in different process scales with frequent alterations in the nature of association between the two. Interestingly, all the first four modes show a strong positive correlation with monsoon rainfall in the recent past (~1997–2010). During 1997–98, the correlation between the two is strongly positive at all scales and some of the past studies also noted the modulation of Indian monsoon by AMO and its link with ENSO (Goswami et al. 2006). In IMF5, there is a long-range negative association between the two.

5 Future Prospects

The multiscale correlation analysis using TDIC method is an effective method for capturing the improved understanding of teleconnections of ISMR with global climatic oscillations. From the TDIC analysis of hydroclimatic teleconnections, it can be concluded that both the strength and nature of dependency between ISMR and climatic oscillation could be different at different process scales. There is a clear evidence of switchovers in the type of association (reversals in nature of correlation from negative to positive or vice versa) between climatic oscillations and ISMR over the period of observation. Further investigation is needed to elucidate the exact reasons or physical processes behind such alterations. Also, the "*lead–lag*" correlation between the correlated series can be captured by performing Time-Dependent Intrinsic Cross-Correlation Analysis. After capturing the non-unique type of association between ISMR and climatic oscillations at different process scales, the efforts for improved prediction of ISMR using decomposition-based hybrid models involving data-driven techniques can be explored in future studies.

6 Conclusions

To investigate the multiscale teleconnections of AISMR with global climate oscillations, the HHT-based TDIC method is found to be a promising approach. The TDIC analysis between orthogonal modes of ENSO and ISMR elucidated the dominant negative association, localized direct associations, and the well-debated positive association during 1997–98 period and the negative association during 2002–03 period. The study recognized that there exists a strong long-range negative correlation between ISMR and EQUINOO along with the respective short spell anti-correlations. In general, both the strength and nature of associations in the link between the climatic oscillations and ISMR vary with timescales and it is not of unique character over the time domain and the exact reasons for such alternations are yet to be investigated. The methodology can be applied between any two relevant time series of concern for understanding the complex dynamics of hydroclimatic variables.

References

Adarsh S, Janga Reddy M (2017) Multiscale characterization and prediction of monsoon rainfall in India using Hilbert-Huang transform and time dependent intrinsic correlation analysis. Meteorol Atm Phys. https://doi.org/10.1007/s00703-017-0545-6

Antico A, Schlotthauer G, Torres ME (2014) Analysis of hydro-climatic variability and trends using a novel empirical mode decomposition: application to Parana river basin. J Geophys Res Atmos 119(3):1218–1233

Azad S, Narasimha R, Sett SK (2008) A wavelet based significance test for periodicities in Indian monsoon rainfall. Int J Wavelets Multi-Res Inf Process 6(2):291–304

Blanford HF (1884) On the connection of the Himalayan snow with dry winds and seasons of droughts in India. Proc Royal Soc London Series A 37:3–22

Chattopadhyay J, Bhatla R (2002) Possible influence of QBO on teleconnections relating Indian summer monsoon rainfall and sea-surface temperature anomalies across the equatorial pacific. Int J Climatol 22(1):121–127

Chen X, Wu Z, Huang NE (2010) The time-dependent intrinsic correlation based on the empirical mode decomposition. Adv Adapt Data Anal 2:233–265

Claud C, Pascal T (2007) Revisiting the possible links between the Quasi-Biennial oscillation and the Indian summer monsoon using NCEP R-2 and CMAP fields. J Clim 20:773–787

Feng S, Hu Q (2008) How the North Atlantic multidecadal oscillation may have influenced the Indian summer monsoon during the past two millennia. Geophys Res Lett 35:L01707. https://doi.org/10.1029/2007GL032484

Gadgil S, Vinayachandran PN, Francis PA, Gadgil S (2004) Extremes of the Indian summer monsoon rainfall, ENSO and equatorial Indian ocean oscillation. Geophys Res Lett 31:L12213. https://doi.org/10.1029/2004GL019733

Goswami BN, Madhusoodanan MS, Neema CP, Sengupta D (2006) A physical mechanism for North Atlantic SST influence on the Indian summer monsoon. Geophys Res Lett 33:L02706. https://doi.org/10.1029/2005GL024803

Grinsted A, Moore JC, Jevrejeva S (2004) Application of the cross wavelet transform and wavelet coherence to geophysical time series'. Nonlin Process Geophy 11(5/6):561–566

Huang Y, Schmitt FG (2014) Time dependent intrinsic correlation analysis of temperature and dissolved oxygen time series using empirical mode decomposition. J Mar Syst 130(2014):90–100

Huang NE, Shen Z, Long SR, Wu MC, Shih HH, Zheng Q, Yen NC, Tung CC, Liu HH (1998) The empirical mode decomposition and the Hilbert spectrum for nonlinear and non-stationary time series analysis. Proc Royal Soc London Series A 454:903–995

Huang Y, Schmitt FG, Lu Z, Liu Y (2009a) Analysis of daily river flow fluctuations using empirical mode decomposition and arbitrary order Hilbert spectral analysis. J Hydrol 454:103–111

Iyengar RN, Raghu Kanth TSG (2005) Intrinsic mode functions and a strategy for forecasting Indian monsoon rainfall. Meteorol Atm Phys 90:17–36

Kripalani RH, Kulkarni A (1997a) Climatic impacts of El Niño/La Nina on the Indian monsoon: a new perspective. Weather 52:39–46

Kripalani RH, Kulkarni A (1997b) Rainfall variability over south East Asia-connections with Indian Monsoon and ENSO extremes: new perspectives. Int J Climatol 17:1155–1168

Krishna Kumar KB, Rajagopalan B, Cane MA (1999) On the weakening relationship between the Indian monsoon and ENSO. Science 284:2156–2159

Kumar KK, Rajagopalan B, Hoerling M, Bates G, Cane MA (2006) Unraveling the mystery of Indian monsoon failure during El Niño. Science 314:115–119

Kumar DN, Reddy MJ, Maity R (2007) Regional rainfall forecasting using large scale climate teleconnections and artificial intelligence techniques. J Intell Syst 16(4):307–332

Liu PC (1994) Wavelet spectrum analysis and ocean wind waves. In: Foufoula-Georgiou E, Kumar P (eds) Wavelets in geophysics, pp 151–166

Maity R, Kumar DN (2006a) Bayesian dynamic modeling for monthly Indian summer monsoon rainfall using El Niño-Southern oscillation (ENSO) and Equatorial Indian Ocean Oscillation (EQUINOO). J Geophys Res Atmos 111:D07104. https://doi.org/10.1029/2005jd006539

Maity R, Kumar DN (2006b) Hydro-climatic association of monthly summer monsoon rainfall over India with large-scale atmospheric circulation from tropical Pacific Ocean and Indian Ocean region. Atmos Sci Lett 7(4):101–107

Maity R, Kumar DN (2007) Hydroclimatic teleconnection between global sea surface temperature and rainfall over India at subdivisional monthly scale. Hydrol Process 21(14):1802–1813

Massei N, Durand A, Deloffre J, Dupont J, Valdes D, Laignel B (2007) Investigating possible links between the North Atlantic oscillation and rainfall variability in northwestern France over the past 35 years. J Geophys Res Atmos 112: 1–10

Massei N, Laignel B, Rosero E, Motelay-Massei A, Deloffre J, Yang ZL, Rossi A (2011) A wavelet approach to the short-term to pluri-decennal variability of streamflow in the Mississippi river basin from 1934 to 1998. Int J Climatol 31:31–43

Nalley D, Adamowski J, Khalil B, Biswas A (2016) Inter-annual to inter-decadal streamflow variability in Quebec and Ontario in relation to dominant large-scale climate indices. J Hydrol 536(2016):426–446

Narasimha R, Bhattacharyya S (2010) A wavelet cross-spectral analysis of solar-ENSO-rainfall connections in the Indian monsoons. Appl Comput Harmonic Anal 28:285–295

Rao AR, Hsu EC (2008) Hilbert-Huang transform analysis of hydrological and environmental time series. Springer, Netherlands

Rao RKS, Lakhole NJ (1978) Quasi-biennial oscillation and summer southwest monsoon. Indian J Meteorol Hydrol Geophys 29:403–411

Rossi A, Massei N, Laignel B, Sebag D, Copard Y (2009) The response of Mississippi River to climate fluctuations of streamflow and suspended sediment load, 1950-1975. J Hydrol 377(3):237–244

Saji NH, Goswami BN, Vinayachandran PN, Yamagata T (1999) A dipole mode in the tropical Indian Ocean. Nature 401:360–363

Shukla J, Paolino DA (1983) The southern oscillation and long-range forecasting of the summer monsoon rainfall over India. Monsoon Weather Rev 111:1830–1837

Torrence C, Webster PJ (1999) Inter-decadal changes in the ENSO–monsoon system. J Clim 12:2679–2690

Torres ME, Colominas MA, Schlotthauer G, Fladrin P (2011) A complete ensemble empirical mode decomposition with adaptive noise. In: IEEE international conference on acoustic speech and signal processing, Prague, 22–27 May 2011, pp 4144–4147

Walker G (1923) Correlation in seasonal variations of weather, VII: a preliminary study of world weather. Mem India Meteorol Depart 24(4):75–131

Wang C, Picaut J (2004) Understanding ENSO physics—a review. In: Wang C, Xie S-P, Carton JA (eds) Earth's climate: the ocean-atmosphere interaction, geophysical monograph series, vol 147. AGU, Washington, D.C., pp 21–48

Urbanisation and Surface Urban Heat Island Intensity (SUHII)

Hiteshri Shastri and Subimal Ghosh

Abstract A complete understanding of urban environment becomes increasingly important in the present world where more than half of the human population resides in the urban areas. A warmer surface temperature exists in the urban area due to change in land cover and population density, indicated as urban heat island (UHI). Traditionally, UHI is linked with air temperature differences between urban and rural areas using pairs of in situ climatology data. The term surface UHI (SUHI) is used to explicitly distinguish UHIs measured using land surface temperature (LST). SUHII is a widely recognized surrogate index indirectly addressing the heat island effect. The urban climate largely differs from the non-urban or rural climate affecting not only the temperature but cloudiness, precipitation, water and air quality as well. The long-term monitoring of UHI is essential to better understand the urban climatology. Understanding the UHI behaviour has a concern primarily for the human health and energy consumption. The factors driving UHI is an important aspect of climate research. Several attempts have been made employing a range of methods to investigate the energy balance of global urban centres. India ranking at second largest urban population of the world is projected to have the highest urban growth rate in the next 30 years. The UHI studies in India have been undertaken mostly within individual cities using varied methodology. This chapter provides a concise compilation different studies characterizing UHI for urban centres of India. The chapter briefly reports results from the first assessment of seasonal and diurnal characteristics of SUHI overall large urban centres of India carried out by Shastri et al. in Sci Rep 7:40178 (2017).

H. Shastri
M. S. Patel Department of Civil Engineering, C.S. Patel Institute of Technology, Charotar University of Science and Technology, Changa, Anand, Gujarat, India
e-mail: shastrihiteshri@hotmail.com; hitehritarak@gmail.com

S. Ghosh (✉)
Interdisciplinary Program in Climate Studies, Indian Institute of Technology Bombay, Mumbai 400076, India
e-mail: subimal@civil.iitb.ac.in; subimal.ghosh@gmail.com

S. Ghosh
Department of Civil Engineering, Indian Institute of Technology Bombay, Mumbai 400076, Maharashtra, India

C. Venkataraman et al. (eds.), *Climate Change Signals and Response*,
https://doi.org/10.1007/978-981-13-0280-0_5

Keywords Urbanization · Urban heat island · India

1 Urbanization

Urbanization refers to the growth of towns and cities due to large proportion of the population living in and moving to the urban areas and its suburban. Urbanization is therefore defined as a population shift from rural to urban areas, the gradual increase in the proportion of people living in urban areas. With the industrial revolution in 1830s and the subsequent increase in the world's population, rapid land use–land cover (LULC) change has been underway. Forests have been altered to support farmland, grassland and other land uses; natural landscapes have been modified to support settlement causing urbanization to rise (World Urbanization Prospects 2014).

Globally, 30% of the world's population resided in urban areas during 1950s and this has been increased to 54% in 2014. By 2050, it is projected to rise up to 66% (World Urbanization Prospects 2014). It is further estimated that 90% of urban growth is taking place in the developing countries (World Urbanization Prospects 2014). The urban growth is particularly pronounced and more rapid in India, than expected (Census of India 2011). During the last 50 years, the population of India has been more than double, whereas the urban population has increased nearly five times over the same period of time (Census of India 2011). At present, India houses three megacities (one with a population of over ten million people) of the world (Mumbai, Delhi and Kolkata). The number of Indian megacities is expected to increase to six by the year 2021 (including Bengaluru, Chennai and Hyderabad), when India will have the largest concentration of megacities in the world (World Urbanization Prospects revision 2014). According to the World Bank's urban challenges studies for India, the estimated urban population of the country will reach 500 million by the year 2017.

2 Urban Climate and UHI Development

The fast-growing urban areas carry out huge anthropogenic activities that burden natural environment and its resources like air–water quality and space. The urban areas have different climatology to their rural surroundings. Urbanization development significantly modifies the moisture, radiation balance, thermal stability and aerodynamic properties at the surface level. The urban climate refers to the climate affected by the presence of a town or city. There exists a distinct urban boundary layer due to modification of the heat balances, fluxes, mass and momentum. The importance of understanding urban weather and climate, and the impacts of urbanization on regional weather and climate, continues to grow. The Intergovernmental Panel on Climate Change (IPCC) (Jones et al. 2007) noted that understanding the role of urban LULC and pollution on climate change is an important topic in future climate research. The urban climate research examines and understands the atmosphere in direct contact with urban land covers (Oke 1988). It mainly refers to understand the difference between the urban and the non-urban climate, the impact of UHI on the regional climate.

The changes in LULC have given rise to significant changes in the ecosystem structure, impacting its functions and induce serious environmental issues threatening sustainable development (Grimm et al. 2008). Different studies have addressed issues relating to urbanization in relation to energy, land use and climate (Grimm et al. 2008). The LULC changes associated with urbanization lead to the changes in climate as evident from enhanced land surface temperatures and formations of UHI (Landsberg 1981). The UHI intensity is conventionally measured as the difference in air temperature between an urban and a nearby rural location (Landsberg 1981). The UHI is observed to have a seasonal and diurnal variability. In addition, UHI patterns vary with the changes of regions (Imhoff et al. 2010). It is of great significance to characterize and monitor the ongoing urbanization processes. At the same time, it is equally important to understand, simulate and project the local and regional environmental effects and feedbacks associated with expanding urban centres (Grimm et al. 2008).

The land–atmosphere interaction is an important component of the earth system. It is largely realized that the urbanization can impact local weather and regional climate due to the land use changes and associated increased emission of greenhouse gases and aerosols. The urbanization impact alters the regional rainfall patterns through the UHI impacts, mesoscale convergences and urban aerosol interactions in a humid and convectively unstable environment. The impact of UHI on global water cycle through the development of clouds and changes of precipitation patterns within and around the urban areas is also widely recognized (Shepherd 2005). The UHI in warm climatic condition is of special concern due to the increased occurrences of heat waves (Basara et al. 2010). Even though there is a long-term increase in the rainfall amounts under global warming, the LULC change can also be an additional factor contributing to regional climate change (Kalnay and Cai 2003; Shepherd 2005).

3 UHI of Global Cities

The urban heat island is first documented by Luke Howard in 1820 in the city of London (Landsberg 1981). The contrast of surface energy exchange between urban and suburban areas is recognized as the major drivers of urban heat island (Oke 1988; Arnfield 2003). The two major sources of energy available to produce UHI is anthropogenic heat flux produced by the dense habitat over an urban region and downward net solar radiation. These sources of energy are converted into sensible heat fluxes, latent heat fluxes, surface heat storage and net heat advection (Arnfield 2003). The partitioning of energy is largely governed by the land cover (urbanization-vegetation). SUHI is a spatiotemporal phenomenon largely governed by the extent of urbanization. The UHI in different regions across the globe is observed to have similar characteristics, with a difference of temperature ranging from 1 to 4 °C between urban and nearby non-urban regions (Peng et al. 2012). The increased temperatures in urban areas may be connected with higher occurrences of heat waves, heat stress-related health problems (Grimm et al. 2008), increase of local convections and heavy precipitation events (Huff and Changnon 1972; Changnon et al. 1981; Shastri et al. 2015; Grimm et al. 2008), with consequent increases in hazard of extremes. High hazard, along with higher vulnerability due to rapidly growing populations and infrastructure, leads to higher risk.

The analysis of 419 global big cities shows positive SUHII with a diurnal variation, as computed from Moderate Resolution Imaging Spectroradiometer (MODIS) data (Peng et al. 2012). The average annual daytime SUHII (1.5 ± 1.2 °C) is reported to be higher than the annual night-time SUHII (1.1 ± 0.5 °C) ($P < 0.001$), with no correlation between the two (Peng et al. 2012). The SUHII in Europe is observed to be dependent on the size of urban regions with seasonal variations (Lei et al. 2014). The analysis of SUHII in the USA reveals dependence of UHI variation on the efficiency of heat convection to the lower atmosphere within urban and rural areas (Lei et al. 2014). An investigation of UHI in Asian megacities illustrates strong negative associations with the urban vegetative cover estimated with positive associations with built-up areas and normalized difference vegetation index (NDVI) (Tran et al. 2006). The study, however, does not investigate the relative contribution of these two factors. On the whole, different studies of global and regional scale suggest warmer urban region as compared to the nearby non-urban regions, with differential long-term trend (Zhou et al. 2015).

4 UHI of Indian Cities

The urban centres in India altogether have the population of 380 million (Census of India 2011). So far, studies of UHI in India have been taken mostly within individual cities using diverse methodology. Table 1 provides a brief listing and findings of UHI studies over India.

5 Diurnal and Seasonal SUHI of Major Indian Cities

A comprehensive analysis of the characteristics of UHI in all major urban centres of India is performed by Shastri et al. (2017). The study presents evaluation of SUHII of 84 cities in India with uniform methodology including both seasonal and diurnal characteristics along with the factors affecting the SUHI development. Amongst the selected 84 urban centres in India, nine have populations of greater than 5 million, 34 have population in the range of 1–5 million and 46 have population in the range of 0.1–1 million. The spatial extent of urban land cover is identified based on the LULC

Table 1 UHI studies for different urban centres of India

S. No.	Urban centre	Conclusion
1	Bangalore (Ramchandra and Kumar 2010)	LST is computed from Landsat thematic mapper (TM), enhanced thematic mapper (ETM) and MODIS LST data for different land use classes of the city. The estimated LST for years 1992, 2000 and 2007 reveals that increased urbanisation has resulted in higher LST due to high level of anthropogenic activities. LST-NDVI relationship is investigated and reported to have a negative correlation
2	Banglore (Ambinakudige 2011)	The UHI is evaluated with the help of Landsat ETM Plus data. A significantly lower mean temperature in the city core than the outgrowth zones is revealed by LST-land cover relationship. The reason behind which is indicated as the presence of vegetation water bodies and in the city's core. Within the different land cover classes, city core temperature is observed to vary from 1 to 7 °C
3	Chennai (Faris and Reddy 2010)	The LST difference between the urban and the surrounding area is assessed with the help of Landsat ETM Plus data for the year 2000. The results indicate that the land surface temperature values hold a positive correlation with dense built up and negative correlation with the vegetation cover
4	Chennai (Devadas and Lilly 2009)	The UHI is assessed through mobile recordings of urban air temperatures covering the major areas of the city for months May 2008 and January 2009. The results indicate UHI with an increasing air temperature in radial fashion from the suburbs towards the city centre. The mean UHI intensity is higher during winter than summer
5	Delhi and Mumbai (Grover and Singh 2015)	The UHI is assessed using the Landsat TM image of April 17, 2010, for the city of Mumbai and May 5, 2010, for the city of Delhi. Due to the mixed land use, the ridge forests and river cutting across the city, and considerable tree cover along roads; the UHI is observed to be not very prominent in the city of Delhi. Mumbai, on the other hand, processes a stronger UHI where the heat is trapped within, the urbanized zones, whereas the sea, creeks and lakes act as heat sinks. The stronger negative LST-NDVI correlation is observed over Mumbai than Delhi
6	Delhi (Mohan et al. 2009)	Surface meteorological observations were taken using multisite ground-based mini-weather stations and meteorological towers for 25–28 May 2008. The UHI intensity is found to be more both in the night and afternoon hours and dominant in areas of dense built up with intense human activity

(continued)

Table 1 (continued)

S. No.	Urban centre	Conclusion
7	Dehi (Mohan et al. 2013)	The UHI is assessed with a network of micrometeorological observational stations across the city. Dense commercial areas were observed to have highest UHI. The UHI based on in situ ambient temperature is compared with the satellite-derived LST, a reasonable comparison is observed between this datasets during night-time; however, the relation was found to be poor during daytime. The impact of land cover was also reflected with the built-up canopies reported largest gradient between air and skin temperature
8	Delhi (Babazadeh and Parvendra 2015)	The UHI is derived from Landsat data over the years 2000–2014. The study reveals that the UHI intensity is high during summer season compared to monsoon and winter seasons. The formation of heat island is controlled by vegetation density. UHI increases during night-time
9	Delhi (Pandey et al. 2012)	The monthly day and night-time UHI is observed with the help of MODIS satellite data during the period 2007–2010. A prominent night-time UHI over central parts of the city is reported almost throughout the year. However, during the daytime an urban cool island is observed for the months of May, June, November and December. The study reports strong negative correlation between UHI intensity and mean monthly aerosol optical depth indicating a significant role played by aerosols in governing the urban thermal structure
10	Hyderabad (Franco et al. 2015)	The growth and LST of urban built-up areas is observed with Landsat 5 TM data of the years 1989, 1999 and 2009. The study shows a 67% loss in water bodies, while the urban built-up areas have grown by 270%. The temperature profile shows a dip in temperature while encountering water bodies, gardens and parks
11	Hyderabad (Badarinath et al. 2005)	The day and night UHI is observed with the help of AATSR satellite data; field campaigns are conducted in synchronous with the satellite overpass for validation purpose. The satellite-derived surface temperature values are found to be within 1 °C from ground measured values. The night-time heat island formation observed with core urban regions showing high temperatures

(continued)

Table 1 (continued)

S. No.	Urban centre	Conclusion
12	Kochi (Thomas et al. 2014)	The UHI intensity is quantified with mobile surveys carried out from January 2011 to March 2013. The pre-dawn UHI is observed to be more intense than early night UHI, also intensity in winter is stronger than in summer. The study area is classified into different local climate zones (LCZ), thermal gradient and cooling rates are observed within the zones and validated with the LCZ classification. The maximum UHI intensity is seen in the central part of the city
13	Nagpur (Kotharkar and Meenal 2015)	Traverse surveys were carried during the summer and winter seasons, for the years 2012–2014, to measure night-time mean canopy UHI intensity. Canopy UHI effects were found to be most prevailing in high building and population density areas. The negative impact of vegetation and positive effect of population density is revealed
14	Pune (Deosthali 2000)	A mobile survey is conducted in April 1997, and the recording of dry bulb temperature and wet bulb temperature is obtained for city. The results indicate that at the sunrise time the core of city appears as heat and dry island, whereas at night-time it appears as both heat and moisture island. There are several morphological variations in hill slope around the city, including dense vegetative cover, barren lands and residential areas. These variations induce a cool airflow of varying intensity. The presence of vegetative cover induces formation of a cool pool resulting in the linearity in the main heat island in downwind direction, whereas warm pockets are developed within the built-up areas. The city processed a basin-like topography, due to which the influence of winds becomes dominant. The thermal circulation system within the city arises from spatial uniformity in the pattern of temperature and humidity
15	Thiruvananthapuram (Ansar et al. 2012)	The variation in air temperature variations was recorded by mobile traverse across the study area. The stationary air temperature recorders are used to record warming and cooling rates in urban and suburban areas. The intensity of UHI at the urban centre is observed to be 2.4 °C. The study also observed significant distinction between the rate of cooling in the urban and rural areas

(continued)

Table 1 (continued)

S. No.	Urban centre	Conclusion
16	Bhopal (Gupta et al. 2009)	The study estimates UHI with the help of Landsat TM thermal bands. The surface temperature. GIS takes into account the spatial scenario. The combination of both provides a model which presents a city map where one can find the buildings which require attention in the sense of energy consumption. This would help in restructuring urban plans, traffic diversions, alternate cooling mechanisms, etc. Therefore, mitigation of thermal pollution would indeed help in planning urban development with minimum environmental degradation
17	Ahmedabad (Joshi et al. 2015)	The study of UHI effect using Landsat ETM satellite data and field measurements is carried out. The field temperature in various zones of the city is measured using IR Gun. As compared to the suburban areas in the city, the surface temperature near dense urban and industrial areas and are reported to be higher
18	Mumbai (Maral and Mukhopadhyay 2002)	The UHI is estimated for the time period 1976–2007 based on the meteorologically observed surface air temperature over the two urban stations of the city Mumbai and two peripheral non-urban stations. The study reveals that UHI is prominent during winter season for both day and night-time than the summer season
19	Noida (Kikon et al. 2016)	The UHI is assessed with the help of field data, meteorological observations and Landsat thermal dataset for year 2000 and 2013. Estimated UHI showed a negative correlation between NDVI, Emissivity and temperature whereas, NDBI, Albedo and temperature showed a positive correlation. The change in temperature is reported mainly due to increase in impervious areas

map from MODIS, during 2008. The city clustering algorithm (CCA) (Rozenfeld et al. 2008) is used to determine the extent of urban area for all the 84 cities. The percentage of urban to the nearby non-urban area is preserved as 50–150%, following earlier studies (Zhou et al. 2015). Figure 1 presents four different urban clusters, namely New Delhi, Hyderabad, Ahmadabad and Varanasi (Fig. 1a–d, respectively) having varied size, and they are from different geographical location in the country.

Here, it is to be noted that the selected non-urban area does not include suburban belts which have built-up areas as dominant land use. The CCA includes peripheral suburban region as a part of the corresponding urban centre as they have urban built up as the dominant LULC.

The spatial average of observed SST over all the grid points for an urban (buffer) region is taken to compute the mean LST over an urban (non-urban) region. For any kth selected urban (non-urban), the mean LST is computed as

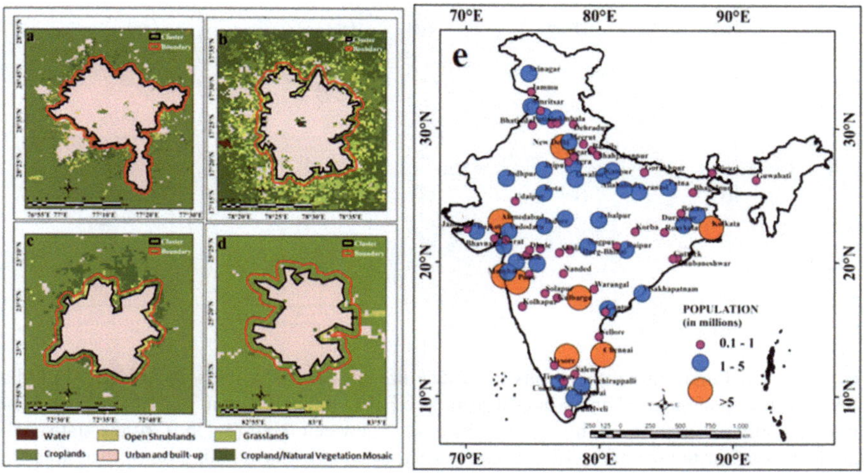

Fig. 1 Selection of urban and surrounding non-urban region **a** identified urban cluster of New Delhi, the pink area represents the cluster for urban centre, while its boundary is outlined with black line. The green areas are other surrounding region near an urban centre, where the land cover is characterized by a large share of crop areas. The boundary of nearby non-urban region is outlined with red line. **b–d** same as (**a**) for the urban centres of Hyderabad, Ahmadabad and Varanasi, respectively. **e** Location of cities with their population

$$T_{(k)}^{U/NU} = \frac{1}{n}\sum_{i=1}^{n} T_{i(k)}^{U/NU} \tag{1}$$

The study evaluates mean LST during 2003–2013 using eight-day composite LST data MODIS-aqua. The difference between LST over an urban and nearby non-urban region is defined as SUHII ($\Delta T_{(k)}$).

$$\Delta T_{(k)} = T_{(k)}^{U} - T_{(k)}^{NU} \tag{2}$$

The study excludes all the grids with LST error more than 3 °C while evaluating mean LST of the urban (non-urban) regions for quality control. The SUHII evaluation with this study for a given day is made only if minimum 50% of the grids LST values are available for both urban as well as nearby non-urban regions. The daytime (early afternoon ~13:30) and night-time (night ~01:30) SUHII is computed separately during winter season (months of December–January–February) and summer season (months of March–April–May). The analysis does not consider the months June to August as summer due to the dominance of summer monsoon rainfall during those months. The SUHII is tested for statistical significance at the 95% significance level with the help of Student's *t* tests. The study makes observation on changes in vegetation condition is made with the help of MODIS-aqua sixteen-day composite normalized differential vegetative index (NDVI) and evapotranspiration (ET) data. The ERA-Interim reanalysis data are accessed to check the background surface-level

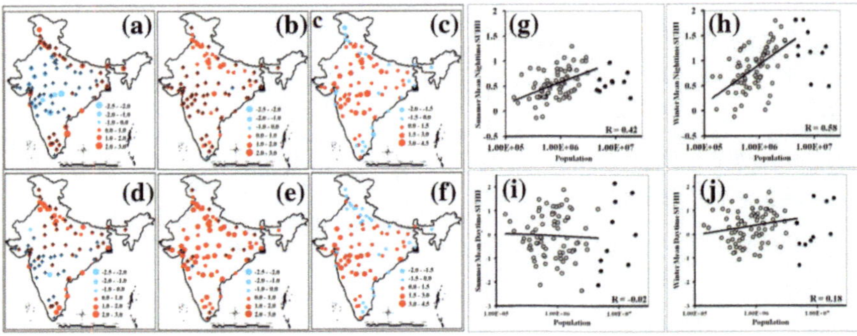

Fig. 2 Seasonal and diurnal variations in surface urban heat island intensity (SUHII) in Indian cities, sensitivity of summer and winter daytime SUHII to the population. SUHII during summer day **a** and summer night-time, **b** with the differences between them (**c**). In **a**, **b**, the statistically significant surface temperature differences between urban and surrounding non-urban areas are shown with "+". **e–g** are same as (**a**), **b**, **c** but for winter season. Sensitivity of summer night-time, **g** winter night-time, **h** summer daytime, **i** and winter daytime, **j** SUHII to the population. Both summer (**g**) and winter (**h**) night-time SUHII are positively correlated with population of the city. Adapted from Shastri et al., 2016 under Creative Commons Attribution 4.0 License

air temperature. The particular reanalysis data are provided by European Centre for Medium-Range Weather Forecasts (ECMWF) (Dee et al. 2011). The study also examine possible role of BC aerosols, with the help of the BC emission data at 0.25° latitude spatial resolution, computed following earlier studies (Sadavarte and Venkataraman 2014; Pandey et al. 2014). The association of SUHII with population, background air temperature, vegetative condition as well the BC emission density is computed with the help of correlation analysis.

The study observes that during summer days, particularly in Gangetic Basin and Central India, majority of the urban centres in India has statistically significant negative SUHII (Fig. 2a). This is completely in contrast to the positive values that are presumed to occur as an impact of urbanization. UHI is prominently observed during summer nights, with SUHII become significantly positive over central India at almost all locations (Fig. 2b). During winter days, positive SUHII is observed in urban regions of the Gangetic Basin. At the same time, southern and west central India negative SUHII is observed (Fig. 2d). During night-time in winter season, leaving just one location all the urban centres observes positive SUHII is (Fig. 2e). The differences between daytime and night-time SUHII are similar during both summer and winter season (Fig. 2c, f). The observed mostly positive differences between night-time and daytime LST in the summer season indicate a strong diurnal characteristic of SUHII within interior of India (Fig. 2c).

The study observes that for both summer and winter season, night-time SUHII is significantly correlated with population in the particular urban centre (Fig. 2g, h). However, the megacities do not reveal the similar correlation, possibly because in megacities the population pattern and built-up areas are different as compared to other urban areas. At the same time, SUHII development during daytime does not reveal any significant correlation with strength of urban population (Fig. 2i, j).

The study investigates and discusses the findings on unexpected negative daytime SUHII in the summer season. The reduced vegetation cover in non-urban regions is found to be responsible for the unusual SUHII during summer daytime. A very low vegetation cover (NDVI) over India is observed during the dry summer (pre-monsoon) compared to winter season (post-monsoon) (Fig. 3a, b). The changes in summer season NDVI between urban and nearby non-urban regions range from low positive to slightly negative (Fig. 3c). This pattern does not occur in winter (Fig. 3d). The difference in NDVI between urban and non-urban regions negatively correlates with SUHII in both summer and winter season (Fig. 3e, f). The differences in vegetative cover (NDVI) between urban and nearby non-urban regions are vastly negative during the winter season, resulting in positive SUHII during the winter season (Fig. 3g, h). The study delineates the types of land use within the non-urban regions. It is observed that for the larger part of non-urban regions, land is used as cropland (Fig. 3i). Low vegetation during the summer (pre-monsoon) results in barren land surfaces that have a lower albedo compared to built-up (urban) areas (Lim et al. 2005). Hence, the observations made by the study support the hypothesis that low vegetation in non-urban regions results in negative SUHII. The SUHII during night-time does not depend on vegetation condition, because in the absence of sunlight, albedo plays a minimal role. This results in night-time positive SUHII.

These observations largely reveal that during summer (pre-monsoon) season, majority of the locations in India behave similar to a dry arid region with low vegetation in the non-urban regions. Therefore, the SUHII characteristics are similar to a dry arid region. Negative UHI has been reported for urban locations in arid or semi-arid regions of north-western China (Quan et al. 2014), western US and central Asia (Peng et al. 2012). The cooling at the urban sites largely attributes to higher evaporation within cities resulting from human consumption of water and the increased ET from the protected vegetative cover in the urban regions. This results in modification sensible and latent heat fluxes at surface level.

The attribution of ET variation amongst urban and nearby non-urban region variation to SUHII during daytime in summer and winter season is examined. The non-urban regions that mostly comprises of croplands and grasslands (Fig. 3i) turn into barren land during the summer (pre-monsoon) period. This reduces the ET in non-urban regions. At the same time, urban regions have comparatively higher ET, largely consequential from consumption of water by humans and the increased ET resulting from the planted vegetation. In the urban regions, this impacts the latent sensible and latent heat flux, resulting into cooler urban places as compared to the nearby non-urban regions. The ET during the two seasons is presented in Fig. 3j, k with the respective SUHI. The plots show negative and statistically significant correlation between the differences in ET and SUHII (Fig. 3n, o). This further establishes the association mentioned above. The positive SUHII during winter season results mainly from highly negative difference in ET between urban and non-urban regions during winter (post-monsoon) season (Fig. 3l, m). The summer daytime temperature within the urban centres is therefore lower than the surrounding non-urban because of

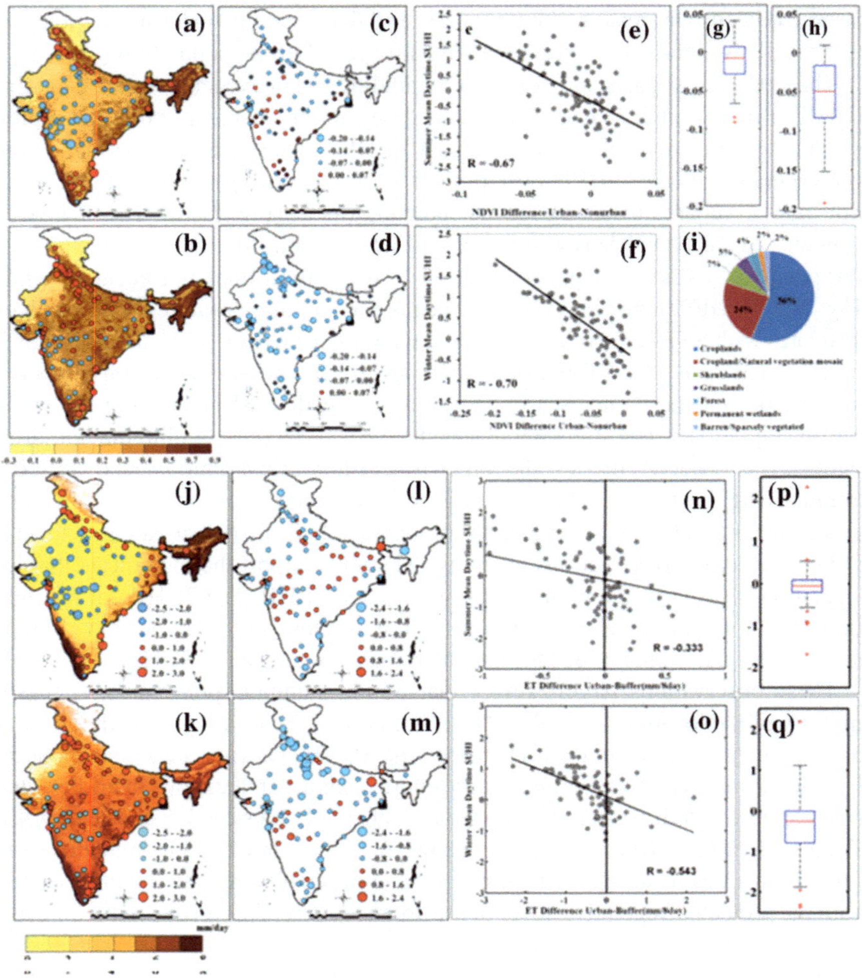

Fig. 3 SUHII association with vegetation cover. The daytime SUHII for summer (**a**) and winter (**b**) season are overlaid on the vegetation cover. The estimated differences in vegetation cover between urban and nearby non-urban regions during summer (**c**) and winter (**d**) season. The SUHII during summer and winter daytime is negatively associated with differences in vegetation cover between urban and nearby non-urban region (**e, f,** respectively). **g, h** presents the differences in vegetation cover between urban and nearby non-urban regions, respectively, for summer and winter season. The nearby non-urban regions largely comprise of cropland (**i**). The UHI indications with red and blue circles are same as Fig. 2j–q same as a–h but for ET. Adapted from Shastri et al., 2016 under Creative Commons Attribution 4.0 License

evaporative cooling of urban area. The ET in non-urban regions increases during winter (post-monsoon) season, resulting into a positive SUHII. Thus, the study explains that changes of sign in the SUHII from dry summer to the winter (post-monsoon).

In north and central India, the opposite seasonal patterns of SUHII are reported between summer and winter season (Fig. 2) with positive winter daytime SUHII, which is in contrast to negative summer daytime SUHII. Daytime land surface temperature could also be influenced by the atmospheric abundance of radiation absorbing constituents, such as black carbon (BC) aerosols, which are pollution particles emitted from incomplete combustion of fuels that strongly absorb radiation over the entire solar spectrum (Bond et al. 2013). Radiation absorption from BC can lead to heating of the atmospheric layers in places where these particles are abundant at instantaneous rates of up to several degrees kelvin per day (Kuhlmann and Quaas 2010). The possible role of BC aerosols is examined with the help of the spatial SUHII plots overlaid on BC emission fluxes. The BC emission fluxes spatially distributed on a 25 km grid are obtained from the recent emissions inventory for India (Sadavarte and Venkataraman 2014; Pandey et al. 2014). The BC spatial distribution in winter revealed larger emissions in the Indo-Gangetic plain and central India, as well as some clusters in the west coast (Ahmadabad-Mumbai belt), east coast and south India, corresponding to the density of users of biomass fuels (Fig. 4a, b). On average, BC emissions and emission density are larger in northern India in the winter months (DJF) than in the summer months (MAM). A BC-mediated mechanism of radiation flux change would only manifest during daytime, from solar radiation absorption by BC. Thus, comparing daytime SUHII values in winter with those in summer, it is observed that SUHII is associated with higher wintertime BC emission density in over 20 sites, where SUHII is positive (red) in winter but negative (blue) in summer (Fig. 4a, b).

This change in SUHII indicates urban land surface temperatures exceeding those in adjoining areas in the winter months, co-located with increases in BC emission density, while falling behind adjoining areas in summer months. Further, at approximately 10 more sites in northwest and central India, negative SUHII values are reduced (smaller blue circles as in Fig. 4a, b) in winter compared to SUHII values in summer, once again indicating that land surface temperatures in urban areas with higher BC emissions increase compared to temperatures in adjoining areas in the winter months. A correlation analysis of point SUHII with BC emission density (Ton grid^{-1} mon^{-1}), however, failed to show a significant association (Fig. 4c, d). This is not unexpected because changes in the radiation balance would be linked to the atmospheric concentration of BC in the surface layer and its vertical distribution rather than to emissions. In winter months, the prevailing meteorology in north India leads to low mixed layer heights and poor ventilation rates (Tripathi et al. 2006), which concentrate pollutants close to the surface. Thus, emission distributions, such as those analysed in this work, reveal a possible influence of BC emissions on SUHII. Further analysis is needed, utilizing such emissions in regional climate models to calculate columnar concentrations of BC and subsequent radiative effects, which could influence the surface radiation balance.

The temperature also observes a strong diurnal and seasonal variability over the Indian sub-continental region. The air temperature over Indian land mass reduces from summer to winter months owing to the differential solar radiative heating (Fig. 5a, b). The air temperature further reduces from day to night with both the

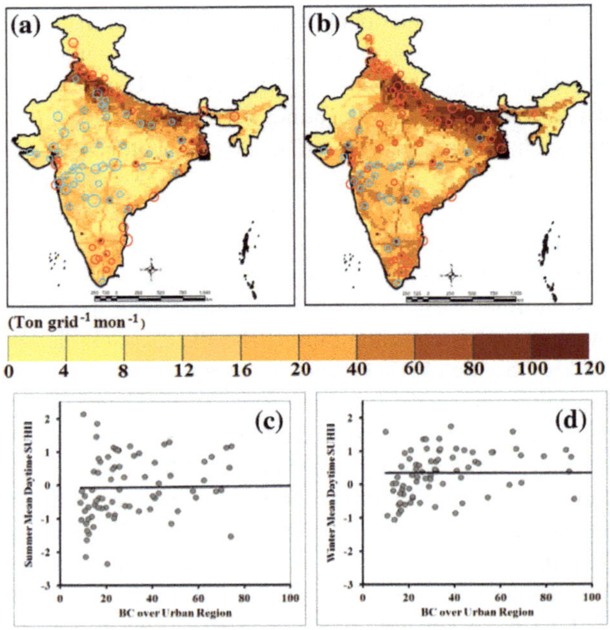

Fig. 4 Effect of seasonality in black carbon (BC) emissions on seasonal variability of SUHII. The SUHII during summer and winter daytime is overlaid on the BC emission colour map (**a**, **b**, respectively). The red and blue circles represent the same as in Fig. 1. Sensitivity of the daytime SUHII to black carbon (BC) emission, **c** dependence of summer (**d**). Winter daytime SUHII on BC. Adapted from Shastri et al., 2016 under Creative Commons Attribution 4.0 License

individual season in absence of direct solar heating (Fig. 5c, d). The observation of surface-level air temperature indicates that the summer daytime high temperature over the peninsular and central highlands mainly results in the negative UHI (Fig. 5a).

The overall surface air temperature is observed to have negative correlation with SUHII during both winter and summer season (Fig. 5e, f). This points out that when the near-surface air temperature is high, the SUHII will be low, which is indicative of are latively lower LST in urban regions. Thus, remarkably the study reveals that in comparison to the other regions around globe the urban heat wave characteristics of India are different.

Further, it is observed that during summer, when temperature reaches at annual maxima the SUHII is negative for more than 50% urban centres (Fig. 5g, h). This observation very importantly leads to a conclusion that in India for majority of urban regions the intensities of heatwaves are lower as compared to the nearby non-urban regions. This finding is in contrast with the general understanding of urban climate, primarily due to low vegetation cover in non-urban regions in the summer season.

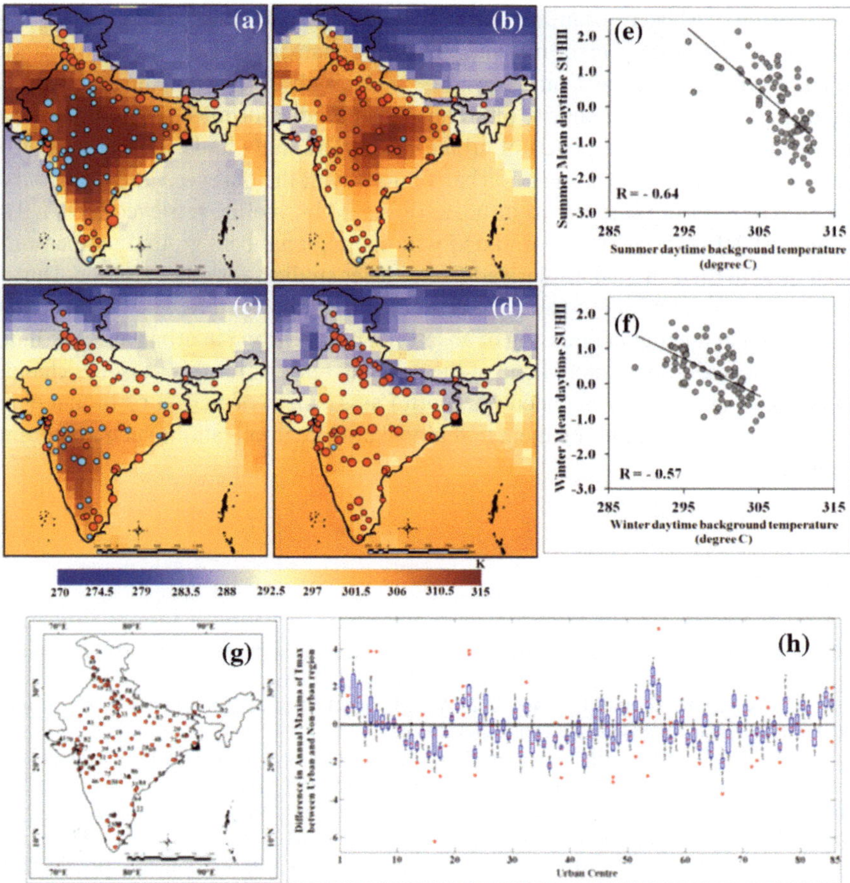

Fig. 5 Association of SUHII in India to the surface temperature. Attribution of SUHI to the surface-level air temperature (T_{air}); Summer daytime **a** night-time, **b** winter daytime, **c** night-time, **d** T_{air} overlaid with the corresponding SUHI, **e** dependence of summer, **f** winter daytime SUHI on T_{air}. **h** Differences of daytime LST between urban and nearby non-urban regions when the temperature reaches annual maxima at individual locations. Each of the boxes represents each city. Adapted from Shastri et al., 2016 under Creative Commons Attribution 4.0 License

6 Summary

Global and regional (USA, Europe, Asian megacities excluding India) studies revealed positive SUHII with seasonal and diurnal variability. The Census of India (2011) reveals an unprecedented urbanization growth over the country over last few decades. The urban population of the country is near to 3.8 million. However, the characteristics of Indian SUHII are highly overlooked. The studies have been undertaken mostly within individual cities using varied methodology, without any definite investigation of SUHII variability during different seasons, different time of the day. Shastri et al. (2017) present the first analysis on the characteristics of SUHII over India with its diurnal and seasonal variations. The analysis observes strong negative SUHII (urban cool island) over majority of urban areas in India during summer daytime, as opposed to the expected impacts of urbanization. The study finds strong association of the low vegetation cover in non-urban regions with the observed unexpected pattern of SUHII during summer daytime, as observed in satellite retrieved NDVI. The daytime SUHII is negatively correlated with the regional background temperature, as opposed to the conclusions derived from earlier studies, for other regions. During summer nights the urban impacts get prominent with positive SUHII. The analysis speculates that the higher emission of absorbing black carbon in urban areas of Indo-Gangetic plain during winter is one of the possible reasons for this seasonal change. This chapter emphasizes the need for careful assessment of SUHII of Indian cities for policy making and linking them to regional climate implications such as extreme precipitation.

References

Ambinakudige S (2011) Remote sensing of land cover's effect on surface temperatures: a case study of the urban heat island in Bangalore India. Applied GIS 71

Ansar S, Dhanya CR, Thomas G et al (2012) A study of urban/rural cooling rates in Thiruvananthapuram Kerala. J Ind Geophys Union 161:29–36

Arnfield AJ (2003) Two decades of urban climate research: a review of turbulence exchanges of energy and water and the urban heat island. Int J Climatol 23(1):1–26

Babazadeh M, Parvendra K (2015) Estimation of the urban heat island in local climate change and vulnerability assessment for air quality in Delhi. Eur Sci J ESJ 11(10)

Badarinath KVS, Chand TK, Madhavilatha K, Raghavaswamy V (2005) Studies on urban heat islands using ENVISAT AATSR data. J Indian Soc Remote Sens 334:495–501

Basara JB, Basara HG, Illston BG, Crawford KC (2010) The impact of the urban heat island during an intense heat wave in Oklahoma City. Advances in Meteorology

Bond TC, Doherty SJ, Fahey DW, Forster PM, Berntsen T, DeAngelo BJ, Flanner MG, Ghan S, Kärcher B, Koch D, Kinne S (2013) Bounding the role of black carbon in the climate system: a scientific assessment. J Geophy Res Atmosp 118(11):5380–552

Census of India 2011. Provisional population totals-India data sheet office of the Registrar General Census Commissioner India Indian Census Bureau. http://censusindia.gov.in

Changnon SA, Semonin RG, Auer AH, Braham RR, Hales J (1981) Metromex: a review and summary. Am Met Soc Met Mono, 40

Dee DP, Uppala SM, Simmons AJ et al (2011) The era-interim reanalysis: configuration and performance of the data assimilation system. Quart J Roy Meteor Soc 125:553–597

Deosthali V (2000) Impact of rapid urban growth on heat and moisture islands in Pune City India. Atmos Environ 3417:2745–2754

Devadas M, DA Lilly (2009) Rose urban factors and the intensity of heat island in the city of Chennai. In: Proceedings of seventh international conference on urban climate Yokohama, Japan

Faris AA, Sudhakar R (2010) Estimation of urban heat Island using Landsat-7 ETM+ imagery at Chennai city—a case study. Int J Earth Sci Eng 33:332–340

Franco S, Mandla VR, Rao KRM et al (2015) Study of temperature profile on various land use and land cover for emerging heat island. J Urban Environ Eng (JUEE) 91:32–37

Grimm NB, Faeth SH, Golubiewski NE et al (2008) Global change and the ecology of cities. Science 319(5864):756–760
Grover A, Singh RB (2015) Analysis of urban heat island (UHI) in relation to normalized difference vegetation index (NDVI): a comparative study of Delhi and Mumbai. Environments 2(2):125–138
Gupta A, Vivek D, Aditya G (2009) Mitigation of thermal pollution to enhance urban air quality through remote sensing and GIS. In: 10th ESRI India User ConfNoida (NCR)
Huff FA, Changnon SA (1972) Climatological assessment of urban effects on precipitation at St Louis. J Appl Meteorol 11:823–842
Imhoff ML, Zhang P, Wolfe RE, Bounoua L (2010) Remote sensing of the urban heat island effect across biomes in the continental USA. Remote Sens Environ 114(3):504–513
Jones PD, Trenberth KE, Ambenje P et al (2007) Observations: surface and atmospheric climate change. IPCC Clim Change, 235–336
Joshi R, Raval H, Pathak M et al (2015) Urban heat island characterization and isotherm mapping using geo-informatics technology in Ahmedabad City, Gujarat State India. Int Geosci 63:274
Kalnay E, Cai M (2003) Impact of urbanization and land-use change on climate. Nature 423(6939):528–531
Kikon N, Singh P, Singh SK, Vyas A (2016) Assessment of urban heat islands (UHI) of Noida City India using multi-temporal satellite data. Sustainable Cities Soc 22:19–28
Kotharkar R, Meenal S (2015) Land use land cover and population density impact on the formation of canopy urban heat islands through traverse survey in the Nagpur urban area India. J Urban Plann Dev 142(1):04015003
Kuhlmann J, Quaas J (2010) How can aerosols affect the Asian summer monsoon? Assessment during three consecutive pre-monsoon seasons from CALIPSO satellite data. Atmos Chem Phys 10:4673–4688
Landsberg HE (1981) The urban climate. Int Geophys Ser, 28
Lei Z, Xuhui L, Ronald BS, Keith O (2014) Strong contributions of local background climate to urban heat islands. Nature l511:216–219
Lim YK, Cai M, Kalnay E, Zhou L (2005) Observational evidence of sensitivity of surface climate changes to land types and urbanization. Geophys Res Lett 32:L22712
Maral SG, Mukhopadhyay T (2002) Signal of urban heat island (UHI) effect: a case study of Mumbai metropolitan region. Signal 551(522):540–53
Mohan M, Kikegawa Y, Gurjar BR et al (2009) Assessment of urban heat island intensities over Delhi. In: Proceeding of 7th international conference on urban climate, pp 1–4
Mohan M, Kikegawa Y, Gurjar BR et al (2013) Assessment of urban heat island effect for different land use–land cover from micrometeorological measurements and remote sensing data for megacity Delhi. Theoret Appl Climatol 112(3–4):647–658
Oke TR (1988) The urban energy balance. Prog Phys Geogr 12(4):471–508
Pandey P, Kumar D, Prakash A et al (2012) A study of urban heat island and its association with particulate matter during winter months over Delhi. Sci Total Environ 414:494–507
Pandey A, Sadavarte P, Rao AB, Venkataraman C (2014) Trends in multi-pollutant emissions from a technology-linked inventory for India: II residential agricultural and informal industry sectors. Atmos Env 99:341–352
Peng S, Piao S, Ciais P et al (2012) Surface urban heat island across 419 global big cities. Env Sci Tech 46(2):696–703
Quan J, Chen Y, Zhan W et al (2014) Multi-temporal trajectory of the urban heat island centroid in Beijing, China based on a Gaussian volume model. Remote Sens Env 149:33–46
Ramachandra TV, Kumar U (2010) Greater Bangalore: emerging urban heat island. GIS Dev 141:86–104
Rozenfeld HD, Rybski D, Andrade JS et al (2008) Laws of population growth. Procd Nat Acad Sci U S A 105(48):18702–18707
Sadavarte P, Venkataraman C (2014) Trends in multi-pollutant emissions from a technology-linked inventory for India: I industry and transport sectors. Atmos Environ 99:353–364

Shastri H, Paul S, Ghosh S, Karmakar S (2015) Impacts of urbanization on Indian summer monsoon rainfall extremes. J Geophys Res Atmos 120(2):496–516
Shastri H, Barik B, Ghosh S, Venkataraman C, Sadavarte P (2017) Flip flop of day-night and summer-winter surface urban heat island intensity in India. Sci Rep 7:40178
Shepherd JM (2005) A review of current investigations of urban–induced rainfall and recommendations for the future. Earth Interact, 9
Thomas G, Sherin AP, Ansar S, Zachariah EJ (2014) Analysis of urban heat island in Kochi India using a modified local climate zone classification. Procedia Environ Sci 21:3–13
Tran H, Daisuke U, Shiro O, Yoshifumi Y (2006) Assessment with satellite data of the urban heat island effects in Asian mega cities. Int J Appl Earth Obs and Geoinf 8:34–48
Tripathi SN, Tare V, Chinnam N (2006) Measurements of atmospheric parameters during Indian Space Research Organization Geosphere Biosphere Programme Land Campaign II at a typical location in the Ganga basin: 1 physical and optical properties. J Geophys Res 111:D23209
World Urbanization Prospects: 2014 Revision. United Nations Population Division Department of Economic & Social Affairs. New York
Zhou D, Shuqing Z, Liangxia Z et al (2015) The footprint of urban heat island effect in China. Sci Rep, 5

Future Heat Wave Projections and Impacts

Kamal Kumar Murari and Subimal Ghosh

Abstract Heat waves are a special class of climatic hazards that have a disastrous impact on many different systems. In the recent past, there has been a considerable improvement in scientific understanding of heat waves; however, most of this understanding is limited to the climatology of developed countries. There is a limited systematic understanding of heat waves and their impact on developing countries including India. This chapter reviews studies in establishing plausible links between climate variations, climate change and the heat waves, particularly in the context of India. The chapter also tries to understand the mechanisms responsible for the occurrence of heat waves. Further, the chapter shows evidence from studies that use climate models with statistical techniques for more accurate characterization of heat extremes and improving projections. Heat waves in India are expected to be intense, more frequent, and to be of longer duration in future due to global warming. This possibility will make the population more vulnerable to the impact of heat waves. The consequence of future heat waves might be severe; therefore, there is an urgent need to prepare a strategy to deal with its likelihood consequences. This is important in the context because the current policy does not consider heat waves as a serious hazard. The content of this chapter aims to provide the general public, policy makers and planners with the kind of effectual information which would enable them to understand and deal with the heat waves as a natural disaster.

Keywords India · Heat waves · Climate variability · Climate change · Hazards Climate models · Climate extremes · Natural disaster

---

K. K. Murari
Centre for Climate Change and Sustainability Studies, School of Habitat Studies, Tata Institute of Social Science, Mumbai 400088, Maharashtra, India
e-mail: kamal.iitd@gmail.com; kamal.murari@tiss.edu

S. Ghosh (✉)
Department of Civil Engineering, Indian Institute of Technology Bombay, Mumbai 400076, Maharashtra, India
e-mail: subimal.ghosh@gmail.com; subimal@civil.iitb.ac.in

C. Venkataraman et al. (eds.), *Climate Change Signals and Response*,
https://doi.org/10.1007/978-981-13-0280-0_6

1 Introduction

One of the most worrisome aspects of global warming is that as the earth becomes warmer overall, other facets of its climate patterns will also change. Extremes may rise everywhere. This is one of the reasons that climatic extremes have received a special attention by the scientific community. The recent Intergovernmental Panel on Climate Change (IPCC) assessment report also stresses upon the observed evidence of increase in climatic extremes and their impacts (Field et al. 2014).

Extreme temperatures are already having impacts on human health (Patz et al. 2005), agriculture (Lobell et al. 2012), forest and ecosystems (Teuling et al. 2010). The projections of increased intensity, duration and frequency also point to a serious concern for ecosystem and society (Field et al. 2014). However, assessment of the impact of extreme temperature on the socio-economic system is not straightforward because of the influence of other stresses such as social deprivation, disease and conflict (Easterling et al. 2000). The future changes in temperature extremes associated with human-induced warming may put some additional challenges by affecting both society and ecosystem (Stott et al. 2004).

All of these will have severe ecological, socio-economic and physical repercussions. It is essential, therefore, to focus on climate extremes such as heat waves and invest in understanding and preparing for them. This requires an understanding of characteristics of extreme temperature episodes, in the past as well as future, to develop a knowledge base about the future course of action. As things stand, however, even as the scientific understanding of human-induced climatic change grows, significant gaps yawn in the scientific understanding of climate extremes. This is largely due to the fact that warming is a global phenomenon, whereas extremes are local ones.

There are several methods of projection global and regional temperatures. General circulation models (GCMs), popularly referred as global climate models, is one such tool to project global and regional temperatures. On the global scale, these are all quite effective for the purposes that they are used, but not for monitoring or understanding localized climate patterns, especially extremes. Statistical methods and other data mining-based approaches would be more effective in this context. However, such methods are applied in the limited sense partially due to lack of availability of appropriate observed data at required spatial scale. Despite significant progress made in recent years, however, our ability to establish credible links between climate variations, climate change and climate extremes is still not robust enough. A great deal of work needs to be done to generate predictive insights and address the knowledge gaps.

There is no doubt that future pattern of heat waves might result in devastating impacts; the poor and vulnerable will be the most impacted (Mazdiyasni et al. 2017). Undoubtedly, India is more vulnerable to the impacts of heat waves, the rationale for this assertion is that it has a large proportion of the population which is poor and has no access to basic amenities such as water, electricity and primary health facility. In addition, a large proportion of population already live in areas where summer is much warmer. What do we know about the past patterns of heat waves? How do we measure them? How will heat waves be changed in the warming world? These are some of the key questions that are centred on the current scientific debate on the issues of climate variability and climate change. This chapter provides an overview of the current scientific literature on the subject and attempts to seek the answers to these questions. This chapter also reviews India-specific studies that utilized more accurate characterization of climate extremes and improving projections by a comprehensive assessment of uncertainties.

In the chapter, first two to six sections are themed around: understanding extreme temperature and heat waves; relevance of heat waves as a climatic hazard, and heat wave as a climatic házard in the Indian context and their projection using climatic models. By the end of this chapter, the reader is expected to have a clear understanding of how heat waves are distinctive with respect to the other climatic hazards. There are, of course, gaps and limitations in the literature, which, if filled, would provide a comprehensive understanding of issues that are discussed in the chapter. The chapter ends with concluding remarks that call for an overview of recent developments where future research on heat waves should focus.

2 Understanding Extreme Temperature and Heat Waves

This section gives a background on an understanding of heat waves. It covers the definition of heat waves and a background on the implication of the use of different definitions of heat waves. In doing so, the section gives an overview of the role of synoptic meteorology, land surface and soil moisture, climate variability and teleconnections, and global warming for the occurrence of heat waves.

2.1 Definition of Heat Waves

Heat waves are defined as a period of anomalously high temperature leading to hotter conditions (Robinson 2001); such episodes are common in the summer months in most places across the world. There is no universally accepted definition of heat waves, the definition of heat waves varies with geography. The definition also varies with the parameters used in the impact assessments by the user community.

The existing definitions of heat wave lead to various estimation of heat wave characteristics and impact (Smith et al. 2013). Heat wave characteristics refer to (1) intensity (maximum temperature during heat wave episodes, also referred as magnitude in some literature), (2) duration (length of heat wave episodes in days), (3) frequency (occurrence of heat wave episodes per year or per season), (4) the date of first occurrence of heat waves and (5) geographical exposure. The implication of the use of different definition of heat waves is that the inference of heat wave

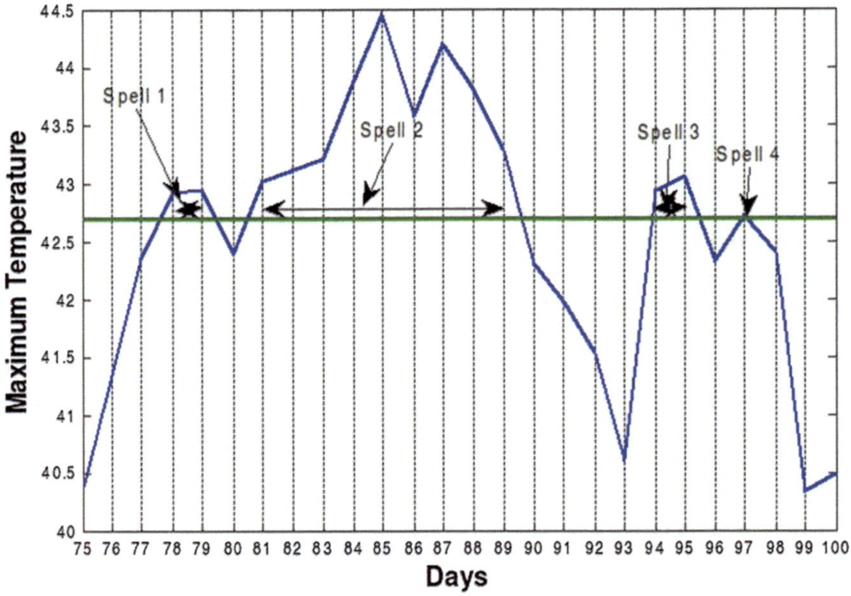

Fig. 1 A schematic illustration of heat wave characteristics. The blue line shows daily maximum temperature profile, and the green line shows the IMD threshold for heat waves in India. Days in the horizontal axis show the days of a year starting from January 1 of that year

characteristics changes with the choice of definition, and therefore, there is a change in impact associated with heat waves.

Figure 1 shows an illustration of heat wave characteristics—intensity, duration and frequency—using a daily maximum temperature data and threshold as suggested by India Meteorological Department (IMD). According to the IMD definition, if a day's temperature is about 6 °C above its long-term average or greater than 45 °C, the days are declared as heat wave day.[1] A heat wave can last for several days if the daily maximum temperature is above the threshold. The consecutive periods of high temperature are referred to clusters or spells. Heat wave clusters or spells are separated if there is at least a single day between them when the maximum temperature is lower than the threshold. Heat wave intensity (or magnitude) is defined as the peak temperature of a cluster or spell, for instance, spell 2 has heat wave intensity of about 44.5 °C. The duration corresponds to the number of days in a cluster or spell, for instance, spell 2 has a duration of 8 days; the frequency is the number of clusters or spells per season, and there are about 4 spells of heat waves in a season as illustrated in Fig. 1.

The choice of definition of the heat wave is an important step for impact studies. Smith et al. (2013) and Perkins and Alexander (2013) provide a comprehensive

[1]See IMD glossary for the definition of heat waves. The glossary can be accessed at imd.gov.in/section/nhac/termglossary.pdf (as on 1 January 2018).

review of various existing definitions of heat waves and their corresponding merits and limitations. According to Perkins and Alexander (2013), there is a wide variation of heat wave definitions within the impact community. Definitions and related indices have been constructed by impact group keeping the sector (such as health, wildlife, agriculture, transport and power) in mind; these are generally too specialized for the individual sector, and therefore, cannot be easily applied to other sectors. This indicates to a point that construction of a common or universal definition of the heat wave is not possible. However, it is important to note that heat waves are abnormal periods of high temperature and their impact on a sector depends upon the susceptibility and resilience of a sector. Therefore, whatever definition or metric we use for a heat wave, it is useful to remember that heat waves are high-temperature conditions leading to warmer atmospheric conditions.

2.2 Physical Drivers for Heat Waves

This section shows evidence on the meteorological background responsible for heat waves. There are four main physical drivers responsible for the occurrence of heat waves.

2.2.1 The Role of Synoptic-Scale Meteorology

The synoptic-scale meteorology is the large-scale atmospheric circulation, which is mainly understood in terms of high- and low-pressure systems. These results in the development of anticyclonic conditions which are responsible for a longer duration high-pressure atmospheric conditions (Matsueda 2011; Dole et al. 2011). Formation of anticyclonic conditions results in the blocking of low-pressure areas that lead to a clear sky in the atmosphere. The clear sky conditions result in more downward solar irradiation and the poor wind circulation, which leads to the warming of the surface and heat waves (Pfahl et al. 2015). Heat waves in Australia are associated with the propagating Rossby waves, which are planetary waves associated with the pressure systems and jet stream. The 2010 heat wave of Russia is explained in terms of the quasi-stationary anticyclonic condition over Western Russia (Barriopedro et al. 2011). While, the occurrence of intense heat waves in Europe in 2003 is associated with the combined influence of large-scale and local-scale drives (Huth et al. 2000). The local-scale drivers are mostly influenced by land surface and soil moisture conditions.

2.2.2 The Role of Land Surface and Soil Moisture

In the context of 2003 heat wave of Europe, a connection of land surface and soil moisture was also explored (Seneviratne et al. 2006). Soil moisture affects the absorption of energy by the earth's surface. The energy that reaches in the land is transformed into latent and sensible heat components. Latent heat is responsible for the evaporation of soil moisture from the land, while sensible heat is responsible for the heating of the land. If antecedent soil moisture is high, latent heat will be the dominant heat flux and will result in increased evapotranspiration from land. High evapotranspiration cools the surface and increases the concentration of water vapour in the air. When the soil is dry, sensible heat is the dominant mode of heat transfer, resulting in heating of the land surface. Soil moisture variability found to be the main driver of temperature variability in Europe in model simulations of both present and future (Seneviratne et al. 2006).

2.2.3 Climate Variability and Large-Scale Teleconnections

Large-scale atmospheric circulation shows a pattern of variability, all of which have effects on the surface climate variation. The global relationship of El Nino Southern Oscillation (ENSO) and the temperature is clear; however, their role for the manifestation of local temperature depends upon its phase such as El Nino or La Nina (Arblaster and Alexander 2012). In Europe, large-scale circulations are associated with heat waves. Della-Marta et al. (2007) suggested that these large-scale circulations were affected by North Atlantic Oscillation (NAO) via Sea Surface Temperature (SST) changes in the Atlantic Ocean. Over Australia, ENSO has been found to influence the seasonal extremes (Nicholls et al. 2005). In case of North America, La Nina patterns were found coincident with heat wave/drought of the year 2011 (Hoerling et al. 2013). Heat waves over India have also been linked to large-scale circulation (Kalsi and Pareek 2001), ENSO (Pai et al. 2013; Murari et al. 2016), and to the variation of SST in Bay of Bengal (Bhadram et al. 2005).

2.2.4 The Role of Global Warming

The previous section recognized the role of natural variability for the occurrence of heat waves. In the context of above discussion, a usual question that needs to be answered is how much more likely global warming is the cause for heat waves? Links between climate change and heat waves are often highlighted in media and other reports, especially aftermath of European heat wave of 2003. Stott et al. (2004), in the first study, applied the formal methodology of detection and attribution and quantified the change in the probability of a heat wave in Europe similar to the one that was experienced in 2003 under the influence of human activity. The study of Stott et al. (2004) covers two sets of model simulation; the first set accounts for the past effect of climate due to both man-made and natural factors (including increasing greenhouse gas concentration), while the second set mimics only natural climatic factors. This study found that the risk of 2003-like extreme European summer heat is attributable to human influences on the climate system.

In another study, using record-breaking statistics by employing Monte Carlo simulation, Rahmstorf and Coumou (2011) suggested that the occurrence of the Russian heat waves had increased by fivefold during the last decade. Further, this suggested that the warming trend in the Russian temperature is responsible for the record-breaking condition of 2010 like heat waves. However, this finding of Rahmstorf and Coumou (2011) in the context of 2010 Russian heat wave was contradicted by Dole et al. (2011), who used an ensemble of multiple climate models to assess the contribution of human activity on Russian heat wave. According to Dole et al. (2011), they detected no trend in Western Russia temperature extremes in the climate model simulations. Dole et al. (2011) suggested that human activity did not contribute to the intensity of 2010 like heat waves in Russia; they indicated that atmospheric mechanism and land-atmosphere feedbacks were the main responsible factor for heat waves in Russia in 2010. There are two contradicting findings for the same heat wave event, making it confusing to understand the causal mechanism. A year later, Otto et al. (2012) suggested that the conclusions of neither of the study are wrong, instead the conclusions should be verified in the context of the question framed. Otto et al. (2012) pointed out that the magnitude of the Russian heat wave is primarily driven by internal dynamics, as outlined by Dole et al. (2011), however, it stressed upon the fact that the probability in the occurrence of this particular event over the region of interest has increased when anthropogenic climate change is accounted for, which is in line with the conclusions of Rahmstorf and Coumou (2011) (Otto et al. 2012). Therefore, global warming plays an important role and might influence the occurrence of a heat wave in the world.

3 Relevance of Heat Waves as Climatic Hazard

Heat waves are an important class of climatic hazard due to its impact on the human health. Changes in the characteristics of heat waves such as intensity, duration and frequency, and geographical exposure have the potential to have some serious impacts. Observations suggest that in some parts of the World such changes are already happening (Meehl and Tebaldi 2004), leading to impact on human health in terms of heat stress and mortality. Heat waves are expected to intensify around the globe in the future, with a potential increase in heat stress and heat-induced mortality. Table 1 shows a comparison of impacts due to various meteorological and climatological hazards from 1985 to 2015. An observation from Table 1 is that the exposure of heat wave does not result in the monetary damages except the case of the 2010 Russian heat wave where Barriopedro et al. (2011) suggested Russian grain harvest suffered a loss of about 30% due to extreme temperature conditions. However, loss of life data from Table 1 suggests that heat waves are a serious climatic hazard. The statistics show that about 1000 lives lost due to exposure to individual heat waves, which is higher in comparison to the lives lost due to other climatological hazards. This suggests that heat waves are the deadliest climate and weather hazards that has caused higher human causalities on per event basis.

Table 1 Exposure and impacts of weather and climate disaster, at a global scale, for the period from 1985 to 2015

Weather and climate extremes	Total events	Persons killed	People affected (in million)	Monetary damage (in billion US$)
Floods	3713	2,08,846	3147.8	644.5
Drought	439	27,150	1635.6	117.6
Cold waves	264	14,574	13.5	10.5
Heat waves	142	1,50,955	4.8	20.5
Tropical cyclones	1355	4,08,926	607.5	666.5

Disasters are defined according to international disaster database, which is a compilation of information from various governmental and non-governmental sources. Further details of the database are available at http://www.emdat.be/

A multi-model analysis suggests that most intense extreme future heat waves are concentrated around densely populated agricultural regions of the Ganges and Indus river basins (Im et al. 2017). This suggests climate change is expected to present a serious and unique risk in South Asia, a region inhabited by about one-fifth of the global human population, due to an unprecedented combination of severe natural hazard and acute vulnerability. This is an important observation on ground of the fact that model analysis suggest that Asia is expected to be highly exposed to heat waves, which is considered to more vulnerable to heat stress on ground of (1) high population density, (2) higher proportion of poorer people and (3) increasing risk of illness and death related to extreme heat conditions (Im et al. 2017).

4 Heat Waves in the Context of India

This section describes heat waves in the Indian context. It gives an overview of studies about temperature patterns and the role of climate variability on heat waves in India.

4.1 Surface Temperature Trend Over India

Heat waves in India are commonly in the summer months (i.e. from March to June and sometimes in the July months). The monsoon arrives in India after the summer months that brings cooler wind and moisture and causes relief in temperature. There are many studies on the observed temperature trends over India. Kothawale and Rupa Kumar (2005) observed that all-India mean annual temperature has a trend of increase of about 0.5 °C per century for the period 1901–2003; however, the 30-year trend for the period 1971–2003 is 2.2 °C per century, which suggest Indian mean annual temperature has an alarming rate of warming in the recent period.

4.2 The Role of Atmospheric Circulation and Heat Wave Characteristics in India

There are limited studies that describe the role of atmospheric circulation on heat waves in India. Jenamani (2012) suggested the possible role of a cyclonic storm in Bay of Bengal that caused westerly flow in the upper atmosphere resulting in blocking of maritime wind over land. This led to a dry atmospheric condition that has triggered 1998 heat waves in Orissa. The 1998 heat wave of Orissa is also linked to El Nino conditions (De and Mukhopadhyay 1998). Kalsi and Pareek (2001) identified the role of formation of anticyclonic circulation on the upper atmosphere that results in subdued convective activity and entering of more heat on the land, this explains 1999 North Indian heat wave. Bhadram et al. (2005) studied the 2003 heat wave of Andhra Pradesh and found that the warmer SST anomalies over North Bay of Bengal were responsible for heat wave-like condition in the state. In addition to these factors, the late onset of Indian monsoon is also understood to be an important factor responsible for the heat waves in India (Pai et al. 2013).

Heat waves in Indian normally occur in the pre-monsoon summer periods, typically during the months from March to May (Murari et al. 2015). This suggests the possibility of the influence of factors affecting monsoon on the heat waves in India. A delay in the arrival of monsoon will result in extending summer period and thereby the exposure of heat waves. Murari et al. (2016) studied the relationship of the El Nino and the onset dates of the summer monsoon in India and its influence on heat wave characteristics. El Niño is associated with an eastward shift in the Walker circulation and breakdown of circulation patterns, resulting in a delay in the onset of the Indian Summer Monsoon (ISM). Because most heat waves in India occur during the pre-monsoon season, Murari et al. (2016) showed that heat waves during El Niño years were longer and hotter, and it is argued that this is related to a delay in the onset of the ISM.

Figure 2 shows the result of Murari et al. (2016) that compares the heat wave characteristics of El Niño and normal years. Most of the positive anomaly in June and July temperatures occurred in El Niño years. Strong El Niño years (i.e. ENSO[2] index ≥ 1) consistently showed warmer temperatures in June and July (Fig. 2a). Figure 2b–d compares distributions of the summer season heat wave characteristics for El Niño and normal years. A large percentage of grid points experienced higher duration and days during El Niño years compared to normal years. This indicates that heat waves were more frequent and lasted longer during El Niño years. Likewise,

[2]ENSO index is defined based on Oceanic Nino Index (ONI), which uses sea surface temperature of Nino 3.4 region. The ONI values are estimated for 3-month window and are used to define ENSO index. See Murari et al. (2016) for details.

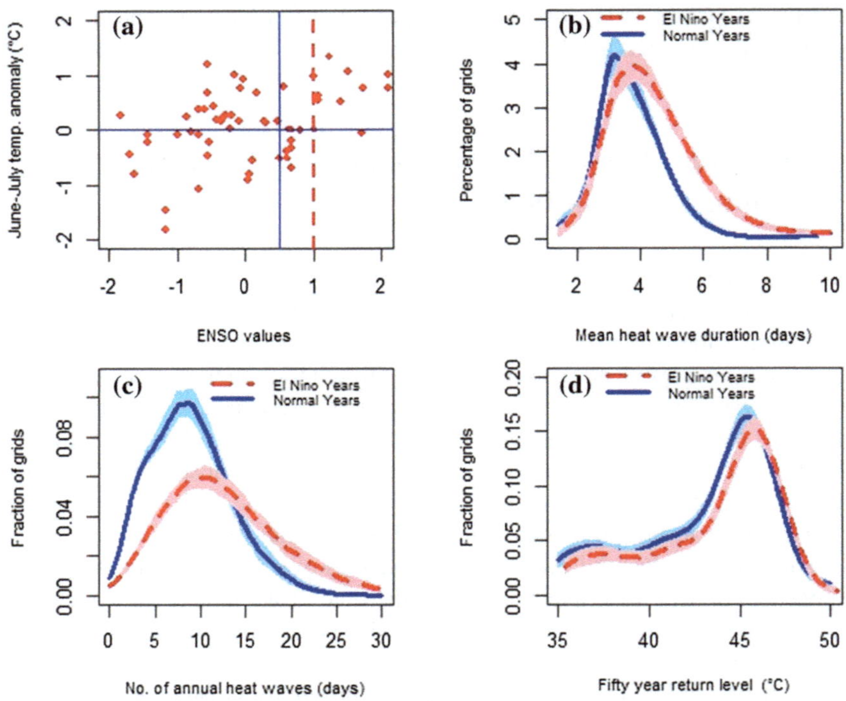

Fig. 2 Comparison of heat wave characteristics between El Niño and normal years. **a** Average June and July temperature anomaly for different values of the El Niño Southern Oscillation (ENSO) index; the vertical continuous and dashed lines show different El Niño conditions. **b–d** Show the fraction of grid points corresponding to different duration, days and T_{max} for El Niño and normal years, respectively. The shaded region shows the 95% confidence intervals for different estimates of duration, days and T_{max} (*Source* Murari et al. (2016))

during El Niño years, a larger number of grid points experienced higher maximum temperature during heat wave episodes (T_{max}), with the curve for El Niño years shifted by about 1 °C with respect to the normal years. These results are consistent with earlier work on extreme temperatures in India (Panda et al. 2014), which suggests that El Niño is related to the number of heat wave days and the maximum temperatures of the pre-monsoon summer heat waves in India. Moreover, the greater number of heat wave days during El Niño years is likely to have resulted in higher mean temperatures.

5 Impacts of Heat Waves in India

There are a number of studies that explored the direct connection between the health and thermal stress (Gosling et al. 2009). The increased stress associated with the heat waves may lead to increased heat-induced mortality in the absence of appropriate

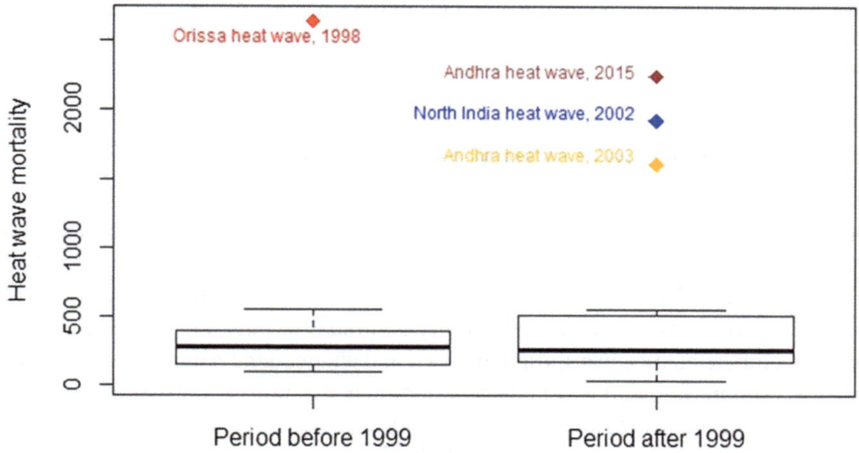

Fig. 3 All-India heat-related mortality for two time periods (*Data source* EM-DAT; www.emdat.be)

adaptation action (Gosling et al. 2009). During heat wave episodes, the exposure to the anomalously high temperature may depress the rate of heat transfer between the human body and the atmosphere. People working outdoors or indoor without basic infrastructures such as electricity and water could be the most vulnerable. In addition to the exposure to high temperature, relative humidity is another important factor that might disturb the thermoregulation of a human body because sweating is the mechanism that controls the body temperature by exchanging with the outside atmosphere. An atmosphere with the high relative humidity results in a decreased rate of sweating and might result in accumulation of heat inside the human body. Extreme conditions, i.e. exposure to high atmospheric temperature and relative humidity, might cause failure of the thermoregulatory system of a human body and might result in skin eruption, heat fatigue, heat cramp, heat exhaustion and heat stroke. Koppe et al. (2004) gave a very good review of the possible implication of exposure to extreme temperature on the human health.

Extreme temperatures in India during summer months often lead to loss of human and animal life and enhanced morbidity and discomfort. Official records of the IMD indicate that there were about 223 heat wave incidences between 1978 and 1999, causing more than 5300 deaths (Choudhary et al. 2000). Figure 3 shows the trend of heat wave mortality in India. The figure divides the mortality data into two periods before and after 1999 and plots the box plots for the mortalities for the two periods. Most of the deadly heat waves in India occurred in the recent period (Fig. 3).

The heat wave of the year 1998 was very much brutal for the most part of India, which is associated with a higher number of deaths, particularly in the states of Delhi and Orissa. Heat waves in India have not, so far, attributed to global warming. Possibly, due to a reason that it is difficult to attribute individual heat waves to climate change, however, their pattern about recent exposure and impact of heat waves gives a first-hand understanding that the effect of global warming on heat waves is already evident in India.

6 Future Projections of Heat Waves in India

The interaction among atmospheric, oceanic and land surface processes that produce prolonged periods of atmospheric conditions of high air temperatures has been explored in the literature (Della-Marta et al. 2007; Dole et al. 2011; Matsueda 2011; Pfahl et al. 2015). While natural variability continues to play a key role in extreme weather, climate change has shifted the odds and changed the natural limits, making heat waves more frequent and intense (Coumou and Rahmstrof 2012). Both record highs and lows are set regularly in a stationary climate. But it is understood that the balance shifts towards a warming climate leading to asymmetry in the temperature distribution (Kodra and Ganguly 2014).

GCMs are good tools available for developing an understanding of how the climate will change under enhanced anthropogenic activity. These models have provided daily data, which are key in the measurement of heat waves. However, a reliable analysis of model outputs depends on the observed data, and therefore for regions where observed records are poorly maintained, such a projection of extremes may lead to insufficient understanding (Field et al. 2014). Regions where good observation records are available, direct uses of GCMs for regional projections are difficult mainly for two reasons. First, GCM data are available at coarse spatial resolution; second, GCM data might be affected by biases because of insufficient knowledge of key physical mechanisms, difficulty with the parameterization and the use of empirical physical laws (Meehl et al. 2007).

In order to employ GCMs for regional- and local-scale studies, downscaling[3] (i.e. generating finer resolution data) and bias correction[4] are essential steps. Application of a suitable bias correction method is an important step for providing regional projections. Murari et al. (2015) used multiple simulations from seven GCMs to understand the future heat wave characteristics in India. Model data are first bias corrected and compared with heat wave characteristics simulated by the IMD gridded data for the observed period. Figure 4 shows the projection of intensity, duration and frequency of heat waves using seven climate model data. The figure indicates the increase in all the three heat wave characteristics for most of India.

---

[3]Downscaling is an approach of obtaining regional high-resolution climatic variables from GCMs. See https://www.gfdl.noaa.gov/climate-model-downscaling/ for further detail on downscaling.

[4]GCMs are typically at coarse spatial and require a high degree of parameterization (i.e. simplification and approximations) of the physical mechanisms of the earth system at the designed resolution of the GCM. This results in error in simulation of climatic variables from the GCM. The error is generally defined with respect to the observed data. This error is referred to biases in the simulation of GCM, which needs to be corrected before using the GCM simulations for impact assessment. See Murari et al. (2015) for bias correction.

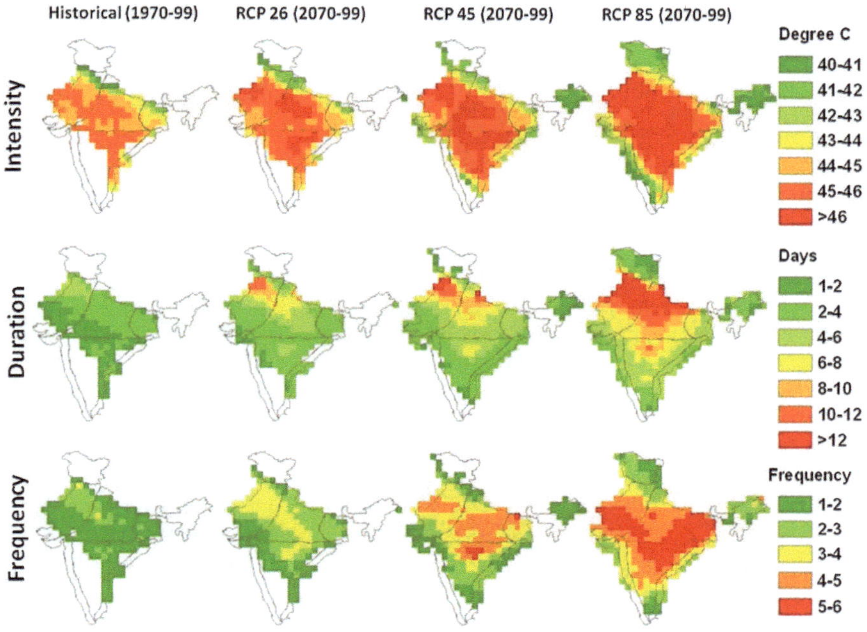

Fig. 4 Intensification of heat wave characteristics in the future

In particular, a large part of Southern India presently not affected by heat waves is projected to be severely affected after 2070. The magnitude of intensification in Northern India is observed to be high compared to other regions of the country. Interestingly, under the RCP 26 scenario, in which most of the climate models are showing the change in global temperature below 2 °C with respect to the average global temperature of the pre-industrial period (van Vuuren et al. 2011), it is observed that intensification in heat wave characteristics is only observed for North India. Southern India, under RCP 26 scenario, does not show intensification; this implicates that the severe heat wave could be avoided for the region if the future forcing is constrained to RCP26.

The first occurrence of heat waves is characterized as the day in the year when a first heat wave hits at a particular location. Figure 5 shows the period during which the first heat wave in summer occurred, where the first occurrence is calculated as the average of the days of the first occurrence in 30-year time window; data are reported for observed data (1970–1999), historical modelled data (1970–1999), and projected data of RCP26, RCP45 and RCP85 (2070–2099). The first heat wave occurs mostly between the second half of May and the first half of June for the observed and historical periods. While towards the end of the twenty-first century, heat waves are expected to occur between April and the first half of May in most parts of India in three scenarios. The impact of early summer heat waves could be severe because of the surprise effect on the population that is not expecting early occurrence of severe

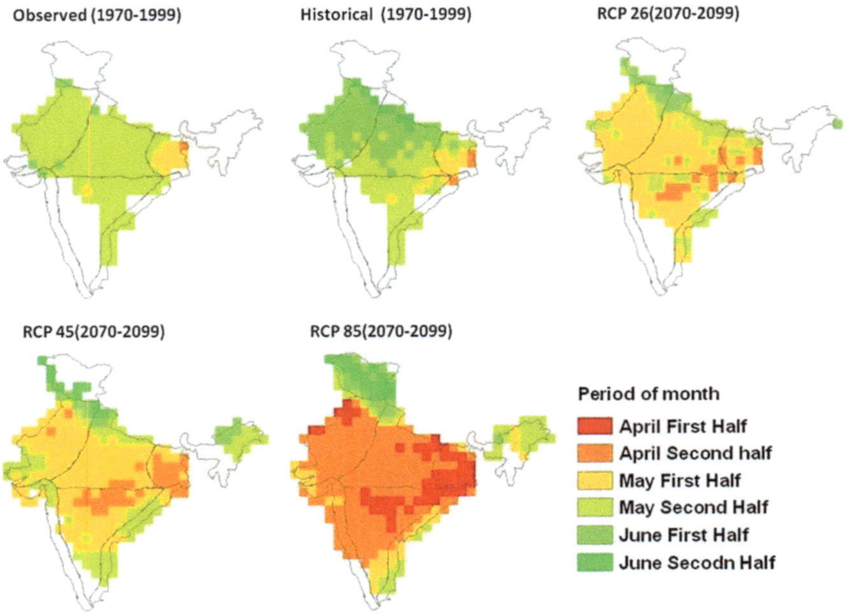

Fig. 5 First occurrence of severe heat waves, expressed as a mean value over a 30-year time window of the day in the year when a severe heat wave first hits a particular location. The figure presents the ensemble mean of the first occurrence values for seven climate models

heat spells (Hajat et al. 2005); this is especially relevant to regions, such as Southern India, that do not experience severe heat wave conditions in the observed data.

These results show that the concerns regarding intensification and early occurrence of a heat wave in North and Central India are serious under all the three scenarios. Unlike Southern India, intensification of heat waves in the remaining parts of India cannot be avoided under any scenario representing less than 2 °C increase in global temperature.

7 Concluding Remarks

A significant amount of work has been done to understand the heat waves and its likely impact due to global warming. This includes understanding heat waves, identifying the physical mechanism responsible for the occurrence of heat waves in various climatological regions, and also the role of global warming on the likelihood of the exposure of heat conditions. This chapter presents an overview of this knowledge. Despite a significant improvement in the scientific understanding of heat waves, there are still limited studies on heat waves, their projections and impact in developing countries, particularly in India. There are studies conducted in a scattered way to understand heat waves in India. This chapter, with a comprehensive review of existing heat wave studies in India, provides a consolidated picture of heat wave studies in India.

Existing studies in India have found various factors that triggered heat waves in different geographical regions. Most of these are related to natural variability in the climatic system. There is no study, so far, that indicates that global warming is responsible for individual heat wave events in India. Such studies are extremely concerning, particularly by impact assessment community, as these provide the basis of likelihood intensification of extreme heat due to global warming. Heat wave community will be significantly benefited by detection and attributional studies. Nevertheless, climate models have been used to understand the future projection of heat extremes in different regions in the world. This chapter presents the findings of Murari et al. (2015), which is perhaps the only study in India that provides a model projection of heat wave characteristics in India.

The results of Murari et al. (2015) suggest that the places which have always experienced heat waves in summer will see heat waves that are greater in intensity as well as duration. This will make the country more vulnerable because a large proportion of the population in India live without sufficient access to water, electricity and primary health facilities. Plants and animals will find it harder to survive through to the end of the summer. A possible consequence of the projected heat waves could be on the health care infrastructure, which might collapse under the strain of overwhelming demand for treatment for ailments as well as emergency services. Moreover, parts of our country that have never been exposed to heat wave conditions will experience conditions that they are unfamiliar with. They will be completely unprepared and unequipped to deal with such situations. Ways of life that have been existed and evolved over countless years will lose the very foundations on which they have been built: the characteristics of the environment which cradled them.

These results point to the necessity for further research on the adverse effects of the heat waves of the future. Policies need to be formulated to cope with their impact on the population. With every passing year, they will occur earlier in the year affect larger areas of India, including Southern India. Worryingly, heat waves are not seriously considered as a disaster, at present.

References

Arblaster JM, AlexanderL V (2012) The impact of the El Nino-Southern Oscillation on maximum temperature extremes. Geophys Res Lett 39:L20702. https://doi.org/10.1029/2012gl053409

Barriopedro D, Fischer EM, Luterbacher J et al (2011) The hot summer of 2010: redrawing the temperature record map of Europe. Science 332(6026):220–224

Bhadram CVV, Amatya BVS, Pant GB (2005) Heat waves over Andhra Pradesh: a case study of summer 2003. Mausam 56(2):385–394

Choudhary SK, Gore JM, Sinha Ray KC (2000) Impact of heat waves over India. Curr Sci 79(2):153–155

Coumou D, Rahmstrof S (2012) A decade of weather extremes. Nat Clim Change 2:491–496

De US, Mukhopadhyay RK (1998) Severe heat wave over the Indian subcontinent in 1998, in perspective of global climate change. Curr Sci 75:1308–1311
Della-Marta PM, Haylock MR, Luterbacher J et al (2007) Doubled length of western European summer heat waves since 1880. J Geophys Res Atmos 112:D15103. https://doi.org/10.1029/2007jd008510
Dole R, Hoerling M, Perlwitz J et al (2011) Was there a basis for anticipating the 2010 Russian heat wave? Geophys Res Lett 38:L06702. https://doi.org/10.1029/2010gl046582
Easterling DR, Meehl GA, Parmesan C et al (2000) Climate extremes: observations, modeling, and impacts. Science 289(5487):2068–2074
EM-DAT: the emergency events database—Université catholique de Louvain (UCL)—CRED, D. Guha-Sapir, Brussels, Belgium. www.emdat.be
Field CB, Barros VR, Dokken DJ et al (2014) Climate change 2014: impacts, adaptation, and vulnerability. Part A: global and sectoral aspects. In: Contribution of working group II to the fifth assessment report of the intergovernmental panel on climate change. Cambridge University Press, Cambridge, United Kingdom and New York, NY, USA, p 1132
Gosling SN, Lowe JA, McGregor GR et al (2009) Associations between elevated atmospheric temperature and human mortality: a critical review of the literature. Clim Change 92:299–341
Hajat S, Armstrong BG, Gouveia N et al (2005) Mortality displacement of heat-related deaths. Epidemiology 16:613–620
Hoerling M, Kumar A, Dole R et al (2013) Anatomy of an extreme event. J Clim 26(9):2811–2832
Huth R, Kysely J, Pokorna L (2000) A GCM simulation of heat waves, dry spells, and their relationships to circulation. Clim Change 46(1–2):29–60
Im E-S, Pal JS, Eltahir EAB (2017) Deadly heat waves projected in the densely populated agricultural regions of South Asia. Sci Adv 3(8):e1603322. https://doi.org/10.1126/sciadv.1603322
Jenamani RK (2012) Analysis of ocean-atmospheric features associated with extreme temperature variation over east coast of India—a special emphasis to Orissa heat waves of 1998 and 2005. Mausam 63(3):401–422
Kalsi SR, Pareek RS (2001) Hottest April of the 20th century over north-west and central India. Curr Sci 80(7):867–873
Kodra E, Ganguly AR (2014) Asymmetry of projected increases in extreme temperature distributions. Sci Rep 4:5884. https://doi.org/10.1038/srep05884
Koppe C, Kovats S, Jendritzky G et al (2004) Heat-waves: risks and responses. Health and global environmental change, series 2. World Health Organization Report
Kothawale DR, Rupa Kumar K (2005) On the recent changes in surface temperature trends over India. Geophys Res Lett 32(18):L18714. https://doi.org/10.1029/2005gl023528
Lobell DB, Sibley A, Ortiz-Monasterio JI (2012) Extreme heat effects on wheat senescence in India. Nat Clim Change 2:186–189
Matsueda M (2011) Predictability of Euro-Russian blocking in summer of 2010. Geophys Res Lett 38:L06801. https://doi.org/10.1029/2010gl046557
Mazdiyasni O, AghaKouchak A, Davis SJ et al (2017) Increasing probability of mortality during Indian heat waves. Sci Adv 3:e1700066. https://doi.org/10.1126/sciadv.1700066
Meehl GA, Tebaldi C (2004) More intense, more frequent, and longer lasting heat waves in the 21st century. Science 305(5686):994–997
Meehl GA, Covey C, Delworth T et al (2007) The WCRP CMIP3 multimodel dataset: a new era in climate change research. Bull Am Meteorol Soc 88(9):1383–1394
Murari KK, Ghosh S, Patwardhan A et al (2015) Intensification of future severe heat waves in India and their effect on heat stress and mortality. Reg Environ Change 15(4):569–579
Murari KK, Sahana AS, Daly E et al (2016) The influence of El Nino Southern Oscillation on heat waves in India. Meteorol Appl 23(4):705–713
Nicholls N, Baek HJ, Gosai A et al (2005) The El Nino-Southern Oscillation and daily temperature extremes in east Asia and the west Pacific. Geophys Res Lett 32(16):L16714. https://doi.org/10.1029/2005gl022621

Otto FEL, Massey N, van Oldenborgh GJ et al (2012) Reconciling two approaches to attribution of the 2010 Russian heat wave. Geophys Res Lett 39:L04702. https://doi.org/10.1029/2011gl050422
Panda DK, Mishra A, Kumar A et al (2014) Spatiotemporal patterns in the mean and extreme temperature indices of India, 1971–2005. Int J Climatol 34(13):3585–3603
Pai DS, Nair SA, Ramanathan AN (2013) Long term climatology and trends of heat waves over India during the recent 50 years (1961-2010). Mausam 64(4):585–604
Patz JA, Campbell-Lendrum D, Holloway T, Foley JA (2005) Impact of regional climate change on human health. Nature 438:310–317
Perkins SE, Alexander LV (2013) On the measurement of heat waves. J Clim 26:4500–4517
Pfahl S, Schwierz C, Croci-Maspoli M et al (2015) Importance of latent heat release in ascending air streams for atmospheric blocking. Nat Geosci 8:610–614
Rahmstrof S, Coumou D (2011) Increase of extreme events in a warming world. Proc Nat Acad Sci 108(44):17905–17909
Robinson PJ (2001) On the definition of a heat wave. J Appl Meteorol Climatol 40:762–775
Seneviratne SI, Luthi D, Litschi M et al (2006) Land-atmosphere coupling and climate change in Europe. Nature 443(7108):205–209
Smith TT, Zaitchik BF, Gohlke JM (2013) Heat waves in the United States: definitions, patterns and trends. Clim Change 118(3–4):811–825
Stott PA, Stone DA, Allen MR (2004) Human contribution to the European heatwave of 2003. Nature 432:610–614
Teuling JA, Reichstein M, Moors E et al (2010) Contrasting response of European forest and grassland energy exchange to heatwaves. Nat Geosci 3:722–727
van Vuuren DP, Edmonds J, Kainuma M et al (2011) The representative concentration pathways: an overview. Clim Change 109:5–31

Decadal Estimates of Surface Mass Balance for Glaciers in Chandra Basin, Western Himalayas, India—A Geodetic Approach

RAAJ Ramsankaran, Ankur Pandit and Avinash Parla

Abstract The Himalayas, which has the highest concentration of glaciers outside the poles, form a huge reservoir of fresh water. These glaciers are not only important because of the contribution they make to our water resources, but also because they are a great help in understanding the effects of climate change. However, monitoring even a few of these glaciers through field techniques is a tedious process due to the tough and unapproachable terrain of the Himalayas. In such situation, remote sensing techniques become a boon. In this article, one such widely used remote sensing-based technique called digital elevation model (DEM). Differencing or geodetic approach available for estimating glacier surface mass balance (SMB) is discussed. However, the accuracy of the estimated SMB using geodetic approach is highly dependent on the accuracy of the DEMs used, which ultimately decides whether this approach can be used for seasonal, annual or decadal studies. To get a better understanding of this approach, this article gives a brief overview of errors present in DEMs and how to correct it before estimating SMB by demonstrating it over 65 glaciers located in the Chandra basin, Lahaul–Spiti district of Himachal Pradesh, India. The obtained results indicate that the Chandra basin has lost 5.01 ± 2.48 Gt of ice mass during the course of 14 years (1999–2013), which gives a clear signal on the effect of climate change occurred over the study area and how climate change could cause imbalance in the health of glacier(s). Considering the advantages of this approach and recent developments in high-resolution DEM data acquisition as well as the need for such accurate mass balance estimates across Himalayas, it is therefore suggested that this

R. Ramsankaran (✉)
Department of Civil Engineering, Indian Institute of Technology Bombay, Room No. 202, Mumbai 400076, India
e-mail: ramsankaran@civil.iitb.ac.in

R. Ramsankaran · A. Pandit
Interdisciplinary Programme in Climate Studies, Indian Institute of Technology Bombay, Mumbai 400076, India
e-mail: ankkurpandit@gmail.com

A. Parla
Water Resources Division, Department of Civil Engineering, Indian Institute of Technology Bombay, Mumbai 400076, India
e-mail: p.avinash283@gmail.com

C. Venkataraman et al. (eds.), *Climate Change Signals and Response*,
https://doi.org/10.1007/978-981-13-0280-0_7

modelling approach should be extended to other basins to get the current status of the health of the glaciers across Himalayas.

1 Introduction

The Hindu Kush Himalayan (HKH) region has the highest concentration of glaciers outside the poles, which makes them a huge source of fresh water for the developing world such as Afghanistan, Pakistan, India, Tibet, Nepal, Bhutan to which they feed. One-third of the humanity on this planet is directly or indirectly dependent on nine major river systems to which headwater is provided by the Himalayan glaciers distributed along the Himalayan arc. Water from these rivers has a significant socio-economic effect on more than 700 million people (Immerzeel et al. 2010). The runoff coming from these glaciers not only affect the drinking water supplies, but also the hydropower projects, biodiversity, industries, agriculture and day-to-day life of dependent individuals among others. HKH glaciers are not only important because of the contribution they make to our water resources, but also because they are widely recognized to be sensitive climatic indicators (Houghton et al. 2001). As compared to the polar region, the HKH glaciers receive more heat because of its location which is closer to the Tropic of Cancer, and hence, they are very sensitive towards climate variability or rising temperature both at regional and global levels.

Out of the entire HKH region, Indian Himalayan region is significantly important because of its rich biodiversity and its influence in Indian summer monsoon, etc. As per one of the latest report by Geological Survey of India (GSI) about 9575 glaciers with an estimated area of 37,466 km^2 exist in Indian Himalayas (Sangewar and Shukla 2009). The three major river systems in India originating from these glaciers, i.e. the Indus, the Ganges and the Brahmaputra, provides approximately 50% of the annual usable water resources (690 km^3) of the country (DST 2012). Out of this, a significant contribution is coming from glacier and seasonal snowmelt especially in Indus basin. Considering the larger extent of water utilization, understanding the changes in these glaciers is vital, especially, for the current and future government policies for the regional water resources management.

The Inter-Governmental Panel on Climate Change (IPCC) in its fourth assessment report (IPCC 2007) highlighted the melting of ice mass and glaciers worldwide as one of the most prominent indicators of climate change. IPCC fifth assessment report on climate impacts (IPCC 2014) says that Himalayan glaciers would shrink by 45% by 2100, if Earth's average surface temperature rose by 1.8 °C. Under a far warmer scenario of 3.7 °C, the reduction would be 68%. IPCC (2007) highlighted that the global mean surface air temperature is increased by 0.74 °C over the last century in the Himalayas. In agreement with the observations of IPCC (2007) and projections of IPCC (2014) reports, the temperature trends profile of the north-west Himalayan region has already shown a significant warming of 1.6 °C (Bhutiyani et al. 2007), whereas 0.98 °C of temperature warming is observed in the western Himalayan regions (Dash et al. 2007). An accelerated warming after

1972 was observed in the north-west and western Himalayan regions (Dash et al. 2007). Another study conducted using microwave satellite measurements from 1979 to 2007 also highlighted that annual mean warming over the whole HKH region is accelerated by 0.21 ± 0.08 °C/decade with maximum warming localized over the western Himalaya, i.e. 0.26 ± 0.09 °C/decade (Gautam et al. 2010). In the Karakoram region, a very interesting contrast is observed by Dimri and Dash (2010) where they found a decrease in winter temperatures in the accumulation zone whereas an increase in winter temperatures in the ablation zone of the Siachen Glacier between year 1984 and 2006. The contrast in temperature is more likely to be directed by local dynamics rather than by regional or global trends (Singh et al. 2011). It is expected that changes in temperature due to climate change might affect the pattern of precipitation. Dimri and Dash (2012) reported decreasing trends in winter precipitation (December–February) in the western Himalaya for 1975–2006. The temperature warming in the western Himalayas has caused an increase in the ablation area of the glaciers by shifting snowline towards the higher elevations. This upward shifting of snowline has gathered pace in the past few years.

Under the climate changing scenario, the most crucial to the survival of a glacier is its mass balance, i.e. the difference between snow/ice accumulation and ablation. A glacier with a sustained negative balance is out of equilibrium and will retreat, whereas a glacier with a sustained positive balance is out of equilibrium and will advance. As climate change may cause variations in both temperature and snowfall, changes in mass balance can control a glacier's long-term behaviour. There is an overwhelming evidence that the Himalayan glaciers especially in western region are not in the best of health and most of them are losing mass (e.g. Vijay and Braun 2016). In another study conducted by Kulkarni et al. (2007) on 466 glaciers of the western Himalayas highlighted that there has been a 21% reduction in the glacierized area, i.e. from 2077 km^2 in 1962 to 1628 km^2 in 2004. Smaller glaciers of less than 1 km^2 have reduced in area by as much as 38% compared to 12% retreat of the larger glaciers. While most of the glaciers are losing mass at a rate similar to glaciers at other places, Karakoram Himalayas emerge out as an exception with some studies suggesting mass gain (Bolch et al. 2012). The region of western Himalayas, in particular, is extremely interesting to investigate because the region is influenced by two major climatic systems, i.e. the mid-latitude Westerlies and the south Asian summer monsoon.

One of the methods to estimate mass balance of a glacier in field is through glaciological method, which requires extensive field surveys over the rugged and tough terrain of the Himalayan glaciers. This traditional stake-based measurement is still carried to collect long-term information regarding the glacier health (Dobhal et al. 2004; Kumar et al. 2009; Azam et al. 2012). In this approach, the rate of data acquisition is slow and expenses for logistics and labour can be high in maintaining observational networks at the higher elevations. Also, extreme weather and terrain conditions restrict the long-term field mass balance measurement in the Himalayan region. In India, there are approximately 25 research institutes/organizations are currently involved in glacier-related research. However, considering the extent of the Himalayan region, field mass balance observations are available only for about 12 glaciers (DST 2012) out of which most of them are not continuous. Till date, most of the glaciers in the Indian Himalayan region are untouched due to their inaccessibility and/or the shortage of manpower with proper training. This largely explains the very limited availability of mass balance data in the Himalaya.

To overcome the dependency of field-based mass balance measurements, a very concerted research effort is required to develop alternate ways of securing this important data. Under such circumstances, remote sensing technology seems to be a very promising tool for temporal monitoring of glacier health. Images acquired from different platforms such as satellite, aircrafts using sensors that operate in different spectral regions such as visible, infrared, microwave have been widely used to study glacier mass balance. (e.g. Gardelle et al. 2012; Kääb et al. 2012; Vincent et al. 2013; Mernild et al. 2013; Rabatel et al. 2016).

Among the available methods for estimating mass balance from remote sensing data, the geodetic method, also known as digital elevation model (DEM) differencing approach has been widely used for large-scale studies by many researchers (e.g. Berthier et al. 2004; Gardelle et al. 2012; Kääb et al. 2012; Vincent et al. 2013). Based on historical topographic maps and/or high-resolution DEMs derived from Advanced Spaceborne Thermal Emission and Reflection Radiometer (ASTER) (e.g. Rivera and Cassassa 1999; Kääb 2007), Satellite Pour l'Observation de la Terre (SPOT) imagery (Berthier et al. 2007), Shuttle Radar Topography Mission (SRTM) (e.g. Surazakov et al. 2006; Racoviteanu et al. 2007), high-resolution Advanced Land Observing Satellite (ALOS) Panchromatic Remote sensing Instrument for Stereo Mapping (PRISM) and Corona (Narama et al. 2009), laser altimetry (Arendt et al. 2002), ERS, Radarsat (Dedieu et al. 2003) and TerraSAR-X Add-On for Digital Elevation Measurement (TanDEM-X) (e.g. Vijay and Braun 2016) glacier surface mass balances can be estimated via estimating glacier thickness change. This method measures the elevation changes over time ($\delta h/\delta t$) from various available DEMs constructed over a glacier surface. Historical topographic information is often used to construct elevations models which are subtracted from more recent DEMs constructed from remote sensing imagery either on a pixel-by-pixel basis or as average elevation change to obtain difference maps. The availability of spaceborne elevation data of glaciers present in the remote location increases the capability to quantify glacier changes. However, the accuracy of the geodetic method has several issues. It highly depends on the accuracy of the DEMs used, discrepancies in the scale of resolution of DEMs and the assumption of the density of lost/gained material (snow/firn/ice). This approach is suitable for the estimation of long-term variation (5–10 years) rather than yearly variation in the glacier elevation because of the low vertical accuracy of generated DEMs. It is important to highlight that many earlier studies use the DEMs without correcting for biases between them as pointed out by, e.g. Rignot et al. (2003), Sund et al. (2009), Muskett et al. (2009). This may lead to wrong estimates of glacier volume changes (Berthier et al. 2010). The geodetic surface mass balance without the correction of potential biases has been compared with the surface mass balance estimated from direct glaciological field measurements in a number of studies, and the differences are often found to be large. (e.g. Krimmel 1999; Østrem and Haakensen 1999; Rolstad et al. 2009). Recently, there are many approaches proposed for correcting the bias in DEMs and tested successfully in various locations (e.g. Schiefer et al. 2007; Van Niel et al. 2008).

Considering the above lacuna in the mass balance information across Himalayas and the recent developments in geodetic approach, the motivation behind the current study is to demonstrate the applicability of bias-corrected DEM Differencing approach to provide an accurate/reliable surface mass balance. For experimental purpose, the DEM differencing approach with appropriate corrections for DEM bias has been demonstrated over 65 glaciers located in the Chandra basin of western Himalayan region in India to get decadal glacier surface mass balance estimates.

Following this introduction (Sect. 1), this chapter discusses details of the study site chosen and the datasets used in the current study in Sect. 2. In Sect. 3, the methodology adopted to investigate the decadal surface mass balance is discussed. Later on, in Sect. 4, the inferences drawn from the current study are discussed followed by the conclusion and recommendations in Sect. 5.

2 Study Area and Datasets Used

The study was conducted over 65 glaciers of Chandra basin located in the western Himalaya, Lahaul–Spiti valley of Himachal Pradesh, India. These 65 glaciers are selected based on their size (<0.5 km^2) because small glaciers have high uncertainty in glacier outlines. The location map of Chandra Basin is shown in Fig. 1.

The basin is having an area of ~2381 km^2 (Pandey et al. 2017). This basin is a sub-basin of Chenab basin and ranging from an elevation range of 2400 m a.s.l. and 6400 m a.s.l. Glaciers from this basin feed the Chandra River—a major tributary of the Chenab river system. Geomorphologically, the area represents highly rugged, inaccessible terrain with high mountains and deeply dissected valleys (Sharma et al. 2016). Bara Shigri Glacier with an area of 136 km^2 is the largest glacier in this basin. This basin falls in the monsoon-arid transition zone and is alternately influenced by Indian Summer Monsoon during summer and mid-latitude westerlies during winter (Wagnon et al. 2007; Bookhagen and Burbank 2010). High wet precipitation is recorded in summer season (July–September), whereas winter season (November–February) experiences a significant amount of solid precipitation due to the influence of the westerlies (Sharma et al. 2013). This basin is characterized by heavy and dry snowfall, fairly cold temperatures with strong wind action (Sharma and Ganju 2000; Negi et al. 2013).

The current study has utilized SRTM C-band DEM and TanDEM-X DEM for estimating decadal surface mass balance of the study area. SRTM is a joint program between NASA and National Geo-spatial intelligence Agency (NGA) to map the globe in three dimensions. Flown on 11–22 February 2000, SRTM had the first space-borne single-pass SAR interferometer. It yielded single-pass interferometry data, C- and X-bands, of land between 60N and 54S for processing into global DEMs. The InSAR antenna configuration (precise measurement of relative antenna location with laser rangers, star trackers and in-orbit Global Positioning System—GPS) resulted

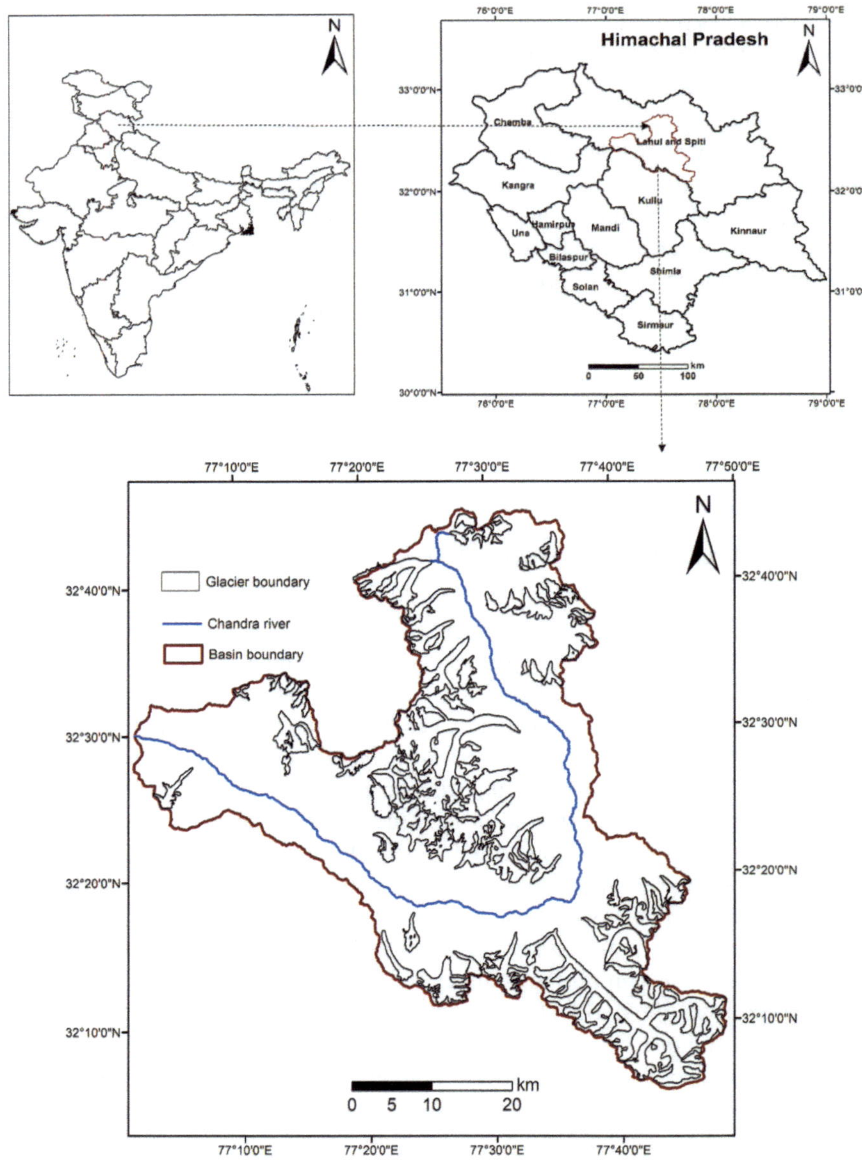

Fig. 1 Map of the study area showing the locations of the selected 65 glaciers in Chandra river basin, Lahaul–Spiti valley, Himachal Pradesh, India

in elimination of mapping errors characteristic of repeat-pass satellite InSAR techniques. The nominal vertical accuracy of SRTM is 6 m relative, 16 m absolute, and the nominal horizontal accuracy is 15 m relative, 20 m absolute, at the 90% confidence level (Muskett et al. 2003). Likewise, high-resolution TanDEM-X (TDX) satellite data obtained under the science proposal XTI_GLAC7043 project has been utilized for generation of another DEM. For this purpose, two raw TDX data was ordered to create a high-resolution DEM (10 × 10 m resolution) of the Chandra basin covering the selected 65 glaciers. The two TDX tiles were acquired on 27th October 2013 in HH polarization with normal baseline of 182.66, 182.72 m and incidence angle of 34.39°, 34.56°, respectively. Both the tiles were acquired on descending pass. SARscape 5.0 licensed software was used for generating DEM from TanDEM-X satellite data. For the current study, the outlines of the study glaciers were obtained from the Randolph Glacier Inventory 6.0 (https://www.glims.org/RGI/) and wherever required, minor corrections have been done using high-resolution Google Earth images.

3 Methodology

The approach adopted for estimating decadal glacier specific Surface Mass Balance (SMB) using bias-corrected DEMs is illustrated in Fig. 2. The biases corrected in this study are (i) the bias due to penetration of microwave signals especially in the C-band of SRTM, (ii) the bias caused due to mis-registration of the two DEMs, (iii)an elevation-dependent bias, which is observed when the DEMs are subtracted and plotted with the altitudes (Berthier et al. 2006, 2007).

Mis-registration bias occurs when two different DEMs of the same terrain surface are not accurately aligned, showing a characteristic bias in elevation difference. However, the reason of the elevation-dependent bias is not clearly known but few studies have suggested some reasons like varying resolution (Paul 2008), more penetration of SRTM C-band at higher elevations (Berthier et al. 2006). Either way, an elevation-dependent bias is extremely significant for estimating glacier volume changes because a glacier and its mass balance vary predominantly with elevation. In this study, we consider that the bias in elevation is caused only due to varying resolutions of the two DEMs used.

The TDX DEM was generated using SAR interferometry technique as given in Pandit et al. (2014) with an overall vertical accuracy of 7.41 m (Ramsankaran et al. 2018). Thus obtained 10-m TDX DEM was then resampled to 30 m to match the SRTM DEM. Due to the fact that C-band radar penetration of SRTM can reach up to 10 m in snow and ice, SRTM DEM thus leading to biased estimates of glacier elevation changes. Therefore, the C-band snow penetration correction as suggested by Kääb et al. (2012) was applied over the SRTM DEM. Then, both the DEMs were further utilized in co-registration algorithm suggested by Nuth and Kääb (2011). For more details about the algorithm, readers can refer to Nuth and Kääb (2011).

Later on, following Gardelle et al. (2012) the bias appeared due to varying DEMs resolution are corrected on and off glacier based on the relation established between the elevation difference and terrain maximum curvature for the stable terrain of the glacier. Maximum curvature is obtained from the high-resolution TDX DEM using its native resolution of 10 m. Once the biases in DEMs are corrected, the ice thickness

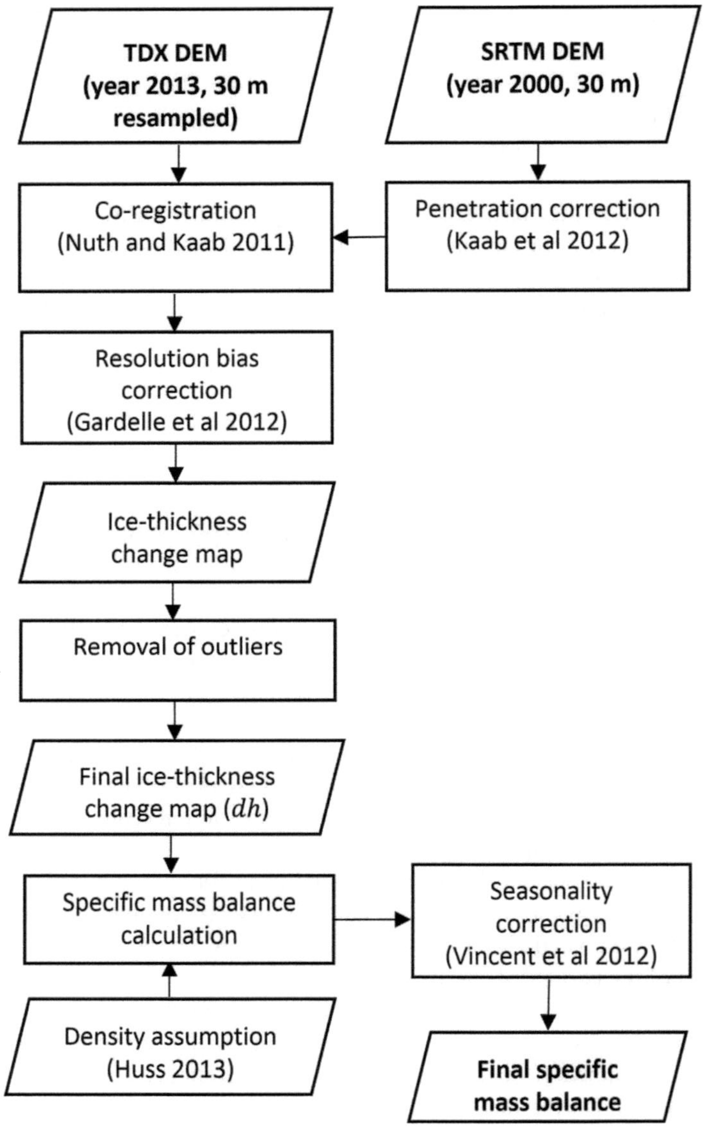

Fig. 2 Methodology flow chart for estimating glacier specific SMB using bias-corrected DEMs

change map for all the 65 glaciers was generated. Then following Berthier et al. (2004), the resultant elevation changes on a glacier that exceed ±100 m are discarded initially, which are surely considered to be outliers. Next, in order to eliminate the further outliers present in the thickness change, we have discarded pixels for which the height difference differed by more than three standard deviations (±100 m) from the mean. Then, using a 3 × 3 averaging filter the excluded pixels were filled. The filled ice thickness map was then converted into specific SMB using Eq. 1 in which the density of ice is assumed to be 850 ± 60 kg/m^3 as adopted by Huss (2013).

$$M = \frac{\int_A dh}{A} \times \frac{\rho_{\text{ice}}}{\rho_{\text{water}}} \tag{1}$$

where M = specific SMB (m w.e), dh = Elevation difference (m), ρ_{ice} = Density of ice (kg/m^3), ρ_{water} = Density of water (1000 kg/m^3), A = Glacier surface area (m^2).

To estimate specific SMB rate seasonal corrections are to be applied which would restore the specific SMB to a value appropriate to the beginning or end of the balance year, or to the date of the other survey. Therefore, a seasonality correction to cover 14 full years from October 1999 to October 2013 was applied between October 1999 and February 2000 as suggested by Vincent et al. (2013). The possible mass changes during this 4-month period were derived from the mean winter mass balance of Chhota Shigri Glacier given in Azam et al. (2016), which comes out to be about 0.15 m w.e per winter month. Finally, uncertainty in the estimated specific SMB rate has also been calculated using Eq. 2 obtained through error propagation method. While calculating the uncertainty in the specific SMB rate, the errors in elevation difference, density (850 ± 60 kg/m^3) and area of the glacier (5% relative uncertainty is assumed) are considered as sources of uncertainty.

$$\frac{\nabla M}{M} = \sqrt{\left(\frac{\nabla \overline{dh}}{\overline{dh}}\right)^2 + \left(\frac{\nabla \rho_{\text{ice}}}{\rho_{\text{ice}}}\right)^2 + \left(\frac{\nabla A}{A}\right)^2} \tag{2}$$

where ∇ represents the uncertainty in estimates.

4 Results and Discussion

Following the methodology specified in Fig. 2, the ice thickness changes and specific SMB rates in the selected 65 glaciers of Chandra basin are calculated. Spatial distribution of the obtained ice thickness change for each of the selected 65 glaciers is presented in Fig. 3. The colour-coded ice thickness change map (Fig. 3) clearly shows that all the glaciers are thinning in most of the parts and a little thickening is observed at accumulation zones of each glacier. The thinning and thickening are found to be varying locally up to 60 and 40 m, respectively.

Specific SMB rate of the selected 65 glaciers of Chandra basin obtained using the bias-corrected DEMs are shown in Fig. 4 along with the area of each glacier. It is found that the specific SMB rate of different glaciers in Chandra basin range from -0.83 ± 0.407 m w.e.y^{-1} to -0.14 ± 0.07 m w.e.y^{-1}. The most negative specific SMB rate of -0.83 ± 0.407 m w.e.y^{-1} is obtained for the G077662E32308N RGI glacier. The mean specific SMB rate of all the glaciers in Chandra basin is about -0.61 ± 0.29 m w.e.y^{-1}. In comparison with the whole basin, it is observed that

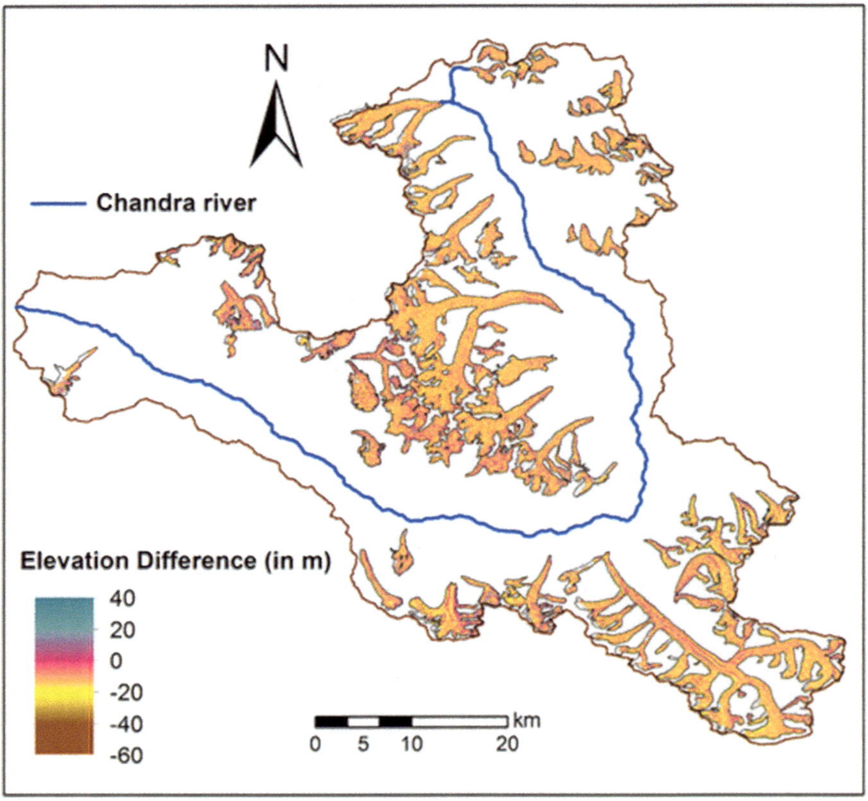

Fig. 3 Map of ice thickness changes of the selected 65 glaciers in the Chandra basin between February 2000 and October 2013

the mass loss rate is lower in Chhota Shigri Glacier (a benchmark glacier), which is estimated to be -0.55 ± 0.26 m w.e.y^{-1}. On the other hand, the surface mass loss rate in large glaciers such as Samudra Tapu and Bara Shigri is found to more than the rest of the region, which are estimated to be -0.69 ± 0.33 m w.e.y^{-1} and -0.66 ± 0.32 m w.e.y^{-1} , respectively.

From the obtained results, it is found that the basin lost around 5.01 ± 2.48 Gt of ice in the past 14 years. Out of which 37.5% was lost from some of the large glaciers such as Bara Shigri and Samudra Tapu glaciers, which indicates that the mass loss is highly dependent on the area of the glacier. To understand how the mass loss is varying in different altitude bands a hypsometric distribution of mass loss is shown in Fig. 5 where mass lost is plotted against the altitude bands along with glaciated area of each altitude band. From Fig. 5, it is inferred that about 3.31 ± 1.66 Gt ice have lost in the elevation band of 5000–5600 m, which is almost 65% of the total loss. The glacier area in lower altitude bands lost less ice mass as their area is less. However, it seen that specific mass loss (defined as mass loss per area) at lower

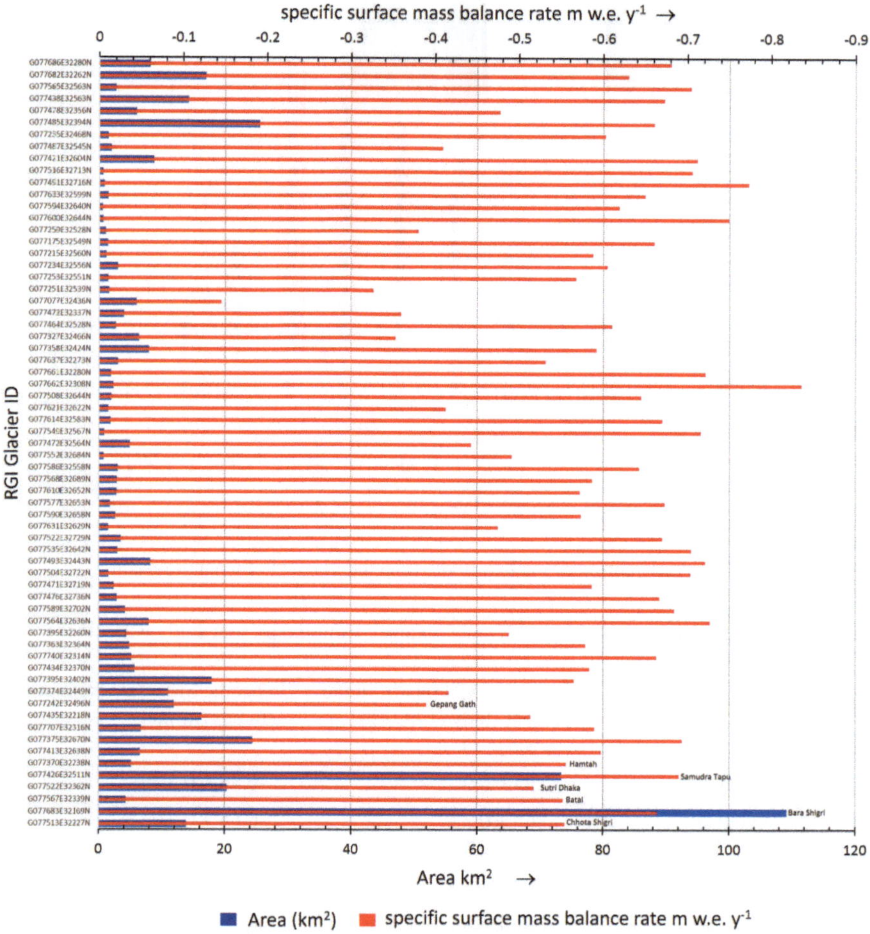

Fig. 4 Specific mass balance rate of the selected 65 glaciers and their area

altitudes is high compared to higher altitudes. This is because the lower elevations are in high-temperature zone and higher altitudes are in lower temperature zone. It is also observed that there is a slight gain in mass at altitude band of 6200–6400 m, which may be because this altitude band is mostly covered by accumulation zones of some of the glaciers.

To quantify the improvements obtained due to DEM bias corrections applied in this study, the specific SMB rate for Chhota Shigri Glacier calculated with bias correction (-0.55 ± 0.26 m w.e.y^{-1}) and without bias corrections (0.18 ± 0.08 m w.e.y^{-1}) is compared with field values (-0.67 m w.e.y^{-1}) obtained from Azam et al. (2016). The comparison indicates that the error has reduced from 127 to 18%. Therefore, one can say that these corrections are very necessary for better estimation of glacier surface mass balance.

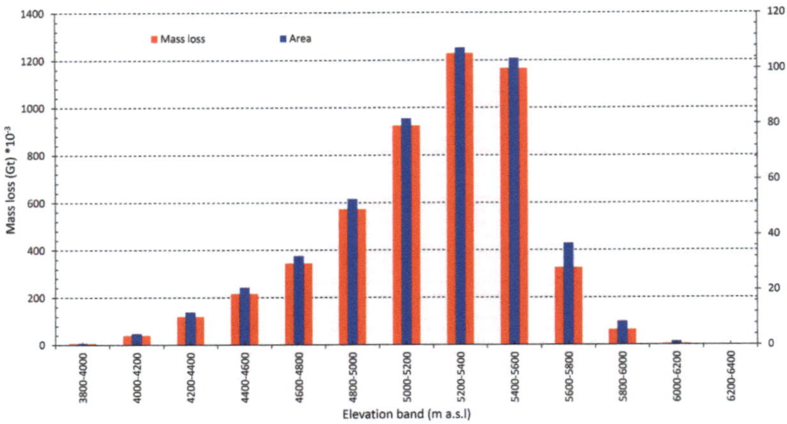

Mass loss (Gt)*10^{-3}	6.7	40.2	120	216.5	344.5	573.6	923.9	1229.3	1165.3	326.9	64.3	3.8	-0.93
Area (km^2)	0.58	4.16	11.98	20.86	32.28	52.75	81.81	107.38	103.57	36.79	8.53	1.26	0.05

Fig. 5 Hypsometric distribution of ice loss in the selected glaciers of Chandra basin

Table 1 Comparison of the specific SMB rate obtained from the present study with other studies in Chandra basin

Study site	Specific SMB rate from the present study (1999–2013) (m w.e.y^{-1})	Specific SMB rate from other studies (m w.e.y^{-1})
Chhota Shigri Glacier	−0.55 ± 0.26	−0.46 ± 0.34[a]
Bara Shigri Glacier	−0.66 ± 0.32	−0.66 ± 0.31[a]
Samudra Tapu Glacier	−0.69 ± 0.33	−0.69 ± 0.31[a]
Hamtah Glacier	−0.55 ± 0.27	−0.45 ± 0.15[b]
Chandra Basin	−0.61 ± 0.29	−0.68 ± 0.15[c]; −0.65 ± 0.04[a]

[a]Vijay and Braun (2016) (2000–2012)
[b]Vincent et al. (2013) (1999–2011)
[c]Gardelle et al. (2013) (1999–2011)

Likewise, a comparison of the mass balance estimates between the present study and other studies given in Table 1. Table 1 indicates that the present study results closely match with the other studies. However, it is observed that the present study estimates for individual glaciers shown in Table 1 are slightly overestimated when compared with other studies. This is possibly due to the difference in the methodology used to correct the penetration of C-band. All other mentioned studies have used SRTM X-band data to correct the C-band penetration, whereas in this study the method given by Kääb et al. (2012) has been used.

It is mainly because SRTM X-band data for the study area is not available but still other studies have used X-band data of nearby basin as a proxy for correction, which may not be appropriate. Specifically, when the results of Chhota Shigri Glacier (Table 1) are compared with the field-based glaciological estimates given by Azam et al. (2016), it is found that the present study results are closer than other studies. On the other hand, it is seen that the results of the whole Chandra basin are underestimated compared to other studies (Table 1). This is because the small glaciers of size <0.5 km^2 are not considered in our study.

5 Conclusion and Recommendations

The current study demonstrates that the geodetic mass balance modelling of glaciers after bias corrections is effective in estimating the long-term surface mass balance. Results of the current study show that the Chandra basin has lost 5.01 ± 2.48 Gt amount of ice mass during the course of 14 years. This is a clear signal indicating the effect of climate change occurred over the study area, which demonstrates how climate change causes imbalance in the health of glacier(s). However, such conclusions are based on very limited number of glaciers which have been studied under this research. Therefore, it is suggested that this modelling approach should be extended to other basins to get the current status of the health of the glaciers across Himalayas. This needs to be taken up on urgent basis considering the fact that many Himalayan glaciers are still untouched and needs to be investigated. In spite of bias corrections, the mass balance calculated from DEM differencing suffers from poor vertical accuracy of the satellite DEM data, thus restricting it to decadal scale studies only. Development of new satellite missions such as Resources at 3S, ICE Sat-2, CryoSat-2, Sentinel-1 can increase the DEM accuracy, providing better future prospects to understand health of the glaciers.

In order to study long-term effects of climate change on the health of glaciers located in the complex Himalayan terrain, continuous surface mass balance data on yearly and seasonal basis are also very much needed. Recent technologies such as Unmanned Aerial Vehicle (UAV), Terrestrial LiDAR are need to be explored which can provide a high-resolution and vertically accurate surface topographic information of the glaciers through which surface mass balance estimates can be obtained on seasonal as well as yearly basis. Such technology can be used to effectively monitor the benchmark glaciers which are presently surveyed using field techniques and also can be used to increase the number of benchmark glaciers studied.

Acknowledgements The lead author and the second author acknowledge the support provided by the Indian Institute of Technology Bombay, through Centre of Excellence in Climate Studies (IITB-CECS) project of the Department of Science and Technology (DST), New Delhi, India. All the authors are thankful to DLR, Germany for freely providing TanDEM-X CoSSC products under TanDEM-X Science proposal XTI_GLAC7043 as well as to USGS for freely providing the SRTM 30 m DEM.

References

Arendt AA, Echelmeyer KA, Harrison WD, Lingle CS, Valentine VB (2002) Rapid wastage of Alaska glaciers and their contribution to rising sea level. Science 297(5580):382–386. https://doi.org/10.1126/science.1072497

Azam MF, Wagnon P, Ramanathan A, Vincent C, Sharma P, Arnaud Y, Linda A, Pottakkal JG, Chevallier P, Singh VB, Berthier E (2012) From balance to imbalance: a shift in the dynamic behaviour of Chhota Shigri glacier, western Himalaya, India. J Glaciol 58(208):315–324. https://doi.org/10.3189/2012JoG11J123

Azam MF, Ramanathan AL, Wagnon C, Vincent AL, Berthier E, Sharma P, Mandal A, Angchuk T, Singh VB, Pottakkal JG (2016) Meteorological conditions, seasonal and annual mass balances of Chhota Shigri Glacier, western Himalaya, India. Ann Glaciol 57(71):328–338. https://doi.org/10.3189/2016AoG71A570

Berthier E, Arnaud Y, Baratoux D, Vincent C, Rémy F (2004) Recent rapid thinning of the "Mer de Glace" glacier derived from satellite optical images. Geophys Res Lett 31(17). https://doi.org/10.1029/2004gl020706

Berthier E, Arnaud Y, Vincent C, Remy F (2006) Biases of SRTM in high-mountain areas: implications for the monitoring of glacier volume changes. Geophys Res Lett 33(8). https://doi.org/10.1029/2006gl025862

Berthier E, Arnaud Y, Kumar R, Ahmad S, Wagnon P, Chevallier P (2007) Remote sensing estimates of glacier mass balances in the Himachal Pradesh (Western Himalaya, India). Rem Sens Environ 108(3):327–338. https://doi.org/10.1016/j.rse.2006.11.017

Berthier E, Schiefer E, Clarke GK, Menounos B, Rémy F (2010) Contribution of Alaskan glaciers to sea-level rise derived from satellite imagery. Nat Geosci 3(2):92–95. https://doi.org/10.1038/ngeo737

Bhutiyani MR, Kale VS, Pawar NJ (2007) Long-term trends in maximum, minimum and mean annual air temperatures across the Northwestern Himalaya during the twentieth century. Clim Change 85(1–2):159–177. https://doi.org/10.1007/s10584-006-9196-1

Bolch T, Kulkarni A, Kääb A, Huggel C, Paul F, Cogley JG, Frey H, Kargel JS, Fujita K, Scheel M, Bajracharya S (2012) The state and fate of Himalayan glaciers. Science 336(6079):310–314. https://doi.org/10.1126/science.1215828

Bookhagen B, Burbank DW (2010) Toward a complete Himalayan hydrological budget: spatiotemporal distribution of snowmelt and rainfall and their impact on river discharge. J Geophys Res Earth Surf 115(F3). https://doi.org/10.1029/2009jf001426

Dash SK, Jenamani RK, Kalsi SR, Panda SK (2007) Some evidence of climate change in twentieth-century India. Clim Change 85(3):299–321. https://doi.org/10.1007/s10584-007-9305-9

Dedieu JP, Rabatel A, Vincent C, Valla F, Thibert E, Arnaud Y (2003) Glacier mass balance determination by remote sensing in the French Alps: progress and limitation for time series monitoring. In: Proceedings of Geoscience and remote sensing symposium, 2003. IGARSS'03, vol 4, pp 2602–2604. https://doi.org/10.1109/igarss.2003.1294523

Dimri AP, Dash SK (2010) Winter temperature and precipitation trends in the Siachen Glacier. Curr Sci, 1620–1625

Dimri AP, Dash SK (2012) Wintertime climatic trends in the western Himalayas. Clim Change 111(3–4):775–800. https://doi.org/10.1007/s10584-011-0201-y

Dobhal DP, Gergan JT, Thayyen RJ (2004) Recession and morphogeometrical changes of Dokriani glacier (1962-1995) Garhwal Himalaya, India. Curr Sci 86(5):692–696

DST (2012) Dynamics of glaciers in the Indian Himalaya: science plan. Prepared by R.K. Midha. Published by the Science and Engineering Board, Department of Science and Technology, New Delhi, Himalayan Glaciology Technical Report No. 2, p 125. Accessed 3 Oct 2017

Gardelle J, Berthier E, Arnaud Y (2012) Slight mass gain of Karakoram glaciers in the early 21st century. Nat Geosci 5:322–325. https://doi.org/10.1038/ngeo1450

Gardelle J, Berthier E, Arnaud Y, Kääb A (2013) Region-wide glacier mass balances over the Pamir-Karakoram-Himalaya during 1999–2011. The Cryosphere 7(4):1263–1286

Gautam R, Christina Hsu N, Lau K.-M (2010) Premonsoon aerosol characterization and radiative effects over the Indo-Gangetic Plains: Implications for regional climate warming. Journal of Geophysical Research 115(D17)

Houghton JT, Ding Y, Griggs DJ, Noguer M, van der Linden PJ, Dai X, Maskell K, Johnson CA (2001) Technical summary of working group 1. In: Climate change 2001: the scientific basis. Contributions of working group I to the third assessment report of the intergovernmental panel on climate change. Cambridge University Press. Accessed 15 Nov 2017

Huss M (2013) Density assumptions for converting geodetic glacier volume change to mass change. Cryosphere Discuss 7(1):219–244. https://doi.org/10.5194/tcd-7-219-2013

Immerzeel WW, Van Beek LP, Bierkens MF (2010) Climate change will affect the Asian water towers. Science 328(5984):1382–1385. https://doi.org/10.1126/science.1183188

IPCC (2007) Summary for policymakers. In: Solomon S, Qin D, Manning M, Chen Z, Marquis M, Averyt KB, Tignor M, Miller HL (eds) Climate change 2007: the physical science basis. Contribution of working group I to the fourth assessment report of the intergovernmental panel on climate change. Cambridge University Press, Cambridge, United Kingdom and New York, NY, USA. Available at https://www.ipcc.ch/pdf/assessment-report/ar4/wg1/ar4-wg1-spm.pdf. Accessed 02 Nov 2017

IPCC (2014) Climate change 2014: synthesis report. In: Pachauri RK, Meyer LA (eds) Contribution of working groups I, II and III to the fifth assessment report of the intergovernmental panel on climate change. IPCC, Geneva, Switzerland, p 151. Available at https://www.ipcc.ch/pdf/assessmentreport/ar5/syr/AR5_SYR_FINAL_SPM.pdf. Accessed 07 Dec 2017

Kääb A (2007) Glacier volume changes using ASTER optical stereo. A test study in Eastern Svalbard. In: Geoscience and remote sensing symposium, 2007. IGARSS 2007, pp 3994–3996. https://doi.org/10.1109/igarss.2007.4423724

Kääb A, Berthier E, Nuth C, Gardelle J, Arnaud Y (2012) Contrasting patterns of early twenty-first-century glacier mass change in the Himalayas. Nature 488(7412):495–498. https://doi.org/10.1038/nature11324

Krimmel RM (1999) Analysis of difference between direct and geodetic mass balance measurements at South Cascade Glacier, Washington. Geografiska Annaler Ser A Phys Geogr 81(4):653–658. https://doi.org/10.1111/1468-0459.00093

Kulkarni AV, Bahuguna IM, Rathore BP, Singh SK, Randhawa SS, Sood RK, Dhar S (2007) Glacial retreat in Himalaya using Indian Remote Sensing satellite data. Curr Sci, 69–74

Kumar R, Areendran G, Rao P (2009) Witnessing change: glaciers in the Indian Himalayas. Pilani, WWF-India and Birla Institute of Technology. Available at http://awsassets.wwfindia.org/downloads/glacier_report.pdf. Accessed 10 Dec 2017

Mernild SH, Pelto M, Malmros JK, Yde JC, Knudsen NT, Hanna E (2013) Identification of snow ablation rate, ELA, AAR and net mass balance using transient snowline variations on two Arctic glaciers. J Glaciol 59(216):649–659. https://doi.org/10.3189/2013JoG12J221P

Muskett RR, Lingle CS, Tangborn WV, Rabus BT (2003) Multi-decadal elevation changes on Bagley Ice Valley and Malaspina Glacier, Alaska. Geophys Res Lett 30(16)

Muskett RR, Lingle CS, Sauber JM, Post AS, Tangborn WV, Rabus BT, Echelmeyer KA (2009) Airborne and spaceborne DEM-and laser altimetry-derived surface elevation and volume changes of the bering glacier system, Alaska, USA, and Yukon, Canada, 1972–2006. J Glaciol 55(190):316–326. https://doi.org/10.3189/002214309788608750P

Narama C, Kääb A, Kajiura T, Abdrakhmatov K (2009) Spatial variability of recent glacier area changes in the Tien Shan Mountains, Central Asia, using Corona (~1970), Landsat (~2000), and ALOS (~2007) satellite data. Glob Plan Change 71:42–54. https://doi.org/10.1016/j.gloplacha.2009.08.002

Negi HS, Saravana G, Rout R, Snehmani (2013) Monitoring of great Himalayan glaciers in Patsio region, India using remote sensing and climatic observations. Curr Sci, 1383–1392

Nuth C, Kääb A (2011) Co-registration and bias corrections of satellite elevation data sets for quantifying glacier thickness change. Cryosphere 5(1):271. https://doi.org/10.5194/tc-5-271-2011

Østrem G, Haakensen N (1999) Map comparison or traditional mass-balance measurements: which method is better? Geografiska Annaler Ser A Phys Geogr 81(4):703–711. https://doi.org/10.1111/1468-0459.00098

Pandey P, Ali SN, Ramanathan AL, Venkataraman G (2017) Regional representation of glaciers in Chandra Basin region, western Himalaya, India. Geosci Front. https://doi.org/10.1016/j.gsf.2016.06.006

Pandit A, Ramsankaran R, Rao YS (2014) Generation and validation of the interferometric SAR DEMs from TanDEM-X data for Gangotri and Hamtah Glaciers of Indian Himalayas. Procedia Technol 16:793–805. https://doi.org/10.1016/j.protcy.2014.10.029

Paul F (2008) Calculation of glacier elevation changes with SRTM: is there an elevation-dependent bias? J Glaciol 54(188):945–946. https://doi.org/10.3189/002214308787779960

Rabatel A, Dedieu JP, Vincent C (2016) Spatio-temporal changes in glacier-wide mass balance quantified by optical remote sensing on 30 glaciers in the French Alps for the period 1983–2014. J Glaciol 62(236):1153–1166. https://doi.org/10.1017/jog.2016.113

Racoviteanu AE, Manley WF, Arnaud Y, Williams MW (2007) Evaluating digital elevation models for glaciologic applications: an example from Nevado Coropuna, Peruvian Andes. Glob Plan Change 59(1):110–125. https://doi.org/10.1016/j.gloplacha.2006.11.036

Ramsankaran R, Pandit A, Azam MF (2018) Spatially distributed ice-thickness modelling for Chhota Shigri Glacier in western Himalayas, India. Int J Rem Sens (in press)

Rignot E, Rivera A, Casassa G (2003) Contribution of the Patagonia Icefields of South America to sea level rise. Science 302(5644):434–437. https://doi.org/10.1126/science.1087393

Rivera A, Casassa G (1999) Volume changes on Pio XI glacier, Patagonia: 1975–1995. Glob Plan Change 22(1):233–244. https://doi.org/10.1016/S0921-8181(99)00040-5

Rolstad C, Haug T, Denby B (2009) Spatially integrated geodetic glacier mass balance and its uncertainty based on geostatistical analysis: application to the western Svartisen ice cap, Norway. J Glaciol 55(192):666–680. https://doi.org/10.3189/002214309789470950

Sangewar CV, Shukla SP, Singh RK (2009) Inventory of the Himalayan Glaciers—a contribution to international hydrological Programme. Special Publication No. 34, Geological Survey of India

Schiefer E, Menounos B, Wheate R (2007) Recent volume loss of British Columbian glaciers, Canada. Geophys Res Lett 34(16). https://doi.org/10.1029/2007gl030780

Sharma SS, Ganju A (2000) Complexities of avalanche forecasting in Western Himalaya—an overview. Cold Reg Sci Technol 31(2):95–102. https://doi.org/10.1016/S0165-232X(99)00034-8

Sharma P, Ramanathan AL, Pottakkal J (2013) Study of solute sources and evolution of hydrogeochemical processes of the Chhota Shigri Glacier meltwaters, Himachal Himalaya, India. Hydrol Sci J 58(5):1128–1143. https://doi.org/10.1080/02626667.2013.802092

Sharma P, Patel LK, Ravindra R, Singh A, Mahalinganathan K, Thamban M (2016) Role of debris cover to control specific ablation of adjoining Batal and Sutri Dhaka glaciers in Chandra Basin (Himachal Pradesh) during peak ablation season. J Earth Syst Sci 125(3):459–473. https://doi.org/10.1007/s12040-016-0681-2

Singh SP, Bassignana-Khadka I, Karky BS, Sharma E (2011) Climate change in the Hindu Kush-Himalayas: the state of current knowledge. International Centre for Integrated Mountain Development (ICIMOD). Available at: http://lib.icimod.org/record/9417/files/icimod-climate_change_in_the_hindu_kush-himalayas.pdf. Accessed 5 Dec 2017

Sund M, Eiken T, Hagen JO, Kääb A (2009) Svalbard surge dynamics derived from geometric changes. Ann Glaciol 50(52):50–60. https://doi.org/10.3189/172756409789624265

Surazakov AB, Aizen VB (2006) Estimating volume change of mountain glaciers using SRTM and map-based topographic data. IEEE Trans Geosci Remote Sens 44(10):2991–2995. https://doi.org/10.1109/TGRS.2006.875357

Van Niel TG, McVicar TR, Li LT, Gallant JC, Yang QK (2008) The impact of 1394 misregistration on SRTM and DEM image differences. Rem Sens Environ, 1395 112(5):2430–2442. https://doi.org/10.1016/j.rse.2007.11.003

Vijay S, Braun M (2016) Elevation change rates of glaciers in the Lahaul-Spiti (Western Himalaya, India) during 2000–2012 and 2012–2013. Rem Sens 8(12):1038. https://doi.org/10.3390/rs8121038

Vincent C, Ramanathan A, Wagnon P, Dobhal DP, Linda A, Berthier E, Sharma P, Arnaud Y, Azam MF, Gardelle J (2013) Balanced conditions or slight mass gain of glaciers in the Lahaul and Spiti region (northern India, Himalaya) during the nineties preceded recent mass loss. Cryosphere 7(2):569–582. https://doi.org/10.5194/tc-7-569-2013

Wagnon P, Linda A, Arnaud Y, Kumar R, Sharma P, Vincent C, Pottakkal JG, Berthier E, Ramanathan A, Hasnain SI, Chevallier P (2007) Four years of mass balance on Chhota Shigri Glacier, Himachal Pradesh, India, a new benchmark glacier in the western Himalaya. J Glaciol 53(183):603–611. https://doi.org/10.3189/002214307784409306

An Experimental Approach Toward Modeling Atmosphere and Ocean Mixing Processes

Sridhar Balasubramanian

Abstract Atmosphere and ocean are host to a variety of submesoscale and mesoscale dynamical processes (e.g., plumes, gravity currents, convection, and baroclinic eddies, to name a few), which trigger episodic turbulent mixing events that govern the variability in weather and climate. The genesis of these processes is attributed to the wind shear interacting with a stably stratified fluid layer, commonly referred to as as a "shear-stratified" flow. In this communication, we consider two variants of shear-stratified flows, namely forced plumes and gravity currents, both of which are commonly encountered in atmospheric and oceanic situations. The dynamics of a forced plume and gravity current are studied with the help of scaled experiments involving simultaneous quantification of fluid velocity and fluid density. The measurements reveal that the mixing is strongly influenced by the shear production flux (P), buoyancy flux (B), and viscous dissipation (ε). The flux Richardson number, $Ri_f = \frac{B}{P}$, which accounts for turbulent motions, is an important parameter used for the characterization of scalar eddy diffusivity (K_ρ), which acts as a proxy for the amount of mixing in the flow. Using the concept of mixing efficiency, Γ, estimates for K_ρ were obtained, which are in agreement with those observed in field conditions. The results documented here would be valuable for dynamical modeling of shear-stratified flows. Additionally, the parameterizations would be beneficial for improvement of numerical models used for weather and climate predictions.

1 Introduction

The equatorial region is a hot spot of variability and host to many small to decadal scale atmospheric disturbances and myriad oceanic processes, the understanding of which remains nascent. It has been widely accepted that the Indian Summer Monsoon (ISM) is an intense phenomena, which affects the livelihoods of more than a

S. Balasubramanian (✉)
Department of Mechanical Engineering and Interdisciplinary Program in Climate Studies, Indian Institute of Technology Bombay, Mumbai 400076, India
e-mail: sridharb@iitb.ac.in

C. Venkataraman et al. (eds.), *Climate Change Signals and Response*,
https://doi.org/10.1007/978-981-13-0280-0_8

billion people in the Indian Ocean rim nations (Gadgil 2003). Therefore, for a tropical climate like that of India, the understanding of atmosphere and ocean dynamics at mesoscale and submesoscale play an important role in monsoon prediction. The influence of the surface energy fluxes are important for understanding the wind patterns, temperature variations, and turbulence profiles over a region, the understanding of which has crucial implications for ISM (Mahadevan et al. 2016). This is because the surface energy balance is directly related to air–sea interactions, height of the atmospheric boundary layer, and the thermal stratification, which in combination are key parameters controlling the sea surface temperature, turbulence, wind profiles, updrafts and downdrafts, and cloud dynamics. Although circulation models incorporate the influence of surface energy and fluxes, the fundamental understanding of mesoscale and submesoscale processes on the ocean and atmosphere dynamics is lacking. In general, the synoptical scale processes are well parameterized in the global models. However, the synoptic scale processes have a strong correlation with the regional scale processes, which are dynamically difficult to model. For example, in both atmosphere and ocean, wind shear and fluid stratification coexist, widely referred to as shear-stratified flows (examples include forced convective plumes and gravity currents) (Thorpe 1969). In these flows, a range of process scales exists (mesoscale and submesoscale), the modeling of which would provide a good basis for improving the crude parameterization during periods of weak and strong stratifications. Additionally, the presence of orography and the forcing induced due to it creates a multitude of time-dependent submesoscale phenomena that contribute to the variability in the weather and climate. Therefore, the process modeling approach proposed in this study would help in improving the understanding at mesoscale and submesoscale levels and provide useful insights into improving the parameterizations used in weather and climate modeling.

2 Forced Plumes and Gravity Currents: A Brief Overview

Forced plumes and gravity currents are two variants of shear-stratified flows that are common occurrences in many geophysical situations. A few examples include hydrothermal vents and oil spills in ocean, rising ash plume from the volcanic eruptions, katabatic wind flows, ocean overflows, and dense water discharges. These flows occur whenever a constant source of buoyancy creates a motion of fluid away from the source. Typically, a forced plume is an initial momentum-dominated flow that has a nonzero density difference between medium and the surrounding ambient fluid, and at some distance becomes dominated by buoyancy (Mirajkar et al. 2015). In most situations, the ambient is in a linear stably stratified state such that the density gradient $\frac{d\rho}{dz} < 0$. The low density plume (ρ_p) fluid intrudes vertically into the linearly stratified ambient leading to complex flow dynamics. An important parameter governing the evolution, growth, and mixing dynamics of a plume is the stratification strength that is characterized by the buoyancy frequency, $N^2 = -\frac{g d\rho}{\rho_0 dz}$, where g is the gravity and ρ_0 is some reference fluid density (generally taken as the ambient density at the source). A forced plume interacting with a stably stratified environment behaves differently compared to a uniform environment, since the mixing near the source diminishes the momentum and buoyancy, thereby rendering the plume to reach a spreading height (Z_s) and spread radially outwards. The intrusion process governs the radial propagation (R_f) of the plume. Past studies on forced plume focussed on scaling arguments and variability of the bulk parameters such as Z_m and R_f and their behaviour with changing buoyancy frequency, N^2 (see Turner 1986; Papanicolaou and Stamoulis 2010; Mirajkar and Balasubramanian 2017 and references therein). Using scaling arguments, Hunt and Kaye (2005) suggested that based on the balance of source momentum flux (M_0), buoyancy flux (B_0), and volume flux (Q_0) at the plume source, different flow regimes for a single-phase plume could be defined using a parameter $\Gamma_0 = \frac{5Q_0^2 B_0}{4\alpha M_0^{5/2}}$, where α is the entrainment coefficient that takes a constant value of $\alpha = 0.08$ for plume-like flows (Turner 1986). Balancing the fluxes at the source, the classifications are as follows: lazy plume ($\Gamma_0 > 1$), pure plume ($\Gamma_0 = 1$), and forced plume ($\Gamma_0 < 1$). Pioneer research on this topic was first done by Morton et al. (1956) to measure the spreading height, Z_s, for the forced plume in the stratified flow using the source conditions and the buoyancy frequency (N^2).

$$B_0 = g\left(\frac{\rho_p - \rho_0}{\rho_0}\right) Q_0 = g' Q_0 \quad (1)$$

$$M_0 = Q_0 U \quad (2)$$

$$Q_0 = \frac{1}{4}\pi D^2 U \quad (3)$$

Here, U is the mean plume velocity, and D is the diameter of the plume. Based on self-similarity arguments, a theoretical formula for Z_s was proposed by Morton et al. (1956) as follows,

$$Z_m = 2.8\alpha^{-0.5} B_0^{0.25} N^{-0.75} \quad (4)$$

The theory by Morton et al. (1956) works well for large-scale flow parameters, but fails to predict the mesoscale and submesoscale plume characteristics (such as turbulent kinetic energy, local momentum and buoyancy fluxes, viscous dissipation, and mixing efficiency). Following the seminal work of Morton et al. (1956), other researchers have also studied the plume dynamics in a stably stratified environment, where the results were extended for fountains, lazy plumes, along with characterisation of bulk quantities and plume entrainment dynamics (see Bloomfield and Kerr 1998; Hunt and Kaye 2005; Devenish et al. 2010; Kaye 2008; Papanicolaou et al. 2008; Papanicolaou and Stamoulis 2010 and references therein). Recently, experiments on forced plume in linear stratification for a varying range of N^2 were performed to study the intrusions from buoyancy-dominated as well as momentum-dominated source conditions (Richards et al. 2014),

Mirajkar and Balasubramanian (2017). However, the results were limited to the characterization of radial propagation of plume, R_f, plume thickness, t_p, and plume spreading height, Z_s. None of the previous studies on forced plume have focused on turbulence and mixing characterization using the kinetic energy budget, which is needed for a better understanding of the mesoscale and submesoscale flow physics. The governing parameter for a forced plume is the bulk Richardson number, given as $Ri_b = \frac{g'D}{U^2}$.

A gravity current is driven by horizontal pressure gradients arising due to density variations between the denser fluid (ρ_1) and the lighter fluid (ρ_2) (Ellison and Turner 1959). The important parameters determining gravity currents propagation are the density difference between the two fluids ($\Delta\rho = \rho_1 - \rho_2$), gravity ($g$), the depth of the gravity current (h), total depth of the fluid layer (H), and the slope of the terrain (α) (Balasubramanian et al. 2015). For a gravity current, the active regions of gravity currents have been well established (Simpson 1982; Simpson and Britter 1979). The interface between two fluids close to the head of a gravity current is a typical frontal zone, that is, a region in which, notwithstanding intense mixing, a high density gradient is present. The frontal zone is immediately followed by the head, which has some fractional depth of the initial height, H, depending on the nature of the gravity current. The head of the current is the region where the fractional depth is $\geq \frac{H}{3}$. The scaling for velocity and depth of two counterflowing gravity currents, produced by lock-exchange, has been well understood (Simpson 1982; Shin et al. 2004; Cantero et al. 2007). For the Boussinesq case, Yih (1965) proposed that the depths of two currents are equal in height, $h = \frac{H}{2}$, along their entire lengths. The speed of both gravity currents are the same and have the value proposed by Benjamin (1968) for energy-conserving gravity currents. Klemp et al. (1994) argued, based on shallow-water theory that idealized energy-conserving gravity currents cannot be realized in a lock-exchange initial-value configuration, as the speed of this current would be faster than the fastest characteristic speed in the channel predicted by the shallow-water theory. Extensive measurements show that, on a horizontal surface, the mean velocity of the current is given by $U = 1.05\sqrt{g'h}$, where $g' = \frac{\rho_1 - \rho_2}{\rho_1}$ is the reduced gravity, and h is the depth of the steady current (Benjamin 1968). These results were mainly obtained from flows occupying about 1/5 of the total depth H, but recent work with lock-exchange flows has shown that U is sensitive to changes in the value of $\frac{h}{H}$ in the range 1/3–1/10 H as proposed by Simpson and Britter (1979). They argue that the inviscid gravity current depth can never be greater than $0.35H$, wherein according to Benjamin's theory (Benjamin 1968), the gravity current has its fastest speed. Therefore, Benjamin's theory for energy-conserving gravity currents is widely accepted, and the velocity and depth of a gravity current are given as,

$$U = 0.4\sqrt{g'H}, \quad h = \frac{H}{2} \tag{5}$$

Most previous studies have focussed on the dynamics of the head, turbulence dissipation and mixing, as well as scaling for the front velocity and fractional depth. The gravity current entrainment as a function of bulk Richardson number, Ri_b was quantified by Ellison and Turner (1959). In their configuration, the bulk Richardson number was variable, since the inertial and buoyancy forces were decoupled. In the present case, however, the governing parameters are g' and H, and in view of Eq. (5) the bulk Richardson number is a constant ($\approx$1). For the configuration of gravity currents considered in our study, the only possible variable is the Reynolds number $Re = \frac{Uh}{\nu} = \frac{UH}{2\nu}$ (Simpson and Britter 1979), which has been consistently used for lock-exchange flows (Shin et al. 2004; Cantero et al. 2007). Similar to the case of a forced plume, none of the previous studies on gravity current have focused on turbulence and mixing characterization near the head of the current using the kinetic energy budget, which is needed for a better understanding of the mesoscale and submesoscale flow physics. Based on this gap, we formulate the problem statement for this communication.

3 Problem Statement

As established from the literature review, most studies have primarily focussed on the bulk characteristics of a forced plume and gravity currents, but not the turbulence and mixing dynamics. The mixing across density interface is a frequent phenomenon in geophysical and engineering flows and there is extensive interest on understanding the turbulent mixing in flows with stable density stratification. For example, the surface wind and temperature advection between ocean and atmosphere drive the upper mixing layer of ocean into stably stratified oceanic pycnocline and this process is important to the dispersion of pollutants. Different to the commonly stable stratification in oceanic flows, the density (or temperature) stratification along gravity direction in the atmospheric boundary layer changes periodically, leading to fundamental differences in the mixing process, where the governing mechanisms are different. The turbulent kinetic energy, shear production flux, buoyancy flux, and viscous dissipation give the local nature of the flow. One key interest is on quantifying the mixing efficiency, which has been studied by in-situ field measurements. The laboratory-based measurements in mixing efficiency is very limited, which prohibits the quantitative characterization of turbulent mixing in density stratified geophysical flows. Accurate quantification of both momentum and scalar diffusivities is imperative given their importance for many practical applications such as air quality prediction, nutrient transport in water bodies and ocean circulation. It is a common practice to quantify turbulent mixing in such flows using a turbulent (eddy) viscosity K_t for momentum and a turbulent (eddy) diffusivity K_ρ for density, which are based on the gradient-diffusion hypothesis (Pope 2000). For a unidirectional shear flow, the momentum eddy diffusivity, K_t, and scalar eddy diffusivity, K_ρ, are defined as

$$K_t = -\frac{\overline{u'v'}}{\frac{\partial \overline{U}}{\partial z}} \tag{6}$$

$$K_\rho = -\frac{\overline{\rho' v'}}{\frac{\partial \overline{\rho}}{\partial z}} \tag{7}$$

In order to characterize turbulence in stratified flows, it is important to understand the evolution of turbulent kinetic energy (TKE). The transport equation of turbulent kinetic energy in stratified flows, $K = \frac{1}{2}\overline{u_i' u_i'}$ is (Pope 2000):

$$\frac{\partial K}{\partial t} + U_j \frac{\partial K}{\partial x_j} + Tr = P - B - \varepsilon \tag{8}$$

Here, Tr is the transport term given as $(\frac{1}{\rho_0}\frac{\partial \overline{u_i' p'}}{\partial x_i} + \frac{\partial \overline{u_j' u_j' u_i'}}{\partial x_i} - \nu \frac{\partial^2 K}{\partial x_j^2})$, $P = -\overline{u_i' u_j'}\frac{\partial U_i}{\partial x_j}$ is the shear production, $B = \frac{g}{\rho_0}\overline{\rho' u_i'}$ is the buoyancy flux and $\varepsilon = 2\nu e_{ij} e_{ij}$, where $e_{ij} = \frac{1}{2}(\frac{\partial u_i'}{\partial x_j} + \frac{\partial u_j'}{\partial x_i})$, is the dissipation of energy due to viscous effects. The term $\overline{u_i' u_j'}$ is the Reynolds stress term, which gives the correlation between the stream-wise and transverse fluctuating velocity components. Major inherent assumptions while using the transport equation for K are that the turbulent flow is statistically stationary and homogeneous. These assumptions are used to simplify the energetics of the turbulent flow field. Under these assumptions, the left-hand side terms of Eq. (8) vanishes, giving a simple balance that yields $P - B - \varepsilon = 0$. Consider, for instance, the model proposed by Osborn (1980) where (under stationary and homogeneous assumptions), the turbulent kinetic energy (TKE) equation can be simplified to obtain the diapycnal diffusivity of density as

$$K_\rho = \frac{Ri_f}{1 - Ri_f}\frac{\varepsilon}{N^2} = \Gamma \frac{\varepsilon}{N^2} \tag{9}$$

$$K_t = \frac{1}{1 - Ri_f}\frac{\varepsilon}{S^2} \tag{10}$$

where Γ is known as the mixing efficiency of the flow, and $S = \frac{\partial U}{\partial z}$ is the mean velocity shear present in the flow. Varying definitions of calculating flux Richardson number, Ri_f, has been defined by past researchers (see Osborn 1980; Venayagamoorthy and Koseff 2016) to measure the amount of TKE that is irreversibly converted to potential energy. The most common definition, under the assumption of stationary and homogeneous flow is as follows,

$$Ri_f = \frac{B}{B + \varepsilon} \tag{11}$$

A value of $Ri_f = 0.2$ was proposed by Osborn based on experiments by Britter (1974), but it is unclear if this value holds good for all genres of shear-stratified flows. Given the fact that the inherent assumptions of statistical stationarity and homogeneity are not always applicable in practice, be it in direct numerical simulations (DNS) or observational studies of geophysical flows, Ivey and Imberger (1991) proposed an alternative definition of Ri_f (denoted by Ri_f^{II}) as

$$Ri_f^{II} = \frac{B}{B+m} \tag{12}$$

where m accounts for contributions from all the terms in Eq. (8). This definition is free from the assumption that turbulence is stationary and homogeneous and hence is a better representation of the flux Richardson number (Ri_f). Venayagamoorthy and Koseff (2016) recently showed that the above definition of Ri_f (Eq. 11) suffers from the effects of counter-gradient fluxes that are common in strongly stratified flows. They proposed another definition based on available potential energy and total dissipation rate. However, to understand the flow energetics, in the present study we stick to the first definition of Ri_f given by Eq. (11).

In order to understand the dynamics of shear-stratified flow (such as a forced plume and gravity currents) at local scales, we need to accurately measure the turbulence quantities such as TKE, Reynolds stresses, production flux (P), buoyancy flux (B), dissipation (ε), flux Richardson number (Ri_f), and mixing efficiency (Γ). Experimentally, this is possible only using simultaneous measurements of velocity and density fields, which is the focus of this present work. Detailed experimental investigations are imminently needed to obtain understanding of the mixing mechanisms for modeling stratified flows, e.g., in the atmospheric boundary layer and the global thermohaline circulation, in particular, to quantify the mixing efficiency and entrainment rate under different stratification and turbulent levels. For measurements of velocity and density fields, we employ particle image velocimetry (PIV) and planar laser-induced fluorescence (PLIF) techniques, respectively. Briefly, Webster et al. (2001) developed simultaneous measurements of the velocity and concentration field using digital particle image velocimetry (DPIV) and planar laser-induced fluorescence (PLIF) for a turbulent jet in the uniform medium to measure the mean velocity, turbulent stresses, mean concentration variance. The results for the stratified case, obtained using DPIV, showed that the mean centreline velocity decreases much more rapidly than the unstratified case, where Reynolds stress profiles never reached a self-similar state, indicating that stratification changes both the overall turbulence characteristics and mixing. Recently, Duo and Chen (2012) observed a horizontal dense jet injected into a lighter stratified solution using combined particle image velocimetry and planar laser-induced fluorescence. They studied flow structure and mixing dynamics of the dense jet in the lighter solution. From the literature, it is clear that simultaneous measurements of velocity and density, using PIV–PLIF technique, for a forced plume and gravity current have not been carried. Below, we will briefly talk about the methodology, followed by results and discussions.

4 Methods

4.1 Experimental Modeling

Due to the complexity involved in measurement of turbulence statistics and related mixing parameter in field observations, it is prudent to study the dynamics of forced plumes and gravity currents using an experimental analogue of the particular geophysical process. By accounting for the unboundedness of the ocean and atmosphere (i.e., shallow-water approximation) through appropriate non-dimensionalization, we gather the rich flow physics embedded in these flows through measurement of various mean and fluctuating quantities. Below, details of the experimental setup along with the important non-dimensional number for each configuration is given.

4.2 Experiments on Forced Plume

The experiments were carried out in a tank facility, whose configuration is illustrated in Fig. 1a. The tank T2 is made of plexiglas, measuring 91 cm long by 91 cm wide by 60 cm high. The second tank (T1), a 60 cm cubical tank, was used as the reservoir for storing the jet fluid. The density of the plume fluid $\rho_p = 998$ kg/m^3 was kept constant in all the experiments. The tank (T2) was linearly stratified using the double bucket technique as discussed in Oster and Yamamoto (1963), Mirajkar and Balasubramanian (2017). The strength of the stratification was maintained at $N = 0.2$ s^{-1}. A portable densitometer (*AntonPaarDMA*35) was used to check the density in the two buckets. Density profile in the stratified tank was checked by collecting the samples at every intervals in the experimental tank to ensure the density profile is linear. A centrifugal pump was used to discharge the jet fluid into the ambient linearly stratified environment using a round jet nozzle fixed at the bottom of tank (T2). The jet nozzle was 160 mm in length with diameter $D = 12.7$ mm. It was made of aluminum and comprised of a diffuser, settling chamber, and a contraction section. A honeycomb was placed in the settling chamber to reduce the flow fluctuations and to generate a stable flow at the nozzle exit. The exit vertical velocity at the nozzle was maintained constant at $U = 17$ cm/s, thereby giving a jet Reynolds number $Re = 2400$, and the initial bulk Richardson number was, $Ri_b = 0.008$. A linear stable stratification with salt–water–ethanol mixture was obtained in T2, such that heavy fluid settles at the bottom and lighter fluid on the top. Once the fluid is filled into tank T2 using the two-bucket technique, it is allowed to stabilize for approximately 2 h to achieve stable uniform linear stratification with height. Upon achieving a stable stratification, the forced plume was injected and the local flow dynamics were captured to understand the flow physics. An experimental image of a forced plume with bulk parameter is shown in Fig. 1a1.

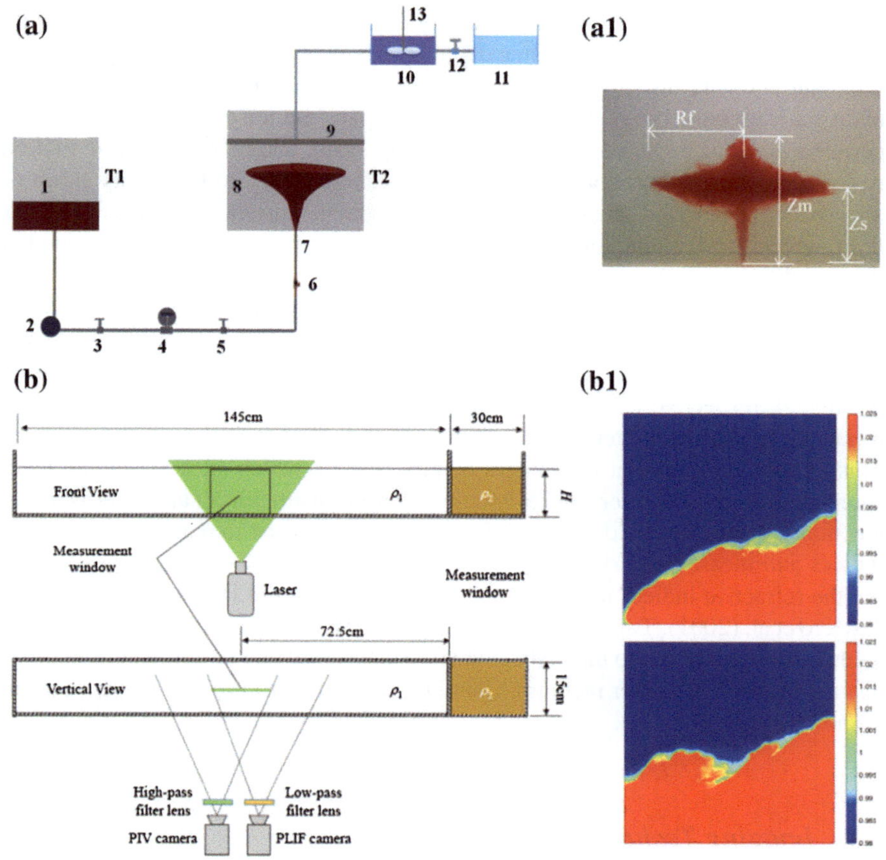

Fig. 1 **a** Schematic of the experimental setup for a forced plume. (1) Reservoir tank (T1), (2) Centrifugal pump, (3, 5, 12) Control valve, (4) Flow meter, (6) Non-return value, (7) Jet nozzle, (8) Stratification tank (T2), (9) Perforated wooden plate, (10) Salt water bucket, (11) Fresh water bucket, (13) Fluid mixer. (a1) Image of a forced plume with representation of bulk parameters. *Note Both (a) and a(1) reproduced with permission from ASCE.* **b** Schematic of experimental setup for a gravity current. (b1). Image of a gravity current at two different instances, (top) at a time instant t, (bottom) at a time instant $t + \Delta t$. Contours represent density. In both the experimental settings, horizontal is the stream-wise coordinate, x, and normal is the vertical coordinate, z

4.3 *Experiments on Gravity Current*

The experiments on gravity currents were conducted in a plexiglass tank equipped with a lock-exchange mechanism. The tank dimensions were 175 cm long, 15 cm wide, and 30 cm high. A schematic of the apparatus is shown in Fig. 1b. The tank was separated into two parts by a lock gate located at 30 cm from right end. The dense fluid ρ_2 in the right slot occupies a predetermined depth H before the gate is released. The dense solution is prepared by adding requisite amount of NaCl to water and mixing it to get a uniform density fluid. The rest of the tank is filled with lighter ambient fluid, ρ_1, to the same depth H, which is separated by the gate. Upon removing the gate instantly, a gravity current is initiated due the difference in the hydrostatic pressure between the two fluids. The quick motion of the gate ensures that perturbations due to gate opening are extremely small, and no relative fluid motion is generated in the direction of the pull. Thus, there are is no secondary flows or disturbances due to the gate release. The denser gravity current undercuts the lighter fluid, which flows in the opposite direction. The measurement section is located at the middle of the tank.

Two different Reynolds numbers were used for the present study, namely $Re =$ 3090 and $Re = 9950$, which covers flow transitioning from weak to strong mixing. The dense and light fluid were created using salt solution and an aqueous solution of ethanol, respectively. This salt–ethanol technique was introduced to match the refractive indices accurately, enabling the use of optical measurement techniques. This method ensured that the images quality is high, which allows accurate measurements. A densitometer (make: Mettle Toledo Densito 30PX) and a refractometer (make: Leica handheld analog refractometer) were used to match the refractive indices and measure the density of the two fluid. Details of the method of matching the refractive indices using salt and alcohol are given in Duo and Chen (2012), Daviero et al. (2001). Two different intensities of gravity currents (represented by Re values) were generated to understand the flow physics of such a dense current propagating in ambient lighter medium. Contour image of a propagating gravity current is shown in Fig. 1b1.

4.4 Imaging Technique

A combination of Particle Image Velocimetry (PIV) and Planar Laser-Induced Fluorescence (PLIF) is used for simultaneous velocity and density measurements as illustrated above in Fig. 1. Before the experiments, the refractive indices of the different fluids in use were matched as explained in the Daviero et al. (2001) and Mirajkar and Balasubramanian (2016). In the present work, the density and the refractive indices were measured using a densitometer and a refractometer (make: *AntonPaar*). A dual-head Nd:YAG pulse laser (532 nm, maximum intensity 145 mJ/pulse) was used for both PIV illumination and PLIF excitation. Through PIV optics, the laser beam is expanded into a 1-mm thick laser sheet illuminating the sample area in the x-z plane along the centre line of the tank. The fluid was uniformly seeded with polyamide tracer particles (median diameter 50 μm, and specific gravity $\rho_{sg} = 1.1$) for PIV measurement. For the PLIF measurement, Rhodamine 6G dye was uniformly mixed to the plume and the gravity current fluid in order for measurement of density. The dye fluoresces at 532 nm and gives an excitation signal at 560 nm. In order to implement the simultaneous PIV/PLIF measurement, the camera lens, PIV filter, PLIF filter, and two cameras are mounted in an optical housing as shown in Fig. 1. The PIV filter (bandpass, 530 nm) blocks most of the fluorescence and passes scattered light from PIV seeding particles. The PLIF filter (high pass in wavelength with cut-off 550 nm) blocks the scattered light and only passes the fluorescence signal. The time delay between the two pulses was set in millisecond. An image acquisition and laser control system synchronized the measurements with a sampling rate of 10 Hz. A two-step processing is applied: 64 × 64 pixels interrogation window and 50% overlap for the first step, and 32 × 32 pixels interrogation window and 50% overlap for the second step. The raw images obtained from PLIF camera were de-warped and then processed using an in-house algorithm to get the density field.

For experiments of forced plume, we observed that intensity of the background image and density was found behave approximately as a linear function. Based on the linear function, the following equation was used to convert intensity to density. The laser light variation was considered while doing this transformation. The final form of the density formula is

$$\rho = \rho_p - \frac{I}{I_1(z,t))}\left[\rho_p - \rho_0(z,t)\right] \tag{13}$$

where ρ_p is the density of the plume fluid. I is the intensity of the evolving plume and $I_1(z, t)$ is the intensity of the background medium, which also takes care of the laser intensity absorption factor in the medium. $\rho_0(z, t)$ is the density of the background image.

For experiments on gravity current, the local R6G concentration can be found from the local gray value, and the local R6G concentration has a linear relationship with the local density (Balasubramanian and Zhong 2018). When the dense fluid (Volume V and density ρ_2) and a lighter fluid (Volume eV and density ρ_1, R6G concentration C_1) are mixed uniformly, the density and the R6G concentration of the mixture are:

$$\rho = \frac{\rho_1 eV + \rho_2 V}{eV + V} = \frac{\rho_1 e + \rho_2}{e+1}$$

$$C = \frac{C_1 eV}{eV + V} = \frac{eC_1}{e+1}$$

Thus, if the local R6G concentration C is known, the local density can be found using:

$$\rho = \rho_2 - \frac{C}{C_1}(\rho_2 - \rho_1) \tag{14}$$

As seen from this equation, if the local concentration, C equals the known concentration of the lighter fluid, C_1, then the local density is same as that of the lighter fluid. This is the initial state, when both the fluids are separated by the gate. Upon the release of the gate, entrainment occurs causing change in the local density.

5 Results and Discussion

The simultaneous velocity and density measurements help in characterization of the turbulence and mixing in shear-stratified flows. Below, we independently discuss the results for the two different cases, namely a forced plume and a gravity current.

5.1 Dynamics of a Forced Plume

The energetics of an evolving forced plume were captured from the source of the plume ($\frac{z}{D} = 0$) to a finite vertical distance such that $\frac{z}{D} = 15$. This finite height corresponds to the fluid layer below the spreading height, Z_s of the plume. The results presented here correspond to the dynamics of moderate stratification strength ($N = 0.2$ s^{-1}). Turbulent kinetic energy, K, is one of the most important statistics in stratified flows, which shows the turbulence distribution in the flow. For 2-D flows, this parameter is given as $K = \frac{1}{2}[\overline{u'^2} + \overline{v'^2}]$. The turbulent kinetic energy profile evolution for the plume was recorded at three different downstream locations $\frac{z}{D} = 4, 8,$ and 12. The normalized turbulent kinetic energy was plotted with the normalized radial coordinate and is shown in Fig. 2a. It is seen that for lower $\frac{z}{D}$ values, i.e., close to the source region, the turbulence kinetic energy is very high and it gradually decreases with increasing $\frac{z}{D}$ values. Such a behavior is expected since the forced plume is losing momentum due to the entrainment between the plume and ambient fluids. The profile of K shows broadening which is attributed to the plume expansion as it moves upwards. Lastly, a double peak structure is seen, which slowly disappears with increasing $\frac{z}{D}$. This trend is attributed to the high shear present near the edges of the plume in comparison to the plume centre. As the $\frac{z}{D}$ increases, the shear reduces and the associated turbulent energy also decreases, thereby reducing the imprint of the double peak in the TKE profiles. The radial Reynolds stress can be plotted to confirm the results from the turbulent kinetic energy profiles. The normalized radial stress plot is shown in Fig. 2b. An off-center peak is seen due to the production of turbulence energy by Reynolds stress working against the mean shear. This further confirms the fact that the flow energetics are predominant in the edges of the plume than at the central region. It is also documented that the Reynolds stress components decrease with increase in the downstream direction, a trend attributed to the reduction in the magnitude of fluctuating components of velocity due to entrainment. The profiles for turbulent kinetic energy and Reynolds stress are in agreement with some of the existing literature for turbulent buoyant jets (Shiri 2010).

The turbulent kinetic energy budget equation given by Eq. (8), under the assumption of stationary and homogeneous turbulence, has three important terms, namely production flux, P, buoyancy flux, B, and viscous dissipation, ε, that govern its evolution. Using the simultaneous velocity and density fields, we can extract all these three terms and study their dynamics. This is done for a forced plume with a stratification strength $N = 0.2$ s^{-1}, and the results are shown in Fig. 3. Evident from Fig. 3a is that

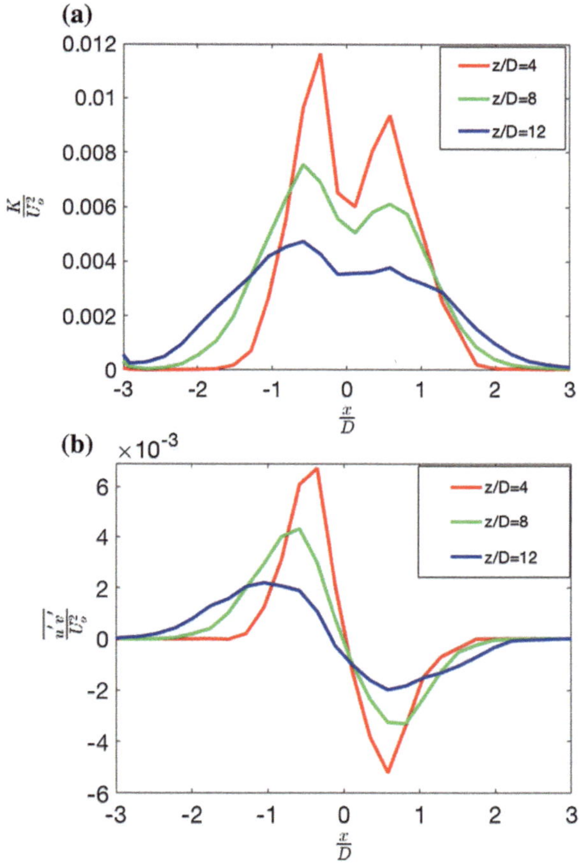

Fig. 2 Profiles of **a** normalized turbulent kinetic energy and **b** normalized radial Reynolds stress, at different normalized vertical locations

the production flux term is always positive indicating mechanical gain of turbulent energy due to the eddies present in the flow. The magnitude of P decreases as z increases due to the plume transitioning from a momentum-dominated to buoyancy-dominated flow. In Fig. 3b, the buoyancy flux is plotted, which has a negative value indicative of unstable stratification. This is a good representation of our flow, since the flow is unstable owing to the upward movement of the lighter fluid. Due to strong density gradients at low z levels, the buoyancy flux is higher. As the plume evolves downstream, the entrainment of the plume fluid with the ambient reduces the density gradient, thereby causing reduction in the buoyancy flux. Nevertheless, the value of B always stays negative, indicating unstable convection in the flow. It should be noted that the magnitude of buoyancy flux is low, which is representative of the low value of bulk Richardson number, $Ri_b = 0.008$, used in this study. Finally, the nature of viscous dissipation is revealed in Fig. 3c. The positive value of ε indicates that turbulent energy is being lost to friction and dissipation of energy from large scales to small scales in the flow. From the kinetic energy budget, we can write $P - B - \varepsilon = 0$,

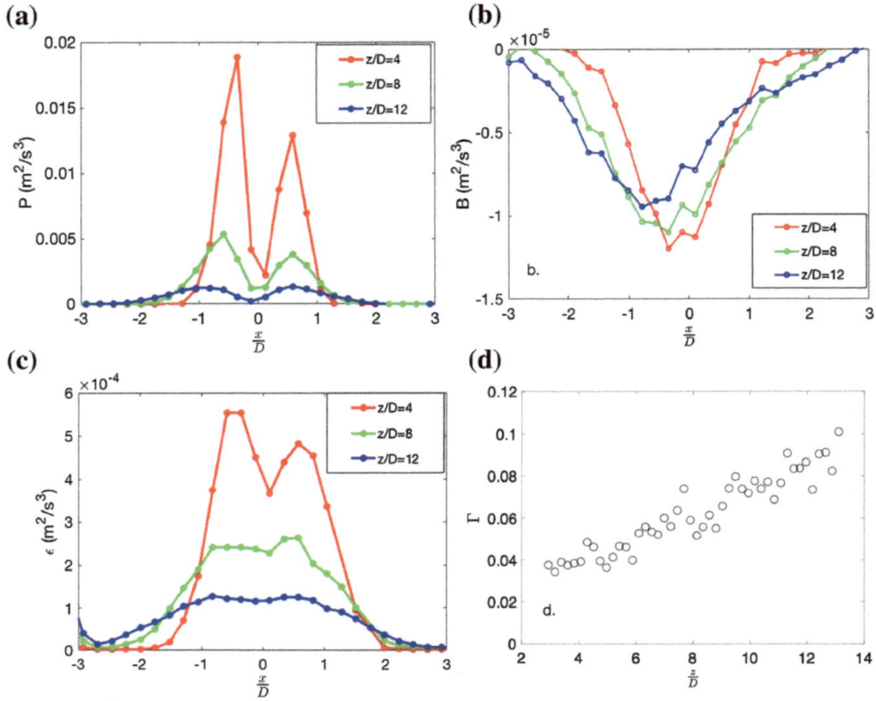

Fig. 3 Turbulent kinetic energy budget terms under stationary and homogeneous assumption. **a** Production flux P, **b** Buoyancy flux, B, and **c** viscous dissipation ε. The mixing efficiency (Γ) as a function of z is shown in (**d**)

which means that the under stationary and homogeneous assumption, the production flux and buoyancy flux must balance dissipation (i.e., $P + B = \varepsilon$). However, this is not evident from Fig. 3 that confirms that the transport term (Tr) also plays an important role in the kinetic energy budget. Despite this, the results from the present work are extremely useful in understanding the dynamics of turbulence and mixing in forced plumes, since it is first of its kind. In Fig. 3d, we present the mixing efficiency, Γ as a function of z. It is observed that Γ increases with z, showing the nature of scalar mixing in the flow. This is expected since dissipation, ε, reduces at a faster rate than the buoyancy flux, B, causing an increase in the value of Γ.

An estimate for K_ρ was also deduced from the experimental results and was found to be of the order of $K_\rho \approx 3 \times 10^{-4}\,\mathrm{m}^2\,\mathrm{s}^{-1}$. This value is seldom seen in field observation depending on the flow conditions (Lozovatsky and Fernando 2012). Therefore, we conclude that the estimate of scalar eddy diffusivity measured in our present study is legit, which gives further confidence in the turbulence characterization of the flow.

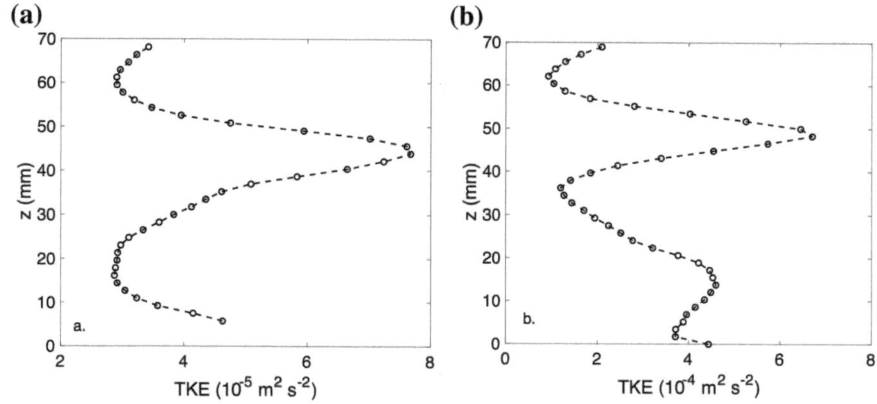

Fig. 4 Turbulent kinetic energy profile at two different values of Reynolds number, **a** $Re = 3090$ and **b** $Re = 9950$

5.2 *Dynamics of Gravity Currents*

The energetics of a gravity current were studied near the head region to understand the turbulence and mixing dynamics of the flow. This was done for two different values of Reynolds numbers, namely $Re = 3090$ and $Re = 9950$. These two values were chosen based on the qualitative differences in the nature of the evolving flow. The turbulent kinetic energy, K, being an important statistics in stratified flows, which shows the turbulence distribution in the flow, it was plotted for these two Re values as shown in Fig. 4. It is clearly seen that the TKE peaks near the central region ($z = 30$ cm to $z = 50$ cm for $Re = 3090$ and $z = 40$ cm to $z = 60$ cm for $Re = 9950$), where the dense and the lighter fluids mix due to strong shear. This is the zone where the energetics of the flow and mixing are dominant. We also notice that the value of TKE is an order of magnitude higher for the case of $Re = 9950$. This is expected since the initial momentum in the flow is large resulting in strong TKE generation.

As given by Eq. (8), the turbulent kinetic energy budget equation, under the assumption of stationary and homogeneous turbulence, is governed by the contributions from production flux, P, buoyancy flux, B, and viscous dissipation, ε. These three terms were plotted as a function of z and for the two Re cases and the results are shown in Fig. 5. An immediate observation from this figure is that P, B, and ε for both values of Re show a peak value in the central z region. This is expected since most of the turbulence and mixing is generated in the central portion allowing the two fluids to mix vigorously. For dissipation, ε, a higher value is also seen near $z = 0$ due to the frictional effect of the wall. For the case of $Re = 3090$, the production flux term is always positive indicating mechanical gain of turbulent energy due to the eddies present in the flow. The buoyancy flux, B, is also positive indicative of stable stratification. This is a good representation of our flow, since the flow is stable owing to the downward movement of the denser fluid, such that it leads to

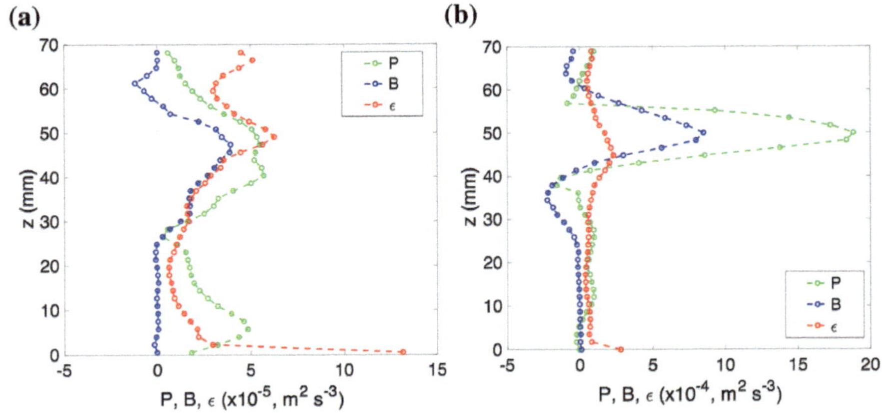

Fig. 5 Turbulent kinetic energy budget terms for two different values of Reynolds number, (left) $Re = 3090$ and (right) $Re = 9950$

dense fluid settling below lighter fluid, thereby giving a stably stratified profile. The positive value of ε indicates that turbulent energy is being lost due to friction and cascade of energy from large scales to small scales in the flow. From the kinetic energy budget, we can write $P - B - \varepsilon = 0$, which means that the under stationary and homogeneous assumption, the production flux must balance buoyancy flux and dissipation (i.e., $P = B + \varepsilon$). However, this is not evident from Fig. 5 (left plot for $Re = 3090$) that again confirms that the transport terms (Tr) may play an important role in the kinetic energy budget. Despite this, the results from the present work are extremely useful in understanding the turbulence and nature of mixing in gravity currents. A very similar picture emerges for $Re = 9950$ (right plot in Fig. 5), the only difference being the magnitude of P, B, and ε are higher than for $Re = 3090$ case due to the higher inertia in the flow.

In Fig. 6, the mixing efficiency Γ as a function of stream-wise location x for the two Reynolds numbers is shown. It is observed that Γ shows a spatial variation showing the random nature of scalar mixing along the stream-wise direction of the gravity current. This is expected due to the chaotic nature of fluid mixing owing to interfacial instabilities. In order to get an estimate of the turbulent diffusivity, a mean value of Γ could be used (shown by the dashed line in Fig. 6). The mean value of Γ is lower for $Re = 3090$ compared to $Re = 9950$, indicated that mixing is vigorous in the high Reynolds number case. Following this, an estimate for K_ρ could deduced from the experimental results using the values of Γ, ε, and N^2. The value of N^2 for the case of gravity current is measured from the vertical density profile that is inherently developed for the particular Re due to the flow dynamics (Balasubramanian and Zhong 2018). The value of K_ρ was found to be of the order of $K_\rho \approx 4 \times 10^{-6}$ m^2 s^{-1} for both $Re = 3090$ and $Re = 9950$. The similar values indicate that the increase in Γ is offset by the corresponding increase in the stability of the density profile yielding higher values of N^2, which appears in the denominator

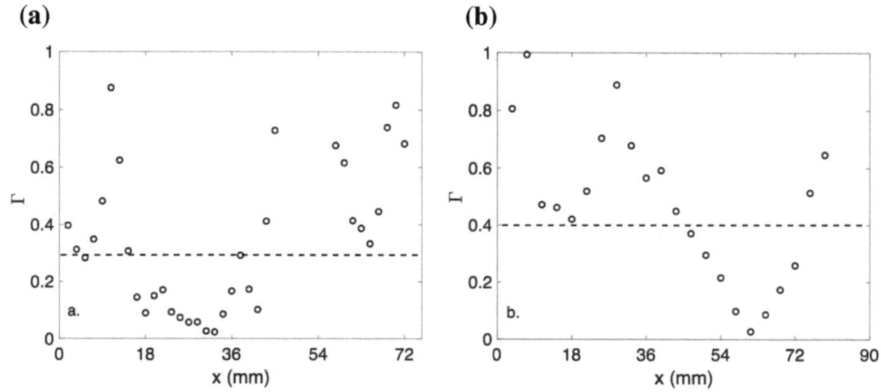

Fig. 6 Mixing efficiency (Γ) as a function of stream-wise distance (x) at two different values of Reynolds number, **a** $Re = 3090$ and **b** $Re = 9950$

of Eq. (9). Similar values of K_ρ were also recorded in field observation depending on the initial flow conditions. Therefore, we conclude that the estimate of scalar eddy diffusivity measured in our present study is legit, which gives further confidence in the turbulence characterization of the flow. Finally, it is interesting to note that the scalar eddy diffusivity value, K_ρ, is lower for gravity currents than for the forced plume case. This could be attributed to the fact that the turbulence is inhibited by stably stratified fluid layers in gravity current. On the other hand, turbulence is augmented for a forced plume due to an unstable configuration.

6 Summary

The dynamics of a shear-stratified flow were studied using experiments by measuring the small-scale flow features and the turbulence statistics. Two variants of such a flow were considered: (a) forced plume evolving in a linearly stratified environment with a low stratification strength of $N = 0.2\,\mathrm{s}^{-1}$ and (b) dense gravity current intruding in a lighter environment for two different Reynolds number namely $Re = 3090$ and $Re = 9950$. The flow evolution was studied using the simultaneous PIV/PLIF measurement technique, which enables capturing velocity and density fields. For a forced plume, the turbulent kinetic energy (K) showed a double peak structure due to vigorous entrainment near the plume edges. It was seen that K was higher near the plume source and the value decreases as the plume moves downstream, indicating decaying nature of the turbulence as the plume evolves. An off-center peak was seen in the radial Reynolds stress plot owing to the production of turbulence energy by the stress working against mean shear. The buoyancy flux, B, confirmed the unstable nature of the plume and its value reduces with increasing $\frac{z}{D}$, as the plume evolves, due to entrainment and reduction in the density gradient between the plume and the ambient fluid. The magnitude of B was small owing to a low value of Ri_b. The production flux, P, and dissipation, ε, also showed decreasing trend with increasing $\frac{z}{D}$. Out of the three budget terms, P was seen to be dominant indicating production of turbulent kinetic energy is mainly due to the stress terms. Due to unstable nature of the flow, the buoyancy flux aids in the TKE production. However, the dissipation acts as a sink for the TKE. A balance between P, B, and ε was not seen indicating that the transport term may also play an important role in governing the flow dynamics. Based on the mixing efficiency, Γ, an estimate for the scalar eddy diffusivity (K_ρ) was found, which had an order of magnitude as that observed in field conditions.

For a gravity current, the turbulent kinetic energy (K) peaks at the central region (between $z = 30$ cm and $z = 50$ cm for $Re = 3090$ and $z = 40$ cm and $z = 60$ cm for $Re = 9950$), where the two fluids mix due to strong shear present in the flow. This region is also known as the mixing layer. The results revealed that K was more for higher Re due to the increased inertia in the flow. The production flux, P, buoyancy flux, B, and viscous dissipation, ε, for both values of Re also show a peak value in the central z region. The buoyancy flux profile revealed the stably stratified nature of gravity current. Unlike for the plume case, the stability of the system acts as a sink for the TKE along with the viscous dissipation. Therefore, for gravity currents, the TKE production is only due to the P term. The terms B and ε act to dissipate this energy through the mechanism of energy cascade and eddy viscosity effects. For $Re = 3090$, all the three terms had similar magnitude, but for $Re = 9950$, the P term was dominant due to increased inertial force. Similar to the plume case, a balance between P, B, and ε was not observed indicating the importance of transport term on the flow dynamics. The value of Γ was more for $Re = 9950$, which translates to more efficient mixing at higher Reynolds number. Using the value of Γ, an estimate for the scalar eddy diffusivity (K_ρ) was found, which had an order of magnitude as that observed in field conditions. The results significantly improve our understanding of the mixing dynamics of shear-stratified flows.

Acknowledgements The author acknowledges funding from Department of Science & Technology, Ministry of Earth Sciences, and IRCC, IIT Bombay (in the form of a start-up grant) for this research work. Sincere thanks to my doctoral students, Mr. Harish Mirajkar and Mr. Partho Mukherjee, who diligently worked on these two problems to collect high quality data.

References

Balasubramanian S, Zhong Q (2018) Entrainment and mixing in lock-exchange gravity currents using simultaneous velocity density measurements. Phys Fluids, 30:015805

Balasubramanian S, Zhong Q, Fernando HJS (2015) Entrainment dynamics in self-adjusting gravity currents using simultaneous velocity-density measurements. APS Meeting Abstracts M 30:005

Benjamin TB (1968) Gravity currents and related phenomena. J Fluid Mech 31:209–248

Bloomfield L, Kerr R (1998) Turbulent fountains in a stratified fluid. J Fluid Mech 358:335–356

Britter RE (1974) An experiment on turbulence in a density stratified fluid. PhD thesis, Monash University, Victoria, Australia

Cantero MI, Lee JR, Balachandar S, Garcia MH (2007) On the front velocity of gravity currents. J Fluid Mech 586:1–39
Daviero GJ, Roberts PJW, Maile K (2001) Refractive index matching in large-scale stratified experiments. Exp Fluids 31:119–126
Devenish B, Rooney G, Thomson D (2010) Large-eddy simulation of a buoyant plume in uniform and stably stratified environments. J Fluid Mech 652:75–103
Duo X, Chen J (2012) Experimental study of stratified jet by simultaneous measurements of velocity and density fields. Exp Fluids 53:145–162
Ellison TH, Turner JS (1959) Turbulent entrainment in stratified flows. J Fluid Mech 6:423–448
Gadgil S (2003) The Indian monsoon and its variability. Ann Rev Earth Planet Sci 31:429–467
Hunt G, Kaye N (2005) Lazy plumes. J Fluid Mech 3:329–388
Ivey GN, Imberger J (1991) On the nature of turbulence in a stratified fluid. Part I: The energetics of mixing. J Phys Oceanogr 21:650–658
Kaye N (2008) Turbulent plumes in stratified environments: a review of recent work. Atmos-Ocean 46:433–441
Klemp JB, Rotunno R, Skamarock WC (1994) On the dynamics of gravity currents in a channel. J Fluid Mech 269:169–198
Lozovatsky ID, Fernando HJS (2012) Mixing efficiency in natural flows. Philos Trans R Soc (A) 371:1–19
Mahadevan A, Gualtiero SJ, Freilich M, Omand MM, Shroyer EL, Sengupta D (2016) Freshwater in the Bay of Bengal: its fate and role in air-sea heat exchange. Oceanography 29:73–81
Mirajkar HN, Balasubramanian S (2016) Dynamics of a buoyant plume in linearly stratified environment using simultaneous PIV-PLIF. International symposium on stratified flows 1:1–8
Mirajkar HN, Balasubramanian S (2017) Effects of varying ambient stratification strengths on the dynamics of a turbulent buoyant plume. J Hydraul Eng 143:1–10
Mirajkar HN, Tirodkar S, Balasubramanian S (2015) Experimental study on growth and spread of dispersed particle-laden plume in a linearly stratified environment. Environ Fluid Mech 15:1241–1262
Morton BR, Taylor GI, Turner JS (1956) Turbulent gravitational convection from maintained and instantaneous sources. Proc R Soc Lond (A) 234:1–23
Osborn TR (1980) Estimates of the local rate of vertical diffusion from dissipation measurements. J Phys Oceanogr 10:83–89
Oster G, Yamamoto M (1963) Density gradient techniques. Chem Rev 63:257–268
Papanicolaou PN, Papakonstantis IG, Christodoulou GC (2008) On the entrainment coefficient in negatively buoyant jets. J Fluid Mech 614:447–470
Papanicolaou P, Stamoulis G (2010) Spreading of buoyant jets and fountains in a calm, linearly density-stratified fluid. In: Environmental hydraulics—Proceedings of the 6th international symposium on environmental hydraulics, vol 1, pp 123–128
Pope SB (2000) Turbulent flows. Cambridge University Press, London, p 771
Richards TS, Quentin A, Sutherland BR (2014) Radial intrusions from turbulent plumes in uniform stratification. Phys Fluids 26:036602
Shin JO, Dalziel SB, Linden PF (2004) Gravity currents produced by lock exchange. J Fluid Mech 521:1–34
Shiri A (2010) Turbulence measurements in a natural convection boundary layer and a swirling jet. PhD thesis, Chalmers University of Technology, Goteborg, Sweden
Simpson EJ (1982) Gravity currents in the laboratory, atmosphere, and ocean. Annu Rev Fluid Mech 14:213–234
Simpson JE, Britter RE (1979) The form of the head of an intrusive gravity currents. Geophys J Res Astron Soc 57:289
Thorpe SA (1969) Experiments on the stability of stratified shear flows. Radio Sci AGU J 4:1327–1331
Turner JS (1986) Turbulent entrainment: the development of the entrainment assumption, and its application to geophysical flows. J Fluid Mech 173:431–471

Venayagamoorthy SK, Koseff JR (2016) On the flux Richardson number in stably stratified turbulence. J Fluid Mech 798:1–10
Webster D, Roberts P, Ra'ad L (2001) Simultaneous DPTV/PLIF measurements of a turbulent jet. Exp Fluids 30:65–72
Yih CS (1965) Dynamics of non-homogenous fluids. Macmillan, New York, p 306

Nearshore Sediment Transport in a Changing Climate

Piyali Chowdhury and Manasa Ranjan Behera

Abstract The impact of changing wave climate on the most important nearshore process, longshore sediment transport (LST), along the central west coast of India is investigated. The main purpose of this study is to provide a better understanding of the meteo-marine climate of the central west coast of India, which is highly influenced by the Arabian Sea and the Indian Ocean. To understand the contemporary evolution of the coastline, hindcast wave climate from ERA-Interim wave data (1979–2016) is used. The annual average significant wave height (H_s), wave period (T_p) and wave direction (α_0) are obtained and used to estimate annual LST. This region receives oblique waves from the W-SW direction which induces a huge gross northerly transport. It experiences two types of waves, swell waves (remotely generated waves that travel thousands of kilometres before hitting the coastline) and wind waves (also known as seas, which are locally generated), both of which are responsible for coastal sediment transport. The swell waves are the major component of a total wave system. It has more strength than the locally generated wind waves and dictates the wave direction and significant wave height at any given point of time. Therefore, the swell wave-induced LST is an order of magnitude higher than the wind wave-induced LST. It was observed that the sediment transport has a seasonal nature due to the influence of monsoonal winds in this region. The total LST in the central west coast of India shows a decreasing trend due to the reduced swell generation in the lower latitudes of the Arabian Sea and the Indian Ocean.

P. Chowdhury · M. R. Behera
Interdisciplinary Program in Climate Studies, Indian Institute of Technology Bombay, Mumbai, India
e-mail: piyali.ceg@gmail.com

M. R. Behera (✉)
Department of Civil Engineering, Indian Institute of Technology Bombay, Mumbai, India
e-mail: manasa.rb@gmail.com

C. Venkataraman et al. (eds.), *Climate Change Signals and Response*,
https://doi.org/10.1007/978-981-13-0280-0_9

1 Introduction

The nearshore region is an ecologically complex and important part of the marine ecosystem. It is a transition zone between land and continental shelf, which includes wetlands, estuaries, beaches, dunes, cliffs, surf zones (regions of wave breaking) and the inner shelf (approximately upto 15 m depth). The nearshore region is a constantly evolving and highly populated region which is under increasing threat from sea-level variability, increased storminess, long-term erosion/accretion processes and anthropogenic activities (Fig. 1). The breaking waves in the nearshore zone produce nearshore currents which transport beach sediments. Sometimes this transport results in a local rearrangement of sand into bars and troughs, or results in extensive longshore displacements of sediments. This transport is the most important nearshore process which controls the beach morphology and determines whether the beaches erode, accrete or remain stable. An understanding of longshore sediment transport is essential to integrated coastal zone management practices.

Climate change-driven rising sea levels, storminess and extreme wave heights have the potential to increase the frequency and magnitude of coastal hazards, increasing risks to coastal community and environment. The latest report by International Panel on Climate Change (IPCC 2014) highlights a lack of information on the potential changes in the wave climate and their impacts. Most of the coastal population live within an elevation of 5–10 m from current mean sea level (ISRO 2011). Long-term

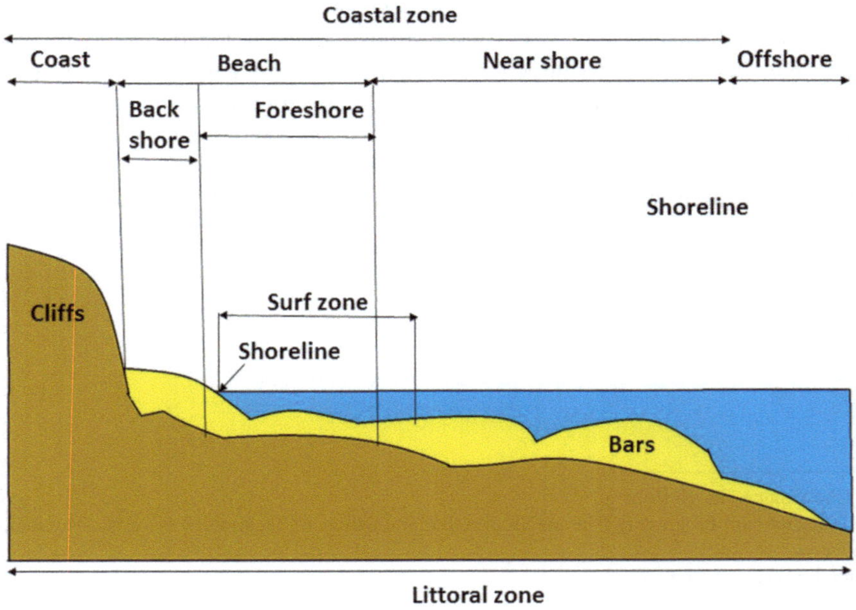

Fig. 1 Schematic of a typical nearshore zone

erosion/accretion may cause changes to the existing ecosystem and infrastructures. Damages worth of millions of rupees every year are seen with increasing number of cyclones. Prediction of changes in nearshore processes under the changing climate scenario is important to effectively manage the constantly evolving yet vulnerable environment.

Coastal ecosystems and infrastructure are threatened by long-term coastal erosion owing to climate change, limited sediment availability and anthropogenic activities. Naturally occurring long-term (10–1000 years) coastal change is a result of the cumulative response of short-term processes that are in turn influenced by surface waves and water levels. Anthropogenic activities in the coastal zone can alter these natural processes (Hapke et al. 2013), potentially inducing change in coastline orientation, which may affect future coastline response (McNamara and Werner 2008). Such two-way interaction processes and feedbacks between coastline dynamics and manmade activities make the coastal zones complicated systems. Understanding future coastal conditions and accurately predicting change over long temporal scales is needed for long-term coastal sustainability.

With growing dependency on coastal areas for urbanization and settlement, it is becoming immensely important to identify the future threats and challenges on these environments. The understanding of the nearshore processes has improved in the last couple of decades. With the availability of climate data and sophisticated wave and current models, it is now possible to predict the future wave climate and its impacts on nearshore processes like sediment transport, wave–wave interactions, wave–current interactions and many more. To have a clear understanding of the evolution of nearshore processes, it is required to investigate the long-term coastal evolution under changing climate and anthropogenic activities, prediction of future waves and storminess, and the physical, chemical and biological processes impacting these processes. In this chapter, we aim to present the hindcast wave climate in the Arabian Sea (1979–2016) and its impacts on coastal sediment transport along the central west coast of India.

1.1 Coastal Geomorphology and Nearshore Sediment Transport

The coastal geomorphology of the western continental shelf of India was studied better after the International Indian Ocean Expedition (1962–1965). The western Indian continental shelf is classified as stable Atlantic-type margin where the shelf is wide near river mouths of north-west India and narrow towards south-west India (ISRO 2011). Transport of sediments and other minerals is influenced by high-energy conditions existing on the shelf. Presence of numerous sedimentary basins in this region indicates the accumulation of sediments since Eocene. This relatively straight shelf of western coast is considered to be a result of an offshore fault extended from south to north-west India. This region forms the nearshore zone of the west coast of India. Most of the coastal processes often take place in this nearshore zone. The most important of which is the nearshore sediment transport, which is responsible for the shoreline evolution and orientation. These processes maintain the topography of beaches by either erosion or accretion processes. Coastal sediment transport is mainly driven by wave-induced or tide-induced currents. The currents interact with seabed and agitate the sediments, which then entrains in fluid flows and gets transported along and/or across the shoreline. There are two types of current systems in the nearshore zone which are responsible for the sediment transport—currents within the surf zone and currents beyond the surf zone. However, the currents in the surf zone are most important for sediment transport. It derives its energy from breaking of waves.

2 Study Area

The study investigates the impact of changing wave climate on the temporal dynamics of longshore sediment transport (LST). LST governs the overall coastal geomorphology of regional coastlines. Shoreline erosion, sedimentation near river mouth, narrowing of deltas and rapid changes in the river mouth configuration are commonly observed problems in the central west coast of India (Nayak et al. 2010). The west coast of India experiences three major seasons: pre-monsoon (JFMAM), monsoon (JJAS) and post-monsoon (OND). The Indian coastline is exposed to significant changes in seasonal to annual wind and wave climate. Winds are south-westerly and significantly high in strength during the Indian summer monsoon (June–September), whereas it is north-easterly during the winter monsoon (November–February). The wave climate is also influenced by the changing wind patterns which makes the wave climate highly seasonal. The study area is located in the central west coast of India (Fig. 2). This region consists of numerous headlands, river deltas, cliffs, bays, sand spits and sandy beaches (Shanas and Kumar 2014). This region is exposed to an energetic wave climate with swell waves travelling from the Arabian Sea and the Indian Ocean. The nearshore wave climate is governed by both wind and swell waves. Wind waves are generated by the local wind conditions and swells propagate from deep ocean. Waves are directed differently in different seasons. Variation in wave direction is an important factor that determines the impact of a changing wave climate on coastal processes, altering the LST direction and rates, and consequently, the overall coastal sediment balance. Beaches in this region are mainly reflective to intermediate in nature (Kumar et al. 2003; Manoj and Unnikrishnan 2009), making it a micro- to meso-tidal region (Short 1991). The presence of several deltas in this region makes the sediment interaction process complicated, especially during monsoon. The varying nature of inflowing sediment complicates the distribution process of sediments in the nearshore zone. These sediments are distributed along the coast by the action of waves and tides. The influx sediment grain size varies from medium to coarse (average $D_{50} = 0.6$ mm) (Veerayya and Varadachari 1975).

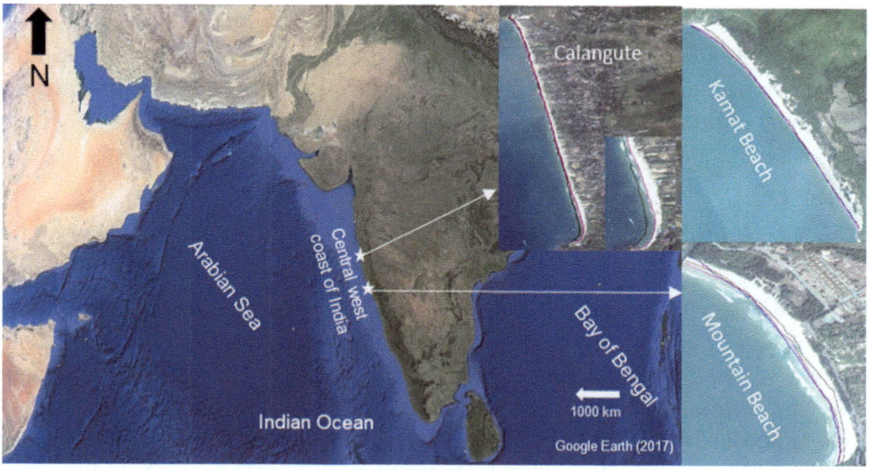

Fig. 2 Selected study regions along central west coast of India

Cyclonic activities in this region are major drivers of the sediment dynamics. It was observed that the Arabian Sea has a bimodal nature of cyclogenesis occurring during the south-west and north-east monsoon periods (Evan and Camargo 2011). However, the sea surface temperature is always above 26.5 °C which facilitates cyclogenesis and high-energy swells. Increased wave heights were observed in regions where the wave climate is subjected to swell propagated from the Arabian Sea and the southern Indian Ocean (Hemer et al. 2010). A large number of extra tropical Indian Ocean swells freely propagate into tropical and subtropical latitudes of the Indian Ocean (Alves 2006) and form an important component of the Arabian Sea and the Bay of Bengal wave climate. However, the tropical north Indian Ocean storms have minimal impact on the global wave climate but can have very rough local wave systems, which impacts coastal processes along the coasts located in this region (Young and Holland 1996). Another type of wave affecting the nearshore wave climate in the Indian region is the locally generated wind waves. The seasonality in the swell and wind waves leads to the seasonal variation of nearshore wave climate and the resulting LST along the west coast of India.

3 Data and Methodology

3.1 Wave Data

Wave parameters like (significant wave height (H_s), wave period (T_p) and wave direction (α_0)) are important inputs in estimating wave-induced LST. ERA-Interim wave data were used to derive these parameters (Dee et al. 2011). The ERA-Interim wave data were extracted on a $0.5° \times 0.5°$ grid with a 6-h-temporal resolution for a period of 38 years. Data were extracted for the grid cell encompassing the study region. Annual and monthly averages of the wave parameters were computed and used in the estimation of LST.

3.2 Methodology

LST is estimated by numerous longshore transport formulae worldwide. Many research works on improving the longshore sediment transport formula are in progress. Also, there is no general rule on the choice of formula. The most commonly used formulae are the Coastal Engineering Research Center (CERC, Shore Protection Manual 1984), Walton and Bruno (1989), Kamphuis (1991), Van Rijn (1989) and the recently developed Kaczmarek et al. (2005) formula. In this study, we used the recently developed Kaczmarek et al. (2005) formula.

Kaczmarek et al. (2005) formula assumes that the volume of LST 'Q' for the entire surf zone is proportional to the wave energy flux (V) and is given as:

$$Q = \begin{cases} 0.023\left(H_b^2 V\right)\ (\mathrm{m^3/s}), & \text{if, } H_b^2 V < 0.15 \text{ (low and medium wave climate)} \\ 0.00225 + 0.008\left(H_b^2 V\right)\ (\mathrm{m^3/s}), & \text{if, } H_b^2 V > 0.15 \text{ (higher waves and storms)} \end{cases} \quad (1)$$

where H_b is the breaker wave height and V is the average longshore current velocity in the surf zone which is determined as:

$$V = 0.25 k_v \sqrt{\gamma_b g H_b} \sin 2\theta_b \ (\mathrm{m/s}) \quad (2)$$

where γ_b is the constant wave breaker parameter (0.78), g is the gravitational acceleration, θ_b is the breaker wave angle and k_v is a site specific constant chosen as 2.9 based on Bertin et al. (2008) for wave-dominated regions with similar grain size features as that of the present study. The breaker wave properties were estimated using the breaker wave predictor formula proposed by Larson et al. (2009) in which deep water wave characteristics were used. The detailed methodology of this model can be found in Chowdhury and Behera (2017) and hence is not repeated here for brevity.

Wind sea and swell heights are computed from the ERA-Interim derived wave data and are used to estimate wind wave-induced LST and swell wave-induced LST. To study the shoreline changes in an interrupted and uninterrupted beach, we identified littoral drift cells that are close to such locations (Calangute Beach, which is an uninterrupted beach and Karwar Naval Beach, which is interrupted by the construction of a naval base since 2005). Landsat imagery (USGS 2014) was used to describe the shoreline evolution on a regional scale. Satellite images with minimal cloud coverage over the area of interest were extracted since 1990. All these images were referenced to the latest 2015 image with clearly visible reference points which are identifiable in the entire period since 1990. This approach is expected not to affect the final result because our ultimate aim is to track relative shoreline position change. Shoreline evolution was computed at both the locations using DSAS tool of Arc-GIS. Model-estimated LST was compared with the Landsat data.

4 Nearshore Sediment Transport in a Changing Climate

4.1 Wave Climate

Annual and monthly average values of H_s, T_p and α_0 were extracted at the grid points encompassing the study region and were used to obtain the nearshore breaker wave characteristics as shown in Fig. 3. The average annual H_s and T_p were found to be 1.32 m and 8.12 s, respectively (Chowdhury and Behera 2017). The annual wave climate experiences dominant waves from the west–south-west direction (~238° clockwise from true north). H_s was the highest during the south-west monsoon (1.4–2.6 m) and low during rest of the year (0.7–1.0 m). The monthly wave data were decomposed to obtain swell and wind wave components using the methodology given by Chen et al. (2002) (Fig. 3). Like swell waves, wind waves also showed seasonal variations, with lower wave heights during October–May and higher wave heights during monsoon months (JJAS). The Arabian Sea has a very calm wave climate in non-monsoon months compared to the monsoon months. This variation was expected to impact the seasonal variability of LST in this region. It was observed that during non-monsoon months the wave activity is low, with monthly breaker height ranging from 0.7 to 1.2 m. With the onset of monsoon in the month of June, a steep increase in the magnitude of breaker wave height is observed with values ranging from 1.2 to 2.3 m.

4.2 Sediment Transport Due to Swells and Wind Seas

The waves in the central west coast of India are oblique (W–WSW) to the coastline. These oblique incident waves generate a large northerly transport. LST is found to be southerly only when the wave direction is between 250–270°. The annual net swell wave-induced LST and wind wave-induced LST are shown in Fig. 4. The average annual swell wave-induced LST is 412,000 m^3/year, which is an order of magnitude higher than the wind wave-induced LST, which is only 25,200 m^3/year. The overall LST shows a decreasing trend of about 5% over the study period of 38 years.

Net monthly LST was also estimated to understand the seasonal variations in LST along the central west coast of India as shown in Fig. 5a. A directional shift in LST from northerly (positive LST values) to southerly (negative LST values) is observed with the onset of monsoon which may be attributed to the directional shift in the winds and longshore current. A similar shift in direction of LST was

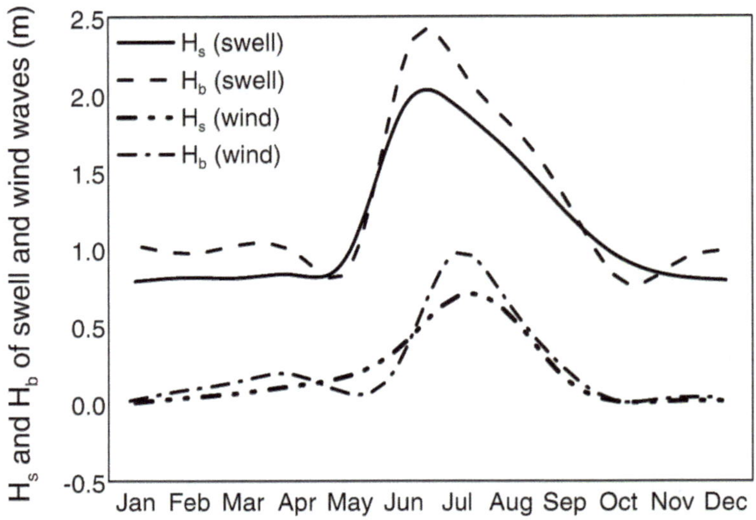

Fig. 3 Significant (H_s) and breaker (H_b) wave height of swell and wind waves (*Source* Chowdhury and Behera (2017))

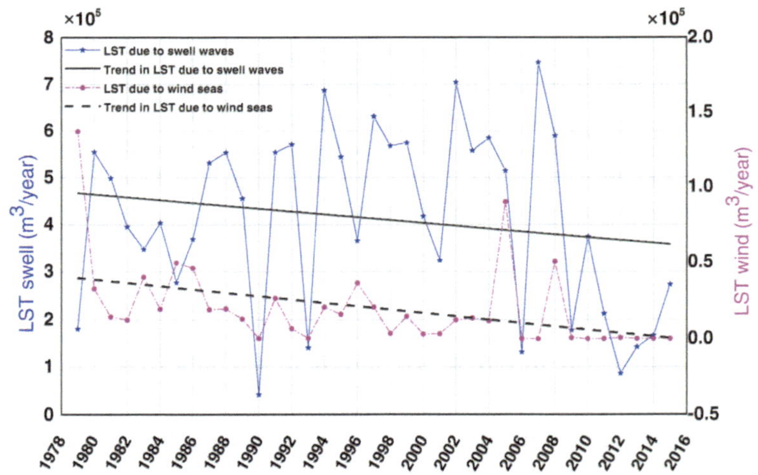

Fig. 4 Swell wave-induced LST (blue) and wind wave-induced LST (magenta) (1979–2016)

again observed towards the end of monsoon season which directed the sediment transport towards the north. The total LST directed towards north is more than that towards south. Also, the monthly LST induced by both swell and wind waves were estimated (Fig. 5b). The wind wave-induced LST is low throughout the year, whereas during monsoon it is higher than the swell wave-induced LST. Intensity of the sea breeze–land breeze system increases during monsoon which in turn amplifies the wave height and hence wind wave-induced LST is the maximum during monsoon.

Fig. 5 **a** Monthly LST and **b** Monthly LST variation induced by swell (blue) and wind waves (magenta) (*Source* Chowdhury and Behera (2017))

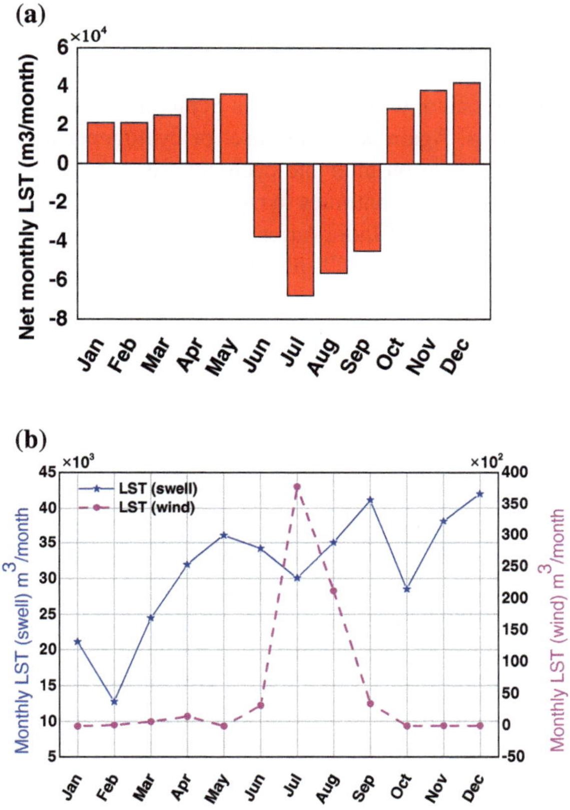

Sea breeze–land breeze system influences wave parameters along the west coast of India (Glejin et al. 2013) and Masselink and Pattiaratchi (1998) reported that the sea breeze—land breeze system has an important influence on LST. Swell wave-induced LST peaks in December due to the cyclogenesis phenomenon in the Arabian Sea. The net swell wave-induced LST during JJAS is only 125,500 m^3/month and that during rest of the year is 250,000 m^3/month, making the total annual transport northerly in nature.

Swells in the Arabian Sea are influenced by cyclonic activities which have a bimodal nature (Evan and Camargo 2011) that peaks in May–June and October–December and are minimum during monsoon. This phenomenon affects the seasonal variation of LST in the west coast of India. Further, Evan and Camargo (2011) state that there is an increase in the cyclonic storm days over the period of 1992–2008 when compared to 1979–1991. This increased trend in cyclonic activity over 1992–2008 is coherent with increase in LST intensity during this period. It is observed that wind blowing over the Arabian Sea is strong during the onset of monsoon which causes strong swells that travels towards the Indian coast and dominates the coastal processes. During monsoon, the wind speed is very high and blows towards the coast causing high rates of swell wave-induced LST with the highest in September (Fig. 5b). During the same period, the wind waves also start growing stronger which is caused by the sea breeze–land breeze system and induces higher rates of LST. With the reversal of wind direction in September, the rate of swell wave-induced LST also peaks up. Again with the onset of cyclogenesis in November, the LST increases. In January, soon after the cyclogenesis, the LST drops down and is the lowest in February. LST continues to rise during March–May. It is concluded that LST along the west coast of India varies seasonally depending upon the wind and wave characteristics. Remote swells also play an important role in determining regional LST by influencing the nearshore wave parameters.

4.3 Shoreline Changes in Interrupted and Uninterrupted Beaches

The evolution of any beach depends on the erosion, and deposition pattern of the beach which in turn depends on the LST rates. Two locations were considered to study their evolution over the past few decades based on LST computed using Kaczmarek et al. (2005) formula, and the same was compared with the shoreline evolution obtained from satellite images. Satellite data from Landsat (USGS 2014) available since 1990, with a resolution of 30 m, are obtained for both the locations (Calangute Beach, which is an uninterrupted beach and Karwar Naval Beach, which is interrupted by the construction of a naval base since 2005) and is used for shoreline change detection. All the satellite images were geo-rectified using a reference image to assign real-world coordinates. These images were projected on to UTM projection system and WGS-84 datum. Shoreline changes from 1990 to 2015 were calculated in Digital Shoreline Analysis System (DSAS) tool of Arc-GIS. Transects at 50-m intervals were drawn along the coastline for determining the shoreline change rate over this period.

Calangute Beach

Coastline evolution over the past decades at Calangute Beach is shown in Fig. 6a. The undisturbed coastline of Calangute Beach shows mild accretion during this period with most of the accretion (+ve value) occurring in the northern part of the beach following erosion (−ve value) in the southern part. The shorelines extracted from Landsat imagery at Calangute Beach since 1990 show an accretion of approx. 2.5 m/year over a period of 26 years, which is similar with the findings obtained from Kaczmarek et al. (2005) formula.

Karwar Naval Base Area (Kamat Beach and Mountain Beach)

Kamat and Mountain beaches were selected for this study as they represent the site condition with anthropogenic activities. The beaches are located on the north and south, respectively, of a newly constructed naval base in Karwar, Karnataka. Shoreline evolution since the construction in 2005 is determined using Landsat images, and it is observed that post-construction of the naval base, the beaches have accreted

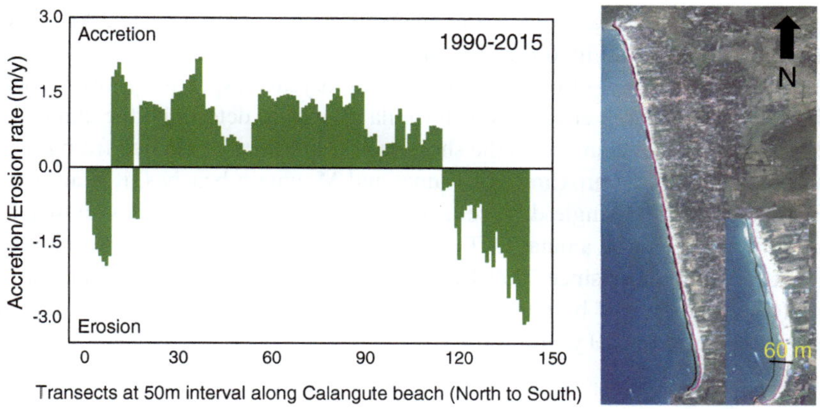

(a)

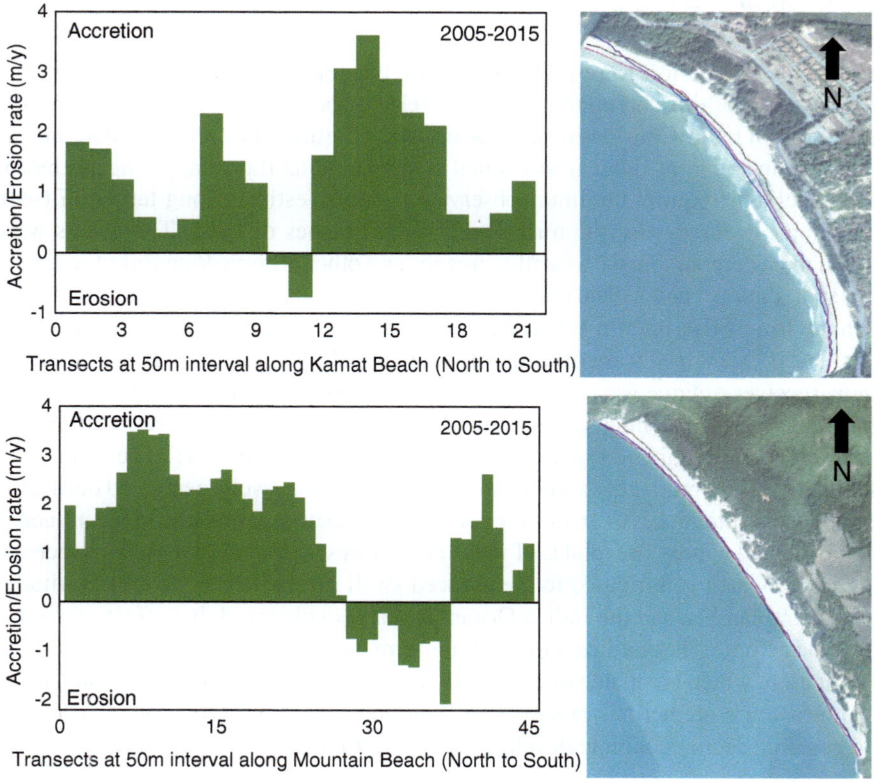

(b)

Fig. 6 Accretion/Erosion rate (m/year) and Shoreline changes in **a** Calangute Beach (1990 (black) and 2015 (magenta)) and **b** Kamat (north of naval base) and Mountain (south of naval base) beaches (2005 (black), 2010 (violet) and 2015 (magenta)) significantly. Using DSAS tool, the annual shoreline evolution since 2005 is computed and presented in Fig. 6b, which shows the rate of accretion along Kamat and Mountain beach to be ~+1.6 m/year and ~+2.8 m/year, respectively, over a period of 10 years. Kaczmarek et al. (2005) formula is used to determine the annual LST in this region and compared with the shoreline evolution rate obtained from Landsat imagery. Owing to the proximity of Kamat and Mountain beaches, it is acceptable to consider them as a single drift cell. It is observed that the beaches show a high amount of variability in annual LST post-construction. However, the variability in annual LST has reduced since 2009. Changes in variability of annual LST indicate the disrupted equilibrium of both the beaches, which has resulted in massive accretion over a small period of 10 years.

5 Conclusions

This study is an effort to provide insights on the contemporary sediment transport regime and resulting shoreline evolution at selected beaches located along the central west coast of India. Considering the scantily available measured sediment data and the difficulty in obtaining them, it was aimed to find the suitability of applying an empirical model (that requires minimum observation data) to estimate long-term LST (with acceptable accuracy) for the micro–meso-tidal beaches of India. The results were found to be comparable with satellite images and other observed/measured sediment transport data. To understand the contemporary evolution of coastline, hindcast wave climate from ERA-Interim was used. The annual average significant wave parameters (H_s, T_p, and α_0) were used to estimate annual LST. The central west coast of India receives oblique waves (swell waves and wind waves) from the W–SW direction which induces a huge gross northerly transport. The swell wave-induced LST is an order of magnitude higher than the wind wave-induced LST. The changes in sediment transport quantity were mainly governed by the swells generated during the cyclogenesis months. LST in this region shows a seasonal nature due to the influence of monsoonal winds. The total LST in the central west coast of India shows a decreasing trend, which is attributed to the reduced swell generation in the lower latitudes of the Arabian Sea and the Indian Ocean. It was also observed that effect of anthropogenic activities is significant on the long-term variation of LST rates and coastline orientation, which is capable of disturbing the equilibrium of beaches. It is suggested that any future predictions of shoreline evolution should be carried out considering the regional socio-economic developments and climate change scenarios.

References

Alves JHG (2006) Numerical modeling of ocean swell contributions to the global wind-wave climate. Ocean Model 11:98–122. https://doi.org/10.1016/j.ocemod.2004.11.007

Bertin X, Castelle B, Chaumillon E, Butel R, Quique R (2008) Longshore transport estimation and inter-annual variability at a high-energy dissipative beach: St. Trojan beach, SW Oléron Island, France. Cont Shelf Res 28:1316–1332. https://doi.org/10.1016/j.csr.2008.03.005

Chen G, Chapron B, Ezraty R, Vandemark D (2002) A global view of swell and wind sea climate in the ocean by satellite altimeter and scatterometer. J Atmos Oceanic Technol 19:1849–1859. https://doi.org/10.1175/1520-0426(2002)019<1849:agvosa>2.0.co;2

Chowdhury P, Behera MR (2017) Effect of long-term wave climate variability on longshore sediment transport along regional coastlines. J Prog Oceanogr 156:145–153. https://doi.org/10.1016/j.pocean.2017.06.001

Dee DP, Uppala SM, Simmons AJ, Berrisford P, Poli P, Kobayashi S, Andrae U, Balmaseda MA, Balsamo G, Bauer P, Bechtold P (2011) The ERA-Interim reanalysis: configuration and performance of the data assimilation system. Q J Royal Meteorol Soc 137:553–597. https://doi.org/10.1002/qj.828

Evan AT, Camargo SJ (2011) A climatology of Arabian Sea cyclonic storms. J Clim 24:140–158. https://doi.org/10.1175/2010JCLI3611.1

Glejin J, Sanil Kumar V, Nair B, Singh J (2013) Influence of winds on temporally varying short and long period gravity waves in the near shore regions of the eastern Arabian Sea. Ocean Sci 9:343–353

Hapke CJ, Kratzmann M, Himmelstoss EA (2013) Geomorphic and human influences on regional shoreline change rates. Geomorphology 199:160–170. https://doi.org/10.1016/j.geomorph.2012.11.025

Hemer MA, Church JA, Hunter JR (2010) Variability and trends in the directional wave climate of the Southern Hemisphere. Int J Climatol 30:475–491. https://doi.org/10.1002/joc.1900

IPCC (2014) Climate change 2014: synthesis report. In: Pachauri RK, Meyer LA (eds) Contribution of working groups I, II and III to the fifth assessment report of the intergovernmental panel on climate change. IPCC, Geneva, Switzerland

ISRO (2011) Coastal zones of India. ISBN No.: 9788190997898. http://www.moef.nic.in/sites/default/files/Coastal_Zones_of_India.pdf

Kaczmarek LM, Ostrowski R, Pruszak Z, Rozynski G (2005) Selected problems of sediment transport and morphodynamics of a multi-bar nearshore zone. Estuar Coast Shelf Sci 62:415–425. https://doi.org/10.1016/j.ecss.2004.09.006

Kamphuis JW (1991) Alongshore sediment transport rate. J Waterway Port Coast Ocean Eng 117:624–640. https://doi.org/10.1061/(asce)0733-950x(1991)117:6(624)

Kumar VS, Anand NM, Chandramohan P, Naik GN (2003) Longshore sediment transport rate—measurement and estimation, central west coast of India. Coast Eng 48:95–109. https://doi.org/10.1016/S0378-3839(02)00172-2

Larson M, Hoan LX, Hanson H (2009) Direct formula to compute wave height and angle at incipient breaking. J Waterway Port Coast Ocean Eng 136:119–122. https://doi.org/10.1061/(ASCE)WW.1943-5460.0000030

Manoj NT, Unnikrishnan AS (2009) Tidal circulation and salinity distribution in the Mandovi and Zuari estuaries: case study. J Waterway Port Coast Ocean Eng 135:278–287. https://doi.org/10.1061/(asce)0733-950x(2009)135:6(278)

Masselink G, Pattiaratchi C (1998) The effect of sea breeze on beach morphology, surf zone hydrodynamics and sediment resuspension. Mar Geol 146:115–135. https://doi.org/10.1016/S0025-3227(97)00121-7

McNamara DE, Werner BT (2008) Coupled barrier island–resort model: 1. Emergent instabilities induced by strong human-landscape interactions. J Geophys Res Earth Surf, 113. https://doi.org/10.1029/2007jf000840

Nayak SR, Hedge VS, Shalini G, Rajawat AS, Girish KH, Jayakumar S, Suryanarayana A (2010) Geomorphic processes in the vicinity of the Venkatapur River mouth, Central West Coast of India: implications on Estuarine sedimentation. J Coast Res 26:925–934. https://doi.org/10.2112/JCOASTRES-D-09-00021.1

Shanas PR, Kumar VS (2014) Coastal processes and longshore sediment transport along Kundapura coast, central west coast of India. Geomorphology 214:436–451. https://doi.org/10.1016/j.geomorph.2014.02.027

Shore Protection Manual (1984) US Army Coastal Eng. Research Center

Short AD (1991) Macro-meso tidal beach morphodynamics: an overview. J Coast Res 7:417–436. http://www.jstor.org/stable/4297847

USGS (U.S. Geological Survey) Landsat: a global land-imaging mission (2014). http://landsat.usgs.gov/

Van Rijn LC (1989) Handbook of sediment transport by currents and waves. Report H461, Delft Hydraulics, Delft, The Netherlands

Veerayya M, Varadachari VVR (1975) Depositional environments of coastal sediments of Calangute, Goa. Sed Geol 14:63–74. https://doi.org/10.1016/0037-0738(75)90017-2

Walton Jr TL, Bruno RO (1989) Longshore transport at a detached breakwater, phase II. J Coast Res 5:679–691. http://www.jstor.org/stable/4297599

Young IR, Holland GJ (1996) Atlas of the oceans: wind and wave climate. Oceanogr Lit Rev 6:640–658

Part II
National and Sub-national Responses to Climate Change

A Comprehensive Social Vulnerability Analysis at a National Scale

H. Vittal and Subhankar Karmakar

Abstract Extreme hydro-climatic events, such as extreme rainfall, droughts, and heat waves, are increasing both in its intensity and frequency over India, leading to corresponding rise in economic losses and human casualties. Moreover, these extreme events continue to accelerate in the future climatic scenarios. Thus, proper understanding of the physical and socioeconomic drivers of these extreme events is essential and eventually improves the adaptation strategies. While the early warning systems of these extreme events have significantly improved, evidences on the evaluation of dynamicity of social vulnerability throughout India are still significantly lacking. With regard to this, there is an urgent need to develop a set of comprehensive vulnerability maps for India, which may be utilized by the national disaster management and policy-making agencies, particularly to manage and recover from disaster events by spatially prioritizing the disaster management. Here, we provide a comprehensive review on the current status of the national-scale vulnerability mapping along with the associated challenges, with the primary focus on the Indian scenario. We also emphasize on the recent advancement in the methodologies in mapping the nationwide social vulnerability. Readers must note that the present chapter serves as a guide for optimal data gathering efforts for future improvements and fine-tuning of social vulnerability maps to develop policy tools to reduce vulnerability, mitigate risks, and increase resilience to hazards.

Keywords Hazards · National scale · Risk · Social vulnerability

---

H. Vittal · S. Karmakar
Centre for Environmental Science and Engineering, Indian Institute of Technology Bombay, Mumbai 400076, India

H. Vittal
Department of Civil Engineering, Manipal Institute of Technology, Manipal 576104, India

S. Karmakar (✉)
Interdisciplinary Program in Climate Studies, Indian Institute of Technology Bombay, Mumbai 400076, India
e-mail: skarmakar@iitb.ac.in

C. Venkataraman et al. (eds.), *Climate Change Signals and Response*,
https://doi.org/10.1007/978-981-13-0280-0_10

1 Introduction

The primary aim of this chapter is to provide a comprehensive literature on the social vulnerability (SoV) mapping to the hazards at a nationwide and also to highlight the challenges associated with the primary focus on the Indian scenario. In addition, we also emphasize on the various definitions of vulnerability and the current status of SoV assessment over India. However, before we proceed, to understand and appreciate the essence of this chapter, the readers should know the definitions of few concepts such as hazard, disaster risk, and vulnerability. The threats that usually have a tremendous potential to harm people and places are termed as hazards. On the other hand, a disaster refers to the singular larger-scale event which makes difficult to the local community to effectively cope/adapt to the event. Risk is defined as the likelihood or probability that some particular type of injuries or loss of infrastructure would result from the hazard/disaster event (NRC 2006; Cutter et al. 2009). It should be noted that normally hazard and disasters are viewed as mainly caused by the interaction between natural environment and human societies, though they may be also caused due to the interaction between society and technology (e.g., nuclear emission from nuclear power plants or any accidental chemical spills from the industries) or within the societies. Often, we give prominence to those situations where interactions between natural environment and societies (earthquakes, droughts, and floods) usually take place since it may trigger a large human displacement across the state/national borders of the country, thereby resulting in huge humanitarian crises (NRC 2006; Cutter et al. 2009).

With regard to the vulnerability, there are multiple definitions available in the existing literature. Table 1 shows the various definitions of vulnerability and the advancement in the conceptual framework.

Table 1 Definitions of vulnerability

Definition	Remarks
Blaikie et al. (1994) defined vulnerability as "the characteristics of a person or group in terms of their capacity to anticipate, cope with, resist and recover from the impact of a natural hazard"	It precisely mentions social system and their characteristics as the subject of analysis, whereas it identifies natural hazards as the source of threat to this social system
Smit et al. (2000) described vulnerability as the "degree to which a system is susceptible to injury, damage or harm"	This definition is very generic. It does not provide information pertaining to hazard events or the subject of analysis
Parry et al. (2007) mentions that according to Intergovernmental Panel on Climate Change (IPCC) the vulnerability can be defined as "the degree to which a system is susceptible to and unable to cope with, adverse effects of climate change, including climate variability and extremes"	This definition is very specific in terms of attributing the hazard, i.e., climate change. However, it provides a broader scope pertaining to the subject of analysis (e.g., economic sector, coupled human–natural system)

Though the definitions of vulnerability provided in Table 1 do have conceptual disparities, it should be noted that all the three concepts provide guidelines for climate change vulnerability assessment.

1.1 Vulnerability Assessment to Natural Hazards

Hazard vulnerability assessment basically explains "who" is exposed to the hazard, and the differential susceptibility aftermath of hazard and finally the impacts of the exposure. Here, the susceptibility mainly includes potential for loss, harm, injury, adverse impacts on livelihoods. In simple words, the main aim of the vulnerability assessment is not only to identify the "who," which is vulnerable, but also to identify the factors which eventually shapes the vulnerability in a respective place (Brikmann 2006). It should also be noted that these vulnerability assessments are either a qualitative or quantitative approach. The quantitative measures are mainly provided by the numerical estimates of the exposures and ranking of vulnerability.

Meanwhile, one should note that the hazard vulnerability assessment essentially consists of three distinct elements, viz. exposure assessment, impact assessments, and damage assessment. Here, the exposure assessment mainly includes the identification of the risk source including estimating the intensity, frequency, magnitude and also analyzing its spatial extent of impact. The impact assessments usually include the consequences of a hazard on the population, and the damage assessment mainly includes analyzing the fatalities and losses to infrastructure and economy. However, the daunting task still persists in integrating these three elements into a comprehensive vulnerability assessment framework. The reason for this may be the difficulties in working across disciplinary or knowledge boundaries (Cutter et al. 2009). Bridging this gap is still a challenging task in the social vulnerability assessment, especially at a national scale.

1.2 Measuring Vulnerability

Developing a universal measurement tool for vulnerability assessment across all disciplines is still a challenging task, provided the discrepancies in the definition in coupled with the dynamicity and changing scale of analysis, i.e., both temporal and spatial scales. Nevertheless, the research on the vulnerability assessment has seen a shift from the qualitative conceptualization and framework to more precise quantitative measure of vulnerability. This dramatic shift is mainly attributed to the advances in the vulnerability science (Cutter et al. 2009).

In conjunction with this, there have been numerous indices related to vulnerability which have been reported in the literature. Some of them, but not limited to, are Environmental Vulnerability Index (Kaly et al. 1999), Human Well-Being Index (Prescott-Allen 2001), Environmental Sustainability Index (Esty et al. 2005), Human

Development Index (Burd-Sharps et al. 2008). Nonetheless, most of these studies are based on the comparison between two or three different nations with course spatial resolutions, but drew little focus towards understanding the vulnerability at sub-national levels. However, it should be noted that there is a significant spatial variation existing within the country in terms of exposure and also the impact.

With regard to climate-related hazards, researchers have developed quantitative indicators. For example, O'Brien et al. (2004) implemented the definition of vulnerability from IPCC, which basically includes exposure, sensitivity, and adaptive capacity. They have used this to map agricultural vulnerability over India to both climate change and globalization at a district level. In this study, the authors have combined three different vulnerability indicators, viz. biophysical, social, and technological indicators which influence the agricultural productivity, specifically for Indian scenario. Similar to O'Brien et al. (2004), Deressa et al. (2008) also conducted an agricultural vulnerability analysis to climate change in seven regions of Ethiopia, as per the definition of IPCC. The authors had developed a vulnerability index which comprises a combination of both biophysical and socioeconomic indicators.

The present-day literature emphasizes more on developing the empirically based indices to assess climate change vulnerability. For example, Hahn et al. (2009) developed a Livelihood Vulnerability Index, wherein assessed the impact of climate change among the individual population residing in two districts of Mozambique with the implementation of several indicators, such as demographic profile, social networks, health, food, water and natural disasters and climate variability. Here, while determining the sensitivity and exposure to climate change impacts, the authors have assigned equal weights to all the selected indicators. On the other hand, Torresan et al. (2008) employed another innovative approach, i.e., Dynamic International Vulnerability Assessment (DIVA), to assess climate change vulnerability along the coast of Venetia, Italy. The major indicators employed are biophysical indicators such as geomorphology, topography, and vegetation.

In addition to these small-scale studies, researchers have also developed indices at the national level to assess the climate change vulnerability. For example, Moss et al. (2002) developed a Vulnerability–Resilience Indicator Prototype (VRIP) model that assessed the ability of different groups to adapt and cope with changing climate in 38 different countries. They considered combination of both environmental and social factors, such as food, water, health, environment, and economics in their lists of indicators. Further, these indicators were scaled against global data to obtain the overall national baselines of vulnerability for each of the considered countries. Further, Sullivan and Meigh (2005) developed a Climate Vulnerability Index comprising six indicators including resource, access, capacity, use, environment, and geospatial dimensions. They argue that the developed vulnerability index has the applicability across various spatial scales ranging from small island developing nations to the national level. Nonetheless, there is no significant theoretical discussion provided on the choice of indicators selected for the study. Though there is a significant potential in research on development of vulnerability indicators and indices, specifically in the area of climate change, the general agreement on measuring vulnerability remains ambiguous (Cutter et al. 2009).

2 Review of National-Level Social Vulnerability Analyses

Using the different conceptual frameworks, as mentioned in the previous section, many researchers have developed the nationwide social vulnerability maps for different countries (Table 2).

3 Need of Social Vulnerability Analysis for India

Studies focused on mapping nationwide social and socioeconomic vulnerability have typically been performed for developed or economically rising countries, such as the USA, China, and European nations. However, few studies have been performed for developing and underdeveloped countries, particularly in India, where a comprehensive social vulnerability map at the national scale has not been attempted yet. Although accurate projection of hydro-climatic extremes over India has received considerable attention, the large population residing within the country remains highly vulnerable as increasing population directly increases the exposure to hydro-climatic extremes, thereby increasing the risk. On the other hand, the efficacy of the disaster governance body has improved over time as demonstrated by comparing two Bay of Bengal cyclone events with similar intensities over Odisha (formerly Orissa) between 1999 and 2014. The impact and number of deaths during the former event was more devastating compared with the latter because of the improved early warning and mass evacuation programs during the latter event, resulting in over a million people evacuating to safe temporary shelters. The capacity of a government to manage a disaster is limited by factors that include geographical size, population, and rapid unplanned urbanization without appropriate infrastructures such as roads and transportation.

In addition, it is reported that due to global warming, the occurrence of extreme events such as precipitation extremes and heat waves is significantly increasing around the world. With regard to Indian perspective, we utilize the data from Indian Meteorology Department (IMD) and analyze the pattern of extremes. Here, we considered two notable extremes, viz. precipitation extremes and heat waves. For precipitation extremes, we utilize 0.25°-resolution precipitation data (Pai et al. 2013). Figure 1a shows the trend of precipitation extreme over India from 1950 to 2000. The results clearly show a significant increase in the intensity of precipitation extremes, which is matching the conclusions derived from the previous study conducted by Goswami et al. (2006), Vittal et al. (2013). Further, we conduct a similar analysis for temperature extremes. As with precipitation extremes, here also we found that the temperature extremes are significantly increasing from post-1950 over India

Table 2 Literature review on nationwide social vulnerability mapping

Author	Region	Indicators used	Remarks
Cutter et al. (2003)	USA	County-level socioeconomic and demographic data is used to develop social vulnerability to environmental hazards	They identified distinct spatial pattern in the social vulnerability. They also noticed that the most vulnerable counties are clustered in metropolitan counties in the East and South Texas and the Mississippi Delta regions
Cutter and Finch (2008)	USA	Country-level socioeconomic status, gender, and housing tenure are used to develop social vulnerability to natural hazards	They observed that initially the vulnerability is more clustered in certain geographic regions and later it has dispersed significantly over time. Moreover, the trend in vulnerability showed a steady reduction in social vulnerability, but there is considerable regional variability, with many counties increasing in social vulnerability during the past five decades
Fekete (2009)	Germany	Three main indicators have been considered to map the social vulnerability such as fragility, socioeconomic condition, and region	It is shown that indeed certain social groups like the elderly, the financially weak or the urban residents are higher-risk groups
Khan and Salman (2012)	Pakistan	Indicators such as population density, lack of knowledge, lack of decent housing, lack of decent standard of living, livestock households, and farm households are considered in this study	The authors have implemented Human Vulnerability Index (HVI) to map the vulnerability for Pakistan. They reported that the adult literacy rate, ownership of livestock, and access to electricity are the three (out of six) key variables that play a critical positive role in recovery after the 2010 floods

(continued)

Table 2 (continued)

Author	Region	Indicators used	Remarks
Lixin et al. (2014)	China	Indicators such as gender, disabled, age, family structure, minority, education, medical service, socioeconomic status, and employment are considered	The results indicate that vulnerability is a location-based regional phenomenon, with the most vulnerable cities being located in the southwest of China and the eastern areas being generally less vulnerable
Siagian et al. (2014)	Indonesia	Population, annual growth rate, population density, population over 65, children under 5, adult illiterate, poor people, Human Development Index	Three main driving factors affecting social vulnerability in Indonesia are found: socioeconomic status and infrastructure, gender, age and population growth and family structure
Frigerio and Amicis (2016)	Italy	Indicators such as family structure, education, socioeconomic status, employment, age, race and ethnicity and population growth are considered	Here, the authors have applied hazard-of-place model approach. The analysis identified different spatial patterns across Italy, providing useful information for identifying the communities most likely to experience negative natural disaster impacts due to their socioeconomic and demographic characteristics
Frigerio et al. (2016)	Italy	Indicators such as age, employment, education, urbanization, quality house, and ethnicity were selected for the study	The study utilizes a geographic information system (GIS) approach which was applied to identify the spatial variability of social vulnerability to seismic hazard. Finally, a qualitative social vulnerability exposure map to an earthquake hazard was produced, highlighting areas with high seismic and social vulnerability levels

(continued)

Table 2 (continued)

Author	Region	Indicators used	Remarks
de Loyola Hummell et al. (2016)	Brazil	Socioeconomic status, gender, race and ethnicity, age, employment loss, urban and rural population, family structure, education, medical services and access, social dependency and special need populations	The result showed a concentration of the most socially vulnerable cities in the north and northeast regions of Brazil, as well as the social vulnerability of metropolitan areas and state capitals in the south and southeast regions. The least vulnerable cities are mainly concentrated in the inland regions of the southeast

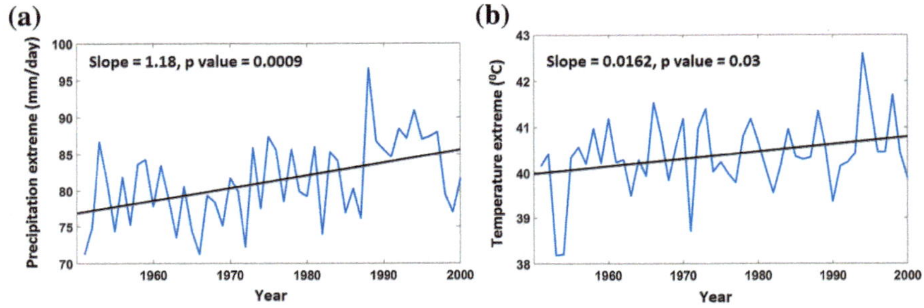

Fig. 1 Trends in the extremes over India. **a** the trends in the precipitation extremes over India from 1950 to 2000. The *P* value provided in the inset of the plot shows that the trend is significant at 5% significance level. **b** Exhibit a temperature extreme over India from 1950 to 2000. As with precipitation extreme, here also we find increasing trend and it is significant at 5% significance level

(Fig. 1b). For this purpose, we utilize 1°-resolution temperature data for all over India (Srivastava et al. 2009). The previous studies also reported that these extremes are further going to increase in the future scenario (Murari et al. 2015).

From our analysis, it is evident that the extreme scenarios are increasing. In conjunction with this, the population in India is also increasing. This situation eventually increases the exposure to the increasing hazards over India. To get more insight into the impact of these hazards on the increasing population, we have analyzed the mortality trends due to different natural hazards over India (Fig. 2). Here, the mortality data is procured from the National Crime Records Bureau (NCRB). Figure 2a shows the mortality trends due to floods from 1967 to 2012. Though the trend is increasing, it is insignificant at 5% significance level. However, the death due to heat waves (Fig. 2b) and cold waves (Fig. 2c) showed statistically significantly increasing trend. Overall, our results suggest that the extreme scenarios are increasing along with the impact in terms of mortality over India. Under such condition, the development of the social vulnerability maps for India is inevitable. Disaster management has been a rapidly evolving discipline in India, particularly over the last decade, and the National Disaster Management Act (2005) provides a mandate for the development of a comprehensive disaster management plan at national, state, and district levels. However, the relatively few studies that exist have focused on mapping social vulnerability to disasters at the city level (Sherly et al. 2015) or basin level (Kelkar et al. 2008) in India, but not on national scale. The nationwide mapping provides valuable vulnerability information that can be used by the central government to develop policies and by regional planners and disaster managers to define target intervention areas based on the required resolution of spatial scales (Fekete 2009).

4 Social Vulnerability Mapping for India

Considering the fact that the social vulnerability mapping is lacking in the literature and also the natural hazards are significantly increasing, the mortality due to these natural hazards is also increasing over India. This motivated us to develop a comprehensive social vulnerability maps for India. To serve this purpose, the demographic data from Census of India for the decade 2011 is procured. Further, from this demographic data, we select both positive and negative vulnerability indicators. Here, the indicators which have a tendency to increase the social vulnerability when it is increased are considered as positive indicators, whereas the indicators which decrease the vulnerability when it is increased are considered as negative indicators.

Based on the previous studies and also the availability of the data from the Census of India, we select total population, female population, main agricultural and cultivator population, children population (population <6 years), rural population, number of households, illiterate population, illiterate female population, SC and ST population (categorized as the backward community by the government of India), SC and ST female population, marginal workers (including cultivators, agricultural laborers, household industry), and non-working population as positive indicators. On the other hand, the indicators such as working population, female working population, literate population, and female literate population are considered as negative indicators.

Further, the vulnerability indicators developed by CoI are standardized mainly to make the indicators dimensionless, which will allow us to compare the different indicators at national scale and also to eliminate anomalies, if any. The method of standardization for indicators (Wu et al. 2002; Karmakar et al. 2010) is provided in the equation below:

$$V_i^{\mathrm{std}} = \frac{V_i - V_i^{\min}}{V_i^{\max} - V_i^{\min}} \tag{1a}$$

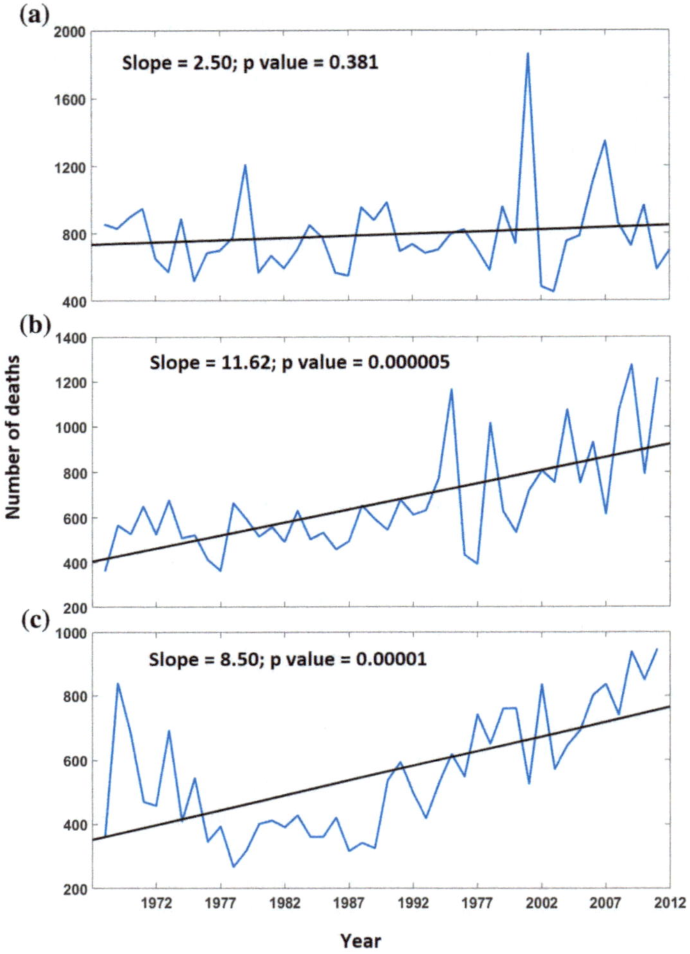

Fig. 2 Trends in mortality due to different natural hazards over India. **a** the trends for floods, whereas the trends for heat waves and cold waves are provided in **b** and **c**, respectively. The slopes and the *p* values are provided in the inset of each subplot. Apart from flood, both heat waves and cold waves showed significantly increasing trend at 5% significance level

where V_i^{std} is standardized vulnerability index of *i*th indicator; $V_i^{\max}$ and $V_i^{\min}$ are the maximum and minimum counts of vulnerability indicator, respectively; V_i is the count pertaining to the *i*th tehsil. This equation considers both maximum and minimum values in the expression and ensures that vulnerability values are within [0, 1] interval (Wu et al. 2002) and always nonnegative (Karmakar et al. 2010). During normalization, the indicators were adjusted for their sign, which indicates whether the indicator contributes positively (+) or negatively (−) to vulnerability. The equation provided in Eq. 1a is generally used for the indicators which have positive influence toward social vulnerability; however, for negative influencing indicators the standardization is usually done with Eq. 1b.

$$V_i^{\text{std}} = \frac{V_i^{\max} - V_i}{V_i^{\max} - V_i^{\min}} \tag{1b}$$

After the standardization of the indicators, the principal component analysis (PCA) is applied mainly to decorrelate and reduce the dimensionality of the indicators. The first principal component (PC1) is able to explain almost 95% of the variability. The PC1 is further fed into the aggregation operation, wherein the vulnerability values are obtained. The present study implements the method developed by Cutter and Finch (2008) to map the vulnerability in which the standard deviation approach is applied to classify the vulnerability ranking. Figure 3 shows the spatial variation of nationwide social vulnerability to the decade 2011 for India. The results show that the social vulnerability to natural hazards is higher in the states of Karnataka, Kerala, Maharashtra, and Uttar Pradesh and lower in the states of Andhra, Orissa, and Telangana.

5 Conclusions

The natural hazards are increasing due to the increasing global warming, and these events are reported to increase in the future scenario as well. These events have huge socioeconomic impacts and also affect the societal well-being. Along with the increase in extreme events, the population around the globe is also rapidly increasing, which eventually increases the exposure to the natural hazards. Under such scenario, the disaster management sector optimally strategizes adaptation options to cope with the increasing hazards. Vulnerability mapping is one of the important tools which provide cartographic information on the regions/areas which are more vulnerable to the hazards. Based on these maps, the regional planners and disaster manager may define target intervention areas.

It should be noted that the comprehensive social vulnerability maps are developed majorly in developed counties, such as the USA, European nations, and China. The maps are even available for the economically backward nations such as Pakistan and Bangladesh. However, such maps are not yet developed for India, even though the natural hazards such as precipitation extremes and heat waves are significantly increasing; moreover, there are evidences that the mortality due to these increasing natural hazards is also significantly increasing. This motivated us to map social vulnerability for India at a fine resolution, i.e., at tehsil level.

We use demographic data as obtained from Census of India for the decade 2011 and implement the method developed by Cutter and Finch (2008) to map the social vulnerability for entire India. For this purpose, we have selected around 20 indicators (both positive and negative) and applied PCA mainly to reduce the dimensionality

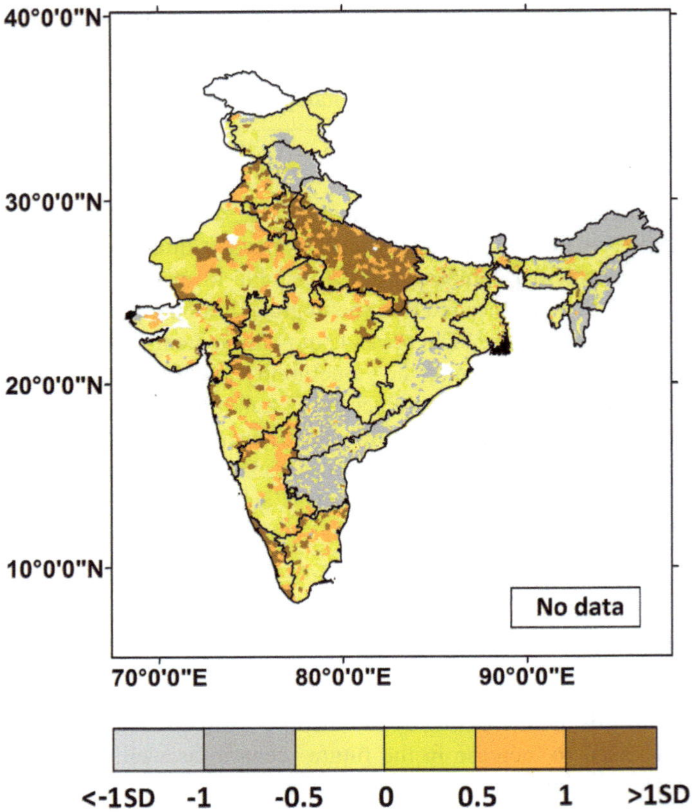

Fig. 3 Spatial variability of the social vulnerability based on 2011 demographic data as obtained from Census of India

and correlation between the indicators. The results show the vulnerability is more in the states such as Uttar Pradesh, Karnataka, Maharashtra and lower in the states of Andhra Pradesh and Telangana.

Though the present study successfully maps the social vulnerability for entire India, these maps can be improved further by the inclusion of certain important indicators such as migration, elder population, medical facilities available. In addition, here we have just implemented the standardization method to obtain the vulnerability; however, in the literature, there are more robust and complex methods which are available in the literature, for example, data envelopment analysis (DEA) which can be implemented, and the present maps can be scrutinized further. In addition, the present study maps the social vulnerability for only one decade, i.e., for 2011. However, the demographic data is also available for the decade 2001. The analysis can be further carried on to map the social vulnerability for these two decades and look into the decadal change in the vulnerability. Finally, in future studies, the risk maps can be generated, which is a combination of hazard and vulnerability components.

References

Birkmann J (2006) Measuring vulnerability to natural hazards: towards disaster resilient societies. United Nations Publications, New York

Blaikie P, Cannon T, Davis I, Wisner B (1994) At risk: natural hazards, people's vulnerability, and disasters. Routledge, London

Burd-Sharps S, Lewis K, Martins EB (2008) The measure of America: American Human Development Report, 2008–2009. Columbia University Press, New York

Cutter SL, Finch C (2008) Temporal and spatial changes in social vulnerability to natural hazards. Proc Natl Acad Sci 105(7):2301–2306

Cutter SL, Boruff BJ, Shirley WL (2003) Social vulnerability to environmental hazards. Soc Sci Q 84(2):242–261

Cutter SL, Emrich CT, Webb JJ, Morath D (2009) Social vulnerability to climate variability hazards: a review of the literature. Final Rep Oxfam Am 5:1–44

de Loyola Hummell BM, Cutter SL, Emrich CT (2016) Social vulnerability to natural hazards in Brazil. Int J Disaster Risk Sci 7(2):111–122

Deressa T, Hassan RM, Ringler C (2008) Measuring Ethiopian farmers' vulnerability to climate change across regional states. International Food Policy Research Institute. Available at: http://www.ifpri.org/pubs/dp/IFPRIDP00806.pdf. Accessed 3 Dec 2017

Esty DC, Levy M, Srebotnjak T, de Sherbinin A (2005) 2005 environmental sustainability index: benchmarking national environmental stewardship. Yale Center for Environmental Law & Policy, New Haven, CT

Fekete A (2009) The interrelation of social vulnerability and demographic change in Germany. Paper accepted for the IHDP OpenMeeting 2009

Frigerio I, De Amicis M (2016) Mapping social vulnerability to natural hazards in Italy: a suitable tool for risk mitigation strategies. Environ Sci Policy 63:187–196

Frigerio I, Ventura S, Strigaro D et al (2016) A GIS-based approach to identify the spatial variability of social vulnerability to seismic hazard in Italy. Appl Geogr 74:12–22

Goswami BN, Venugopal V, Sengupta D et al (2006) Increasing trend of extreme rain events over India in a warming environment. Science 314(5804):1442–1445

Hahn MB, Riederer AM, Foster SO (2009) The livelihood vulnerability index: a pragmatic approach to assessing risks from climate variability and change—a case study in Mozambique. Glob Environ Change 19(1):74–88

Kaly U, Briguglio L, McLeod H et al (1999) Environmental vulnerability index (EVI) to summarise national environmental vulnerability profiles. SOPAC Technical Report 275. South Pacific Applied Geoscience Commission, Suva

Karmakar S, Simonovic SP, Peck A, Black J (2010) An information system for risk-vulnerability assessment to flood. J Geogr Inf Syst 2(03):129

Kelkar U, Narula KK, Sharma VP, Chandna U (2008) Vulnerability and adaptation to climate variability and water stress in Uttarakhand State, India. Glob Environ Change 18(4):564–574

Khan FA, Salman A (2012) A simple human vulnerability index to climate change hazards for Pakistan. Int J Disaster Risk Sci 3(3):163–176

Lixin Y, Xi Z, Lingling G, Dong Z (2014) Analysis of social vulnerability to hazards in China. Environ Earth Sci 71(7):3109–3117

Moss RH, Malone EL and Brenkert AL (2002) Vulnerability to climate change: a quantitative approach. Joint Global Change Research Institute. http://www.globalchange.umd.edu/data/publications/Vulnerability_to_Climate_Change.PDF

Murari KK, Ghosh S, Patwardhan A et al (2015) Intensification of future severe heat waves in India and their effect on heat stress and mortality. Reg Environ Change 15(4):569–579

National Research Council (NRC) (2006) Facing hazards and disasters: understanding human dimensions. The National Academies Press, Washington DC

O'Brien K, Leichenko R, Kelkar U et al (2004) Mapping vulnerability to multiple stressors: climate change and globalization in India. Glob Environ Change 14(4):303–313

Pai DS et al (2013) Development and analysis of a new high spatial resolution (0.250 × 0.250) long period (1901–2010) daily gridded rainfall data set over India. National Climate Centre, Asheville

Parry M, Canziani O, Palutikof J (2007) Climate change 2007: impacts, adaptation and vulnerability, vol 4. Cambridge University Press, Cambridge

Prescott-Allen R (2001) The wellbeing of nations. Island Press, Washington DC

Sherly MA, Karmakar S, Parthasarathy D et al (2015) Disaster vulnerability mapping for a densely populated coastal urban area: an application to Mumbai, India. Ann Assoc Am Geogr 105(6):1198–1220

Siagian TH, Purhadi P, Suhartono S, Ritonga H (2014) Social vulnerability to natural hazards in Indonesia: driving factors and policy implications. Nat Hazards 70(2):1603–1617

Smit B, Burton I et al (2000) An anatomy of adaptation to climate change and variability. Clim Change 45(1):223–251

Srivastava AK, Rajeevan M, Kshirsagar SR (2009) Development of a high resolution daily gridded temperature data set (1969–2005) for the Indian region. Atmos Sci Lett 10(4):249–254

Sullivan C, Meigh J (2005) Targeting attention on local vulnerabilities using an integrated index approach: the example of the climate vulnerability index. Water Sci Technol 15(5):69–78

Torresan S, Critto A, Valle MD et al (2008) Assessing coastal vulnerability to climate change: comparing segmentation at global and regional scales. Sustain Sci 3(1):45–65

Vittal H, Karmakar S, Ghosh S (2013) Diametric changes in trends and patterns of extreme rainfall over India from pre-1950 to post-1950. Geophys Res Lett 40(12):3253–3258

Wu SY, Yarnal B, Fisher A (2002) Vulnerability of coastal communities to sea-level rise: a case study of Cape May County, New Jersey, USA. Clim Res 22(3):255–270

Review of Indian Low Carbon Scenarios

Pankaj Kumar, Trupti Mishra and Rangan Banerjee

Abstract An increase in global temperatures at rapid rate is attributed to greenhouse gas emissions mainly from our energy system. This paper performs a comparison and meta-analysis of low carbon scenarios based on integrated assessment models to provide insights on India's greenhouse gas mitigation potential. This review compares the range of scenario formulation methods, analytical models, energy demand estimation methods, baseline scenarios, mitigation drivers, and low carbon scenario assumptions. A meta-analysis was conducted to provide insights on findings from these scenarios by analyzing trends in energy mix, electricity production, and emissions for 2050. The analysis of mitigation scenarios shows a clear decline in energy and carbon intensity for year 2050. The studies use a range of policies and targets for analyzing mitigation potential for India. This paper highlights trends in past studies where earlier studies have used parametric changes to create energy scenarios while recent studies have also used structural changes and policy options. The policy scenarios suggest a reduction in energy intensity by 70% with emission intensity declining by more than 90% by 2050. The emissions in policy scenarios stabilize in the range of 1300–2600 million tonne CO_2 by 2050 averaged at 2118 million tonnes. The policy scenarios also suggest a phaseout of coal and its substitution with gas combined with carbon capture, nuclear, and solar. The study further discusses emerging avenues for future research and implications for policy modeling in Indian context.

P. Kumar (✉) · T. Mishra · R. Banerjee
Interdisciplinary Program in Climate Studies, Indian Institute of Technology Bombay, Mumbai 400076, India
e-mail: pankaj.iitb14@gmail.com

T. Mishra
Shailesh J. Mehta School of Management, Indian Institute of Technology Bombay, Mumbai 400076, India
e-mail: truptimishra@iitb.ac.in

R. Banerjee
Department of Energy Science of Engineering, Indian Institute of Technology Bombay, Mumbai 400076, India
e-mail: rangan@iitb.ac.in

C. Venkataraman et al. (eds.), *Climate Change Signals and Response*,
https://doi.org/10.1007/978-981-13-0280-0_11

Keywords Low carbon scenarios · Energy modeling · Renewable energy Decarbonization · Carbon emissions · Meta-analysis

1 Energy and Emissions in Indian Context

Policymakers in Indian energy sector are informed frequently by scenarios of energy models to make decisions for energy security and system decarbonization. These energy system models differ in their scope, theoretical frameworks, scope and other parameters. Understanding these factors and their effect on results is necessary to enable comparison across scenarios and obtain insights for informed policy decisions.

With a population of 1295 million, India accounted for 17.8% of world's population in 2015 (IEA 2015). The country also accounted for 6.2% of global emissions and 3.9% of energy consumption in 2014 which is poised to increase with its rapidly growing population projected to reach 1.7 billion in 2050 (UN 2018). In spite of low per capita energy consumption (0.65 toe/capita) and per capita emissions (1.58 t CO_2/capita) (IEA 2015), India forms an important part of Paris agreement due to its large population and rapidly growing economy.

Within last decade, many modeling studies have been published on India's greenhouse gas (GHG) mitigation potential for medium- and long-term scenarios. Several major forums (Energy Modeling Forum, Intergovernmental Panel on Climate Change Working group III, the United States Climate Change Science Program) along with the previous literature have explored and compared energy scenarios and integrated assessment models in global context. Energy Modeling Forum's study (Luderer et al. 2013) of EMF 27 scenarios analyzed the role of renewable energy in mitigation of climate change and found the significance of climate policies in deployment of renewable energy technologies. Asia Modeling Exercise (Calvin et al. 2012) involved a comparison exercise of 23 energy-economy and integrated assessment models and exploration of the role of Asia in mitigation of climate change.

However, very limited studies (CPR 2015) have reviewed low carbon scenarios for India. This review is structured in two parts. First, we compare and provide insights on scenario formulation methods, models, energy demand estimation methods, baseline scenarios, mitigation drivers, and low carbon scenario assumptions. Second, a meta-analysis is conducted to provide insights on findings from these scenarios by analyzing trends in energy mix, electricity production, and emissions for 2050.

This study is structured as follows. Section 2 explains the literature specific to low carbon scenarios in Indian context. Section 3 presents comparison and meta-analysis of low carbon scenarios. Section 4 assesses the implications for future low carbon scenarios. Section 5 is the summary.

2 Existing Low Carbon Scenarios

To conduct the meta-analysis, we collected data from selected studies which provided results for medium- or long-term scenarios (i.e., 2030 or 2050) and met the following criteria: (i) a study based on detailed top-down, bottom-up, or hybrid model assessment with interventions for mitigation, excluding those which have Indian component as a part of global scenarios and (ii) a publication in peer-reviewed journal. We filtered the studies published between 2006 and 2017.

As a result, we filtered nine studies in total from a set of 22 which are in the best of our knowledge as summarized in Table 1. We assess the qualitative and quantitative attributes of each study. Here the qualitative attributes include objectives, analytical models, modeling framework, and assumptions. The quantitative attributes include demand projections, carbon prices, energy and electricity mix, and emissions.

The study further also reveals gaps in the existing literature and highlights avenues for future research which will be critical for Indian energy system transition. Here meta-analysis of scenarios contributes to the scholarship by assessing how these differences in policies and targets affect results and their implications for policymaking.

Comparisons of energy scenarios have been made in the previous literature to assess the range of energy futures and their implications (Luderer et al. 2013; Calvin et al. 2012). The motivation for recent low carbon scenarios for India varies and includes climate change mitigation, energy security, and potential role of renewables.

Shukla (2006) created GHG and non-GHG scenarios for India using IPCC SRES framework for energy sector without any climate policy intervention but with different endogenous emission drivers. Shukla et al. (2008) created low carbon society scenarios by visualizing social, technological, and economic transition for a low carbon future. Parikh et al. (2009) took a top-down approach with an activity analysis model (Integrated Energy System Model) to find out extremes of feasible energy technology options. Massetti (2011) used a hybrid World Induced Technological Change Model (WITCH-IAM) to study the impacts of carbon taxes in order to reduce greenhouse gas emissions. Shukla and Chaturvedi (2011) created a conventional and a sustainability scenario to analyze the implications of carbon prices and technology push policies in transforming energy system. Shukla and Chaturvedi (2012) used Global Change Assessment Model (GCAM) to apply a targets approach for pushing low carbon technologies like solar, nuclear, and wind in electricity sector from 2005 to 2095. Anandarajah and Gambhir (2014) analyzed the role of renewables using The Integrated MARKAL-EFOM System (TIMES) integrated assessment model in meeting India's climate change mitigation targets with two scenarios, with and without carbon capture and storage (CCS). Gambhir et al. (2014) explored potential benefits of entering a 2 °C mitigation target for India by analyzing long-term mitigation options using a TIMES integrated assessment model. Mittal et al. (2016) assessed the implications of renewable energy deployment targets for India's Intended Nationally Determined Contributions (INDCs) and their associated GDP and welfare loss using Asia-Pacific Integrated Model (AIM) model under climate constrained scenarios. Here the scenarios with very high renewable energy penetration may be highly

Table 1 Studies selected for comparison of scenarios

Author(s)	Timeline	Model(s)	Model assessment method	Scenario type	Scenarios
Shukla (2006)	2000–2030	MARKAL + AIM + ERB (Loulou et al. 1997; Kainuma et al. 2002; Edmonds and Reilly 1985)	Bottom-up (national + local), GIS interface	Explorative	1 baseline and 3 low carbon scenarios
Shukla et al. (2008)	2000–2050	AIM-CGE + ANSWER-MARKAL + AIM-Snapshot + End Use Demand Model + AIM SDB (NIES 2006; Hibino et al. 2003)	Top-down recursive dynamic (energy environment impacts), bottom-up (optimization), end use demand modeling	Explorative	1 reference and 2 low carbon scenarios
Parikh et al. (2009)	2000–2031/32	IESM	Bottom-up, optimization, linear programming	Explorative	1 base case (coal based) and 11 low carbon scenarios
Massetti (2011)	2005–2050	WITCH-IAM (Bosetti et al. 2006)	Top-down, hybrid, optimization	Explorative	1 baseline and 4 low carbon scenarios
Shukla and Chaturvedi (2011)	2005–2095	GCAM-IIM	Top-down, recursive dynamic, partial equilibrium	Normative	1 reference and 1 low carbon scenario
Shukla and Chaturvedi (2012)	2005–2095	GCAM-IIM (Edmonds and Reilly 1983)	Top-down, recursive dynamic, partial equilibrium	Normative	2 baseline and 2 carbon price scenarios
Gambhir et al. (2014)	2010–2050	TIAM UCL (Loulou and Labriet 2007)	Bottom-up, partial equilibrium, technology explicit	Normative	1 reference, 3 low carbon scenarios
Anandarajah and Gambhir (2014)	2005–2050	TIAM UCL	Bottom-up, partial equilibrium, technology explicit	Normative	1 reference, 2 low carbon scenarios
Mittal et al. (2016)	2005–2030	AIM/CGE	Top-down recursive dynamic (energy environment impacts)	Normative	12 scenarios: RE deployment (high, low, and medium) x emission constraints (none, high, strong, weak)

ambitious since they do not include system operations and complexity to match supply with demand at peak and nonpeak hours. This supply and demand balance may require modifications in transmission grid, storage capacity, and operations which are difficult to model. Yet the scenarios provide useful insights on future low carbon transition pathways.

3 Results

3.1 Models

The analytical models used in conducting analyses vary in spatial resolution, solution mechanisms, and degree of foresight which differ with their assumptions of future growth. Models used in these studies have analytical structures framed so as to represent real-world phenomena (like market behavior and resource endowments) with mechanisms to impose restriction on use of resources as well as environmental quality. Few models also have parameters which can be estimated to change over time (e.g., technology learning). The choice for analytical model for analysis appears to reflect the study purpose. Although these models possess analytical strengths for analyzing measures for low carbon transition, they possess limitations due to their reliance on mainstream economic theories and inability to account behavioral and socioeconomic characteristics.

3.2 Energy Demand

We observe that assumptions of population, macroeconomic factors like economic growth and energy consumption have varied in studies and developed with time. These assumptions are critical in shaping future scenarios and supply-side factors associated with them. The variation in assumptions can be attributed to data sources and methodologies used for energy demand projection. Table 2 summarizes the data and methods used where it is observed that variations in exogenously input assumptions are legitimate and adopted from official sources or quantified from justified methodologies.

Table 2 Data and methods for energy demand estimation

S. No.	Study	Energy demand
1	Shukla (2006)	The energy use drivers like GDP growth rate and population are taken same as SRES scenarios
2	Shukla et al. (2008)	GDP projections from historical data (Ministry of Finance 2007). Logistic regression for projecting sectoral demands using elasticity of sectors and gross value added
3	Parikh et al. (2009)	Commercial energy demand assumed by estimating elasticity with respect to GDP using log-linear regressions. Non-commercial energy demand by projecting per capita expenditure and GDP with expenditure classes
4	Massetti (2011)	GDP, energy use, electricity use, emissions from World Bank Indicators (World Bank 2018)
5	Shukla and Chaturvedi (2011)	GDP growth from 2005 to 2095 assumed at 5.16% per annum. Energy intensity assumed in line with the scenarios
6	Shukla and Chaturvedi (2012)	Targets for electricity generation constructed for model for short, medium, and long run in line with policy announcements by govt. [GOI documents (Planning Commission 2006), and expert opinion]
7	Gambhir et al. (2014)	Residential and commercial energy demand projections till 2036 from TERI methodology (TERI 2006) and extended further. Industrial activity assumed to increase 20 fold by 2050 in comparison to 2005 levels. Road transport increases with economic growth. Passenger and freight demand increase with an elasticity of 0.8 w.r.t GDP. Domestic and international travel elasticity w.r.t GDP assumed as 0.9
8	Anandarajah and Gambhir (2014)	
9	Mittal et al. (2016)	Population and GDP growth till 2030 in line with SSP2 SRES (O'Neill et al. 2013; Riahi et al. 2015)

3.3 *Baseline and Low Carbon Scenarios*

3.3.1 Baseline Scenarios

Energy demands are often exogenously determined, which combined with energy mix changes determine total emissions. The studies present an increasing trend of baseline emissions in baseline scenarios. A prime source of variation among assumptions is assumed growth rate of GDP per capita (which ranges between 9 and 4.5% in scenarios (Parikh et al. 2009; Massetti 2011), while population projections are incorporated from UN median projections in six studies as shown in Table 3 (UN 2018).

The model projections also indicate increase in GDP and energy efficiency gains over modeling period (which vary from 2030 to 2100). The carbon intensity of energy systems is the function of energy efficiency gains and substitution effect of coal fleet from other technologies. The studies use different metrics for setting up the targets in atmospheric CO_2 concentration trajectory, carbon budgets, or energy targets as shown in summarized in Table 3. Most of the studies have relied on a coal-dominated baseline scenario with no explicit carbon price with baseline CO_2 concentration trajectory reaching 770–740 ppmv by 2100.

3.3.2 Low Carbon Scenarios

The relative amount of emission reduction from a baseline scenario is established by the mitigation potential of a policy intervention. Five studies (Massetti 2011; Shukla et al. 2008; Shukla and Chaturvedi 2012; Gambhir et al. 2014; Anandarajah and Gambhir 2014) employ an explicit carbon price which drive mitigation by enabling economic viability of low carbon technologies. Two studies (Shukla 2006; Gambhir et al. 2014) also account for interlinkages of markets, commodities, and emission credits. Many studies differ in technology database and mitigation options used for analysis which have been summarized in Appendix. Shukla (2006) explored the impact of factors like economic growth, population profile, energy resources, technological change, market integration, policies, and institutions on emission trajectories. However, Shukla et al. (2008) take a co-benefits approach with carbon price as key instrument enabling fuel switching and use of renewables, while change of preferences and demand reduction drive mitigation in sustainability scenarios.

This inconsistency in technology options is attributed to model choices by studies where bottom-up models have detailed representation of systems while top-down models lack such detail. Such a variation in technology options also poses significant changes in incremental costs for low carbon transition. In addition, the models have less representation of non-CO_2 gases and aerosols responsible for air quality.

Energy system optimization models like MARKAL and TIMES use a bottom-up accounting approach and give optimistic estimates of the best performance of system while ignoring market barriers. Top-down models consider production possibilities frontier for technologies with representation of market behavior but lack in robust technological detail. Thus, top-down models give pessimistic estimates on best performance of performance of energy system. Models like TIMES, MARKAL, and CGE have price-based analytical structure, i.e., they optimize technology selections within a given number of constraints.

Table 3 Baseline scenarios

S. No.	Study	Base year	GDP growth and population	Baseline scenario
1	Shukla (2006)	2000	7.5, 5.5, 6.5, 4.5% assumed in scenarios (Planning Commission 2002; UNPD 2003)	IA2: mixed economy (ref scenario) (A: centralization, 2: fragmented markets)
2	Shukla et al. (2008)	2005	8.1% (2005–30), 5.9% (2030–50), 7.1% (2005–50) (UN 2018)	650 ppmv stabilization (resource-intensive path, BAU improvements in energy intensity and renewables)
3	Parikh et al. (2009)	2000	Assumed 8 and 9% till 2031–32	S1: coal-based development scenario
4	Massetti (2011)	2005	Assumed 6.9%(2005–20), 5.5% (2020–35), 4.3% (2035–50)	No policy intervention. BAU energy efficiency gains. Emissions increase from 1.41 GT CO_2 in 2005 to 5.15 GT CO_2 by 2050
5	Shukla and Chaturvedi (2011)	2005	7.62% (2005–2035), 5.2% (2035–65), 2.71% (2065–95), Overall 5.16% (2005–95) (UN 2018)	BAU (baseline scenario with no mitigation interventions) 740 ppmv CO_2 stabilization by 2095
6	Shukla and Chaturvedi (2012)	2005	7.73% (2005–35), 5.22% (2035–65), 2.71% (2065–95) overall 5.2% (2005–95) (Planning Commission 2006; UN 2018)	BAU (baseline scenario without carbon price or electricity targets) 770 ppmv CO_2 stabilization by 2095
				BAU-T (baseline scenario without carbon price but exogenously specified electricity targets) 770 ppmv CO_2 stabilization by 2095
7	Gambhir et al. (2014)	2005	Assumed 7.7% till 2030. 5% from 2030 to 2050. (UN 2018)	No emission constraint. Emissions increase to 8 GT CO_2 by 2050
8	Anandarajah and Gambhir (2014)	2005	Assumed 7.7% till 2030 and 5% till 2050 (UN 2018)	No emission constraint. Emissions increase to 8 GT CO_2 by 2050
9	Mittal et al. (2016)	2005	Assumed 6.33% growth till 2030	No emission cap scenarios with high, medium, and low RE deployment (Riahi et al. 2015; O'Neill et al. 2013)

A comparison of all scenarios presented in Tables 1 and 4 shows a methodological shift with time where earlier studies focus on explorative scenarios while recent studies focus on normative scenarios. Here, the explorative scenarios refer to those which focus on the implications of certain strategies on future outcomes (emissions/energy mix). The normative scenarios answer how an energy system can be transformed to attain an emission target by policy or structural interventions. Since scenarios are developed using energy models which are quantitative in nature, almost all the studies have created scenarios using changes in parameters or outcomes. However, parametric uncertainties for scenario formulation become a limited choice since it does not allow structural changes (for instance scenarios like no CCS, no nuclear). The prior studies have relied more on parametric changes to create baselines and policy scenarios while recent studies have focused more on structural changes in energy system (refer Table 1).

All the above scenarios assume stable economic growth and oil prices; however, none of the scenarios has captured uncertainty in oil prices. The development of CCS and uncertainties around its commercial viability has led the recent studies to create with and without CCS scenarios. Parallels in this regard can also be drawn with countries having considered no nuclear scenarios.

3.4 Trends in Energy and Electricity Mix

3.4.1 Energy Mix

The fossil fuel demand in reference scenarios rises through 2030 with a coal-dominated energy mix and increasing oil and gas demand. It is also observed that coal-based energy will slow down or decline after 2030 in most low carbon scenarios, with oil and gas demand rising till 2050. There is wide divergence in studies for coal as part of energy mix as illustrated in Fig. 1 where coal in reference scenarios increases by a multiplier of about six to four as compared to 2010 levels [six in case of Gambhir et al. (2014) and four in case of Shukla et al. (2008)]. Coal in energy mix varies by a factor of 0.2–2 times the 2010 levels in low carbon scenarios [0.2 for Anandarajah and Gambhir (2014); 2 for Shukla et al. (2008)].

There is wide divergence seen in oil and gas with a factor of 1.6–4.1 times 2010 levels in oil (Shukla et al. 2008; Gambhir et al. 2014). All scenarios suggest a future with increasing gas dominance with a factor of 3.7 between maximum and minimum projections for gas in 2050. The studies involving macroeconomic soft-linked modeling frameworks ride on broad assumptions for sectors to achieve cumulative mitigation.

Table 4 LCS assumptions in studies

S. No.	Study	Emission target	Scenario
1	Shukla (2006)	Evaluated as deviation from 550 ppmv trajectory	IA1: globalization and high growth (A: centralization, 1: high market integration)
			IA2: mixed economy (ref scenario) (A: centralization, 2: fragmented markets)
			IB1: sustainable development (B: decentralization, 1: high market integration)
			IB2: self-reliance (B: decentralization, 2: fragmented markets)
2	Shukla et al. (2008)	550 ppmv (480 ppmv for CO_2 + 70 ppmv for non-CO_2)	Carbon tax scenario with greater reductions in energy intensity and increase in commercial renewable energy
		550 ppmv	Sustainability scenario: sustainability rationale in line with IPCC SRES B1 global scenario. Economic growth decoupled from resource-intensive conventional path. High regional cooperation and energy and electricity trade
3	Parikh et al. (2009)	No targets	S2: maximize nuclear
			S3: full hydro potential
			S4: maximize nuclear and hydro
			S5: natural gas induction
			S6: demand-side management
			S7: higher efficiency coal plant scenario
			S8: DSM and high-efficiency coal plants
			S9: higher railways freight share
			S10: higher vehicle frequency
			S11: accelerated renewables

(continued)

Table 4 (continued)

S. No.	Study	Emission target	Scenario
4	Massetti (2011)	No emission targets for three scenarios. 535 ppm pathway for CT4	CT1: carbon tax of $10/ton
			CT2: carbon tax of $30/ton
			CT3: carbon tax of $50/ton
			CT4: carbon tax in line with 535 ppm trajectory
5	Shukla and Chaturvedi (2011)	370 ppmv stabilization trajectory by 2095	Conventional stabilization: carbon price scenario
			Renewable push stabilization: govt. policies focused toward reducing life cycle costs of wind and solar
6	Shukla and Chaturvedi (2012)	384 ppmv stabilization trajectory for carbon price by 2095	CP-noT (carbon price with no electricity targets)
			CP-T (carbon Price with electricity targets)
7	Gambhir et al. (2014)	1.3 T CO_2 per person per year by 2050	LC1: emission constraint for 2050. 1.3 T CO_2/person/yr for 2 °C pathway
		1.3 T CO_2 per person per year by 2050	LC2: emissions trading. Results in 1.7 GT CO_2 by 2050
8	Anandarajah and Gambhir (2014)	LC1: 2.4 GT CO_2 (2020–2050)	LC1: Emission constraint from 2020–2050
		LC2: 2.4 GT CO_2 (2050)	LC2: 2.4 GT CO_2 stabilization by 2050
9	Mittal et al. (2016)	Based on strength of emission cap and RE deployment target	9 Scenarios: emission cap (strong, medium, weak) x RE target (high, medium, low)

3.4.2 Electricity Mix

Indian electricity supply accounted for 65.4% of national emissions in 2007 (MOEF 2007). The scenarios predict higher electricity supply over coming decades with a likely higher role of gas, nuclear, and renewables. There is an observable trend that electricity supply will increase through 2030 and 2050. Even in reference scenarios, electricity consumption is projected to be in ranges of nearly 6–8 times the 2010 levels for year 2050. The reference scenarios also depict an absolute increase in coal

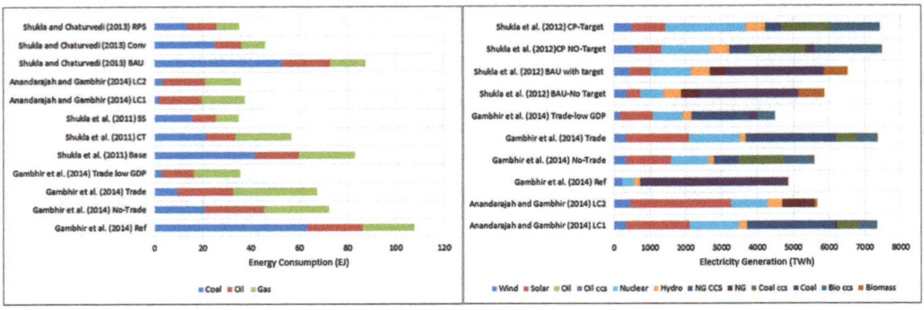

Fig. 1 Comparative trends in energy and electricity mix

use ranging from 42 to 84% (Shukla and Chaturvedi 2012; Gambhir et al. 2014) of total electricity mix in 2050 as compared to 67% in 2010 (IEA 2018). This divergence in baseline scenarios is attributed to assumptions of coal fleet and renewable energy utilization with least action.

Coal thermal capacity in the policy scenarios is assumed to be phased out and substituted with gas CCS, nuclear, and solar energy. The studies project additional 797–2214 TWh generation from nuclear power in policy scenarios [Gambhir et al. (2014)—Trade low GDP, Shukla et al. (2008)—CP no target] over 2010 levels of 0.123 EJ which offers a very big contrast from past trends. Most low carbon scenarios predict a very large role of solar energy in the range of 8–49% [Shukla and Chaturvedi (2012)—BAU with targets, Anandarajah and Gambhir (2014)—LC2] which is in line with current government policies and India's proposed INDCs for Paris agreement. The share of renewable energy in electricity generation varies from 8 to 33% among the baseline scenarios, whereas the share varies from 38 to 66% in policy scenarios. The scenarios with the highest renewable penetration assume a future without carbon capture and storage with a higher viability of solar and wind energy [Anandarajah and Gambhir (2014)—LC2]. The majority of policy scenarios are clustered around 30–50% electricity share from renewables. These policy scenarios provide a very divergent picture due to differences in policy assumptions, coal fleet retirement, technology dissemination, and emission targets.

3.5 Trends in Emissions and Costs of Mitigation

3.5.1 Trends in Emissions

The findings from all studies show a clear divergence in energy use and emissions between baseline and low carbon scenarios. The baseline scenarios across studies project emissions to reach 6.6–8 GT CO_2 by 2050 (Shukla et al. 2008; Gambhir et al. 2014; Anandarajah and Gambhir 2014).

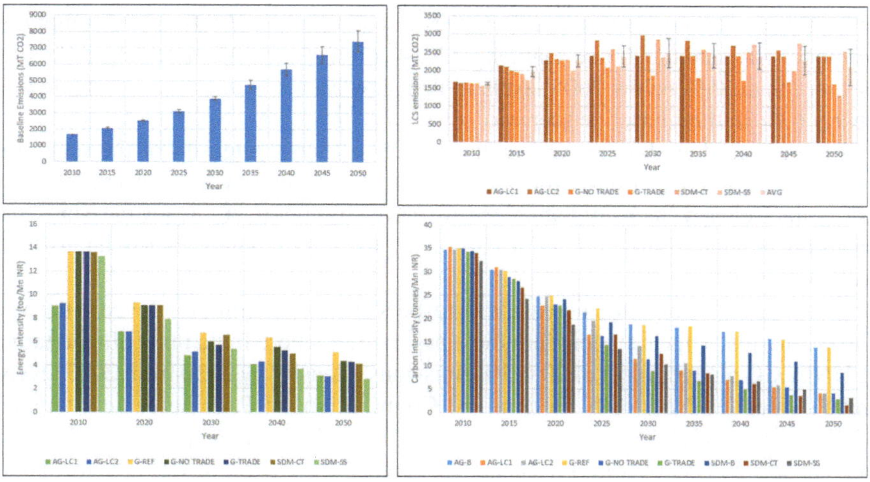

Fig. 2 Trends in emissions in baseline and policy scenarios, energy intensity and carbon intensity (AG-B: Anandarajah and Gambhir-BAU; G-ref: Gambhir et al.-reference scenario; SDM-B: Shukla et al.-BAU; AG-LC1: Anandarajah and Gambhir-LC1; AG-LC2: Anandarajah and Gambhir-LC2; G-NO TRADE: Gambhir et al.-no trade; G-TRADE: Gambhir et al.-trade; SDM-CT: Shukla et al.-carbon tax; SDM-SS: Shukla et al.-Sustainability Scenario; AVG-averaged trend of all scenarios)

Figure 2 illustrates the baseline and low carbon scenarios with likely bounds for future carbon emissions. Together the policy scenarios represent upper and lower ranges for 2050 energy and emissions. The baseline scenarios are framed with assumptions containing high GDP growth rates (7–8% till 2030 and 5% from 2030 to 2050) and increasing share of coal in primary energy. Baseline scenarios project emissions in 2050 with a multiplier of 4.7–3.5 times 2010 levels (Massetti 2011; Anandarajah and Gambhir 2014). In contrast, policy scenarios project emissions in the range of 1.5–0.8. The policy scenarios target 2300–2550 MT CO_2 stabilization range by 2050 (Fig. 2).

The lowest emission scenario for 2050, i.e., carbon tax scenario from Shukla et al. (2008) mitigates about 62.6 BT CO_2 from baseline scenario where majority of mitigation takes place in power sector. This scenario assumes more than 90% decline in carbon intensity and nearly 70% decline in energy intensity in comparison to 2010 levels.

The decoupling of energy and emissions takes place only after 2030 due to the presence of a large existing stock of fossil energy-based infrastructure and large investments promised for the same in near future. This lock-in effect will be the reason for lags before impacts of carbon prices materialize. This is also the prime reason why emission peak does not occur before 2030 in low carbon scenarios.

The energy intensity declines from an average 12.1 toe/Million INR in 2010 to 3.6 toe/Million INR in 2050 as shown Fig. 2. The trends in carbon intensities show a decline of more than 90% below 2010 levels (Anandarajah and Gambhir 2014; Gambhir et al. 2014; Shukla et al. 2008). However, there are wide ranges in energy intensity starting points due to differences in base years and intrinsic assumptions in models. Anandarajah and Gambhir (2014) project a carbon intensity reduction of more than 85% below 2005 levels in their low carbon scenarios as compared to nearly 63% in their baseline scenario.

Figure 2 further shows a higher divergence in energy intensity of the scenarios as compared to carbon intensities. This divergence in energy and carbon intensity arises since energy intensity reductions are driven by energy efficiency measures and changing economic structure while carbon intensity is driven by energy intensity as well as energy mix changes.

The total emissions will increase with respect to absolute emissions even in the best case scenarios. This is evident in India's past targets where India in its official target to UNFCCC pledged a cut of 20–25% emissions intensity by 2020 from 2005 levels. India's Intended Nationally Determined Contributions at Paris agreement target a 33–35% reduction in emission intensity of GDP by 2030 with respect to 2005 levels. With nearly 20 GW of solar PV capacity by Jan 2018, a target of 100 GW for INDC will be a challenge, but a bigger challenge will be providing the storage and flexibility provisions along with effective demand forecasting to ease integration of high wind and solar in the grid. Even with renewable energy capacity addition, the fossil-based energy is likely to stay high by 2030. GDP in most of these studies has been assumed with India's current growth potential and hence ignore macroeconomic impacts of carbon taxes.

3.5.2 Costs of Mitigation

An assessment of mitigation cost is vital for informing policy decisions but not all studies focus on cost. Intermodal comparison of mitigation costs is difficult to perform due to differences in underlying assumptions for technologies. The studies provide mixed details of costs of mitigation. Gambhir et al. (2014) find that welfare and GDP cost of mitigation can be minimized by selling carbon credits in international markets. Further, it is also found that deployment of CCS lowers the incremental costs to energy system in comparison with no CCS scenario with overall cost of meeting Copenhagen targets at US$1.123–1.162 trillion (2005) (Anandarajah and Gambhir 2014). Stabilizing at 50% emissions above 2005 baseline for India would require 1% of GDP for low carbon transition, and any carbon tax beyond that level can pose adverse economic impacts for India (Massetti 2011). The costs of mitigation are comparatively lower if developmental pathways are aligned with climate policy (Shukla 2006; Shukla et al. 2008). While some of these studies address mitigation costs to some extent, it is also important to address sectoral investments required for prioritization of mitigation technologies. Past studies as summarized in Table 4 use carbon tax as key instrument for policy scenarios; however, IEA (2015) notes that capital grants, tax breaks, subsidies, and performance standards have been more successful forms of intervention. While these models perform least cost analysis for technology selection in the presence of climate constraints, in reality the decisions by policymakers are also influenced by co-benefits like traffic congestion, energy security, air quality.

4 Implications for Future Policy Modeling in Indian Context

Although the studies analyze cost-effectiveness of carbon tax for renewables, potential environmental effectiveness of carbon tax and co-benefits of mitigation, none of them focuses specifically on current policy regimes under implementation. There is little emphasis on challenges and costs involved in energy system transformation by addition of alternate energy technologies like wind, solar and bioenergy. Many low carbon scenarios discussed in past studies envision large-scale variable renewable energy deployment for emission reductions which has implications on stability of electricity grid. In order to establish stability and reliability of power supply while adding renewable energy, the operational systems need to be integrated with large-scale storage and alternative power for rapid ramp-up during evening peak demand hours. The current Indian electricity grid is not capable of handling two-way variations in electricity flow to handle variable renewable energy. Thus, the incremental costs incurred in developing key infrastructure like smart grid and storage capacity need to be analyzed in future studies. It will be more practical to create regional and subregional disaggregated energy scenarios to obtain a more comprehensive picture for low carbon transition of the grid.

Carbon capture and storage is a technology which can contribute significantly toward mitigation of GHG emissions by 2050. However, its commercial availability in India depends on its successful implementation in developed countries. The implementation of CCS will incur a huge pressure on the country's water resources since its water footprint is much larger than conventional technologies. While carbon tax and cap and trade form effective policy instruments for generating price signals, the carbon revenues can be used for subsidizing green energy or as general funds for reducing budget deficits. Most of studies in past do not account for recycling of carbon revenue for subsidizing green spending. It would be interesting to analyze future policy scenarios with revenue recycling mechanisms for large-scale renewable energy integration.

5 Summary and Discussion

The review reveals some common trends and variations in total emissions, energy, and electricity mix. The baseline scenarios among studies are constructed with no or minimal policy actions and predict an increase in total emissions ranging from 5.13 to 8 GT CO_2 by 2050. The total emissions in policy scenarios target a stabilization range of 1300–2600 million tonne CO_2 averaged at 2118 million tonne CO_2 in 2050. The mitigation scenarios consider normative policy assumptions like emissions trading, carbon pricing, with and without CCS cases, and societal action. The associated carbon and energy intensity decline in all baseline and policy scenarios. The carbon intensity of GDP declines by about 90% while energy intensity of GDP declines by about 70% from 2010 to 2050 in policy scenarios. This divergence is attributed to the fact that energy intensity reductions are driven by energy efficiency measures and changing economic structure while carbon intensity reduction is driven by energy intensity reduction as well as energy mix changes. The use of coal in energy mix of baseline scenarios increases by a multiplier of six to four times by 2050 as compared to 2010 levels while in policy scenarios, it varies by a factor of 0.2–2. Almost all scenarios favor gas in energy mix ranging from 3.7 to 13.8 times the 2010 levels by 2050. The reference scenarios depict an increased role of coal ranging up to 84% of electricity generation by 2050. On the contrary, policy scenarios assume a phaseout of coal and its substitution with gas CCS, nuclear, and solar which play a dominant role in decarbonizing energy system.

We observe a methodological shift in studies with time where earlier studies focus on exploratory scenarios while recent studies focus on normative scenarios. It is also observed that earlier studies have used parametric changes to create scenarios while recent studies focus more on structural changes and policy options for energy system. Among the policy options, carbon price across models have been used as a prevalent price signal for low carbon scenarios instead of capital grants, subsidies, and tax breaks which have been more successful in past. There are other few common trends across studies, for example, a number of studies have some dominant technologies for India (wind, solar, nuclear) while there are some whose deployment is doubtful due to commercial viability (CCS) or energy security issues (gas). Some studies have performed hybrid analyses with soft-linked approaches to account for a comparatively larger set of variables.

With varying cost assumptions in energy-economy models, it is difficult to perform an accurate intermodal comparison of incremental costs involved in policy scenarios. It is also observed that very high carbon prices in Indian context in order to achieve rapid decarbonization involve large costs and are infeasible. However, decreasing fuel consumption, credits selling, and co-benefits with renewables like improved air quality and energy security can provide government economic benefits while relieving them from budgetary constraints.

India's INDC at Paris agreement targets a 33–35% decline in emission intensity of GDP by 2030, but it does not detail out the sectoral plans and gases covered. Achieving the current targets under INDCs would require high investments in solar, wind, and nuclear capacity which will have implications on load balancing and grid stability. Effective grid integration of large-scale renewables will require flexibility in coal-fired power plants, maintaining storage capacity, robust wind and solar predictions along with ancillary services from the grid.

Acknowledgements We would like to thank Department of Science and Technology, Climate Change Programme for supporting fellowship of students at Centre of Excellence in Climate Studies at IIT Bombay.

Appendix A: Catalogue of Studies with Low Carbon Scenarios for India

Sr. no.	Studies	Scenario years
1	Shukla (2006)	2000–2100
2	Shukla et al. (2008)	2005–2050
3	Parikh et al. (2009)	2006–2031
4	Massetti (2011)	2005–2050
5	Shukla and Chaturvedi (2011)	2005–2095
6	Shukla and Chaturvedi (2012)	2005–2095
7	Gambhir et al. (2014)	2010–2050
8	Anandarajah and Gambhir (2014)	2005–2050
9	Mittal et al. (2016)	2010–2030
10	TERI (2006)	2001–2031
11	Shukla et al. (2010)[a]	2005–2050
12	Johansson et al. (2014)	2010–2050
13	Lucas et al. (2013)	2010–2050
14	Van Sluisveld et al. (2013)	2000–2100
15	Shukla et al. (2017)	2016–2030
16	Calvin et al. (2012)	2005–2050
17	Planning Commission (2014)	2007–2030
18	NCAER (2009)	2003/04–2030/31
19	TERI (2013)	2001–2051
20	Shukla et al. (2015)	2005–2050
21	CSTEP (2015)	2012–2030
22	World Bank (2011)	2007–2031

[a]The study was not included due to limitations in access

Appendix B: Mitigation Drivers (Measures Deployed Are Marked with x)

S. No.	Study	Energy mix changes	Energy efficiency measures	CCS	Demand-side factors	Credits trading
1	Shukla (2006)	x	x	x	x	x
2	Parikh et al. (2009)	x	x			
3	Massetti (2011)	x	x			
4	Shukla et al. (2008)	x	x	x	x	
5	Shukla and Chaturvedi (2012)	x		x		
6	Shukla and Chaturvedi (2011)	x	x	x	x	
7	Gambhir et al. (2014)	x	x	x	x	x
8	Anandarajah and Gambhir (2014)	x	x	x		
9	Mittal et al. (2016)	x	x	x	x	

References

Anandarajah G, Gambhir A (2014) India's CO_2 emission pathways to 2050: what role can renewables play? Appl Energy 131:79–86. Available at: https://www.sciencedirect.com/science/article/pii/S0306261914006084. Accessed 30 Jan 2018

Bosetti V, Carraro C, Galeotti M, Massetti E, Tavoni M (2006) WITCH—a world induced technical change hybrid model. Energy J 27(Special Issue). https://www.jstor.org/stable/23297044. Accessed 31 Jan 2018

Calvin K, Clarke L, Krey V, Blanford G, Jiang K, Kainuma M, Kriegler E, Luderer G, Shukla P (2012) The role of Asia in mitigating climate change: results from the Asia modeling exercise. Energy Econ 34:S251–S260

CPR (2015) Informing India's energy and climate debate: policy lessons from modelling studies. http://cprindia.org/sites/default/files/Informing%20India%27s%20Energy%20and%20Climate%20Debate_CPR-IIASA.pdf. Accessed 30 Jan 2018

CSTEP (2015) A sustainable development framework for India's climate policy. Centre for Study of Science, Technology and Policy, Bangalore. http://www.cstep.in/uploads/default/files/publications/stuff/30683f0adbd6f81820ab9a37eb55c7b0.pdf. Accessed 25 Jan 2018

Edmonds J, Reilly J (1983) A long-term global energy-economic model of carbon dioxide release from fossil fuel use. Energy Econ 5(2):74–88. Available at: https://www.sciencedirect.com/science/article/pii/0140988383900142. Accessed 30 Jan 2018

Edmonds J, Reilly J (1985) Global energy: assessing the future, 1st edn. Oxford University Press, New York

Gambhir A, Napp T, Emmott C, Anandarajah G (2014) India's CO_2 emissions pathways to 2050: energy system, economic and fossil fuel impacts with and without carbon permit trading. Energy 77:791–801. Available at: https://www.sciencedirect.com/science/article/pii/S0360544214011116. Accessed 30 Jan 2018

Hibino G, Matsuoka Y, Kainuma M (2003) AIM/common database: a tool for AIM family linkage. Climate policy assessment, 1st edn. Springer, Tokyo

IEA (2015) India energy outlook. World energy outlook special report. International Energy Agency, Paris, France. Available at: https://www.iea.org/publications/freepublications/publication/IndiaEnergyOutlook_WEO2015.pdf. Accessed 30 Jan 2018

IEA (2018) International Energy Agency. https://www.iea.org/. Accessed 30 Jan 2018

Johansson D, Lucas P, Weitzel M, Ahlgren E, Bazaz A, Chen W, Den Elzen M, Ghosh J, Grahn M, Liang Q, Peterson S, Pradhan B, Van Ruijven B, Shukla P, Van Vuuren D, Wei Y (2014) Multi-model comparison of the economic and energy implications for China and India in an international climate regime. Mitig Adapt Strat Glob Change 20(8):1335–1359. Available at: https://link.springer.com/article/10.1007/s11027-014-9549-4#citeas. Accessed 27 Jan 2018

Kainuma M, Matsoka Y, Morita T (2002) Climate policy assessment—Asia-pacific integrated model (AIM) applications, 1st edn. Springer, Tokyo

Loulou R, Labriet M (2007). ETSAP-TIAM: the TIMES integrated assessment model part I: model structure. Comput Manag Sci 5(1–2):7–40. Available at: https://link.springer.com/article/10.1007/s10287-007-0046-z#citeas. Accessed 30 Jan 2018

Loulou R, Shukla P, Kanudia A (1997) Energy and environment strategies for a sustainable future: analysis with the Indian MARKAL model, 1st edn. Allied Publishers, New Delhi

Lucas P, Shukla P, Chen W, Van Ruijven B, Dhar S, Den Elzen M, Van Vuuren D (2013). Implications of the international reduction pledges on long-term energy system changes and costs in China and India. Energy Policy 63:1032–1041. Available at: https://www.sciencedirect.com/science/article/pii/S0301421513009506. Accessed 28 Jan 2018

Luderer G, Krey V, Calvin K, Merrick J, Mima S, Pietzcker R, Van Vliet J, Wada K (2013) The role of renewable energy in climate stabilization: results from the EMF27 scenarios. Climatic Change 123(3–4):427–441. Available at: https://link.springer.com/article/10.1007/s10584-013-0924-z. Accessed 30 Jan 2018

Massetti E (2011) Carbon tax scenarios for China and India: exploring politically feasible mitigation goals. Int Environ Agreements: Polit, Law Econ 11(3):209–227. Available at: https://link.springer.com/article/10.1007/s10784-011-9157-7. Accessed 30 Jan 2018

Ministry of Finance (2007) Economic survey. Government of India, New Delhi. Available at: http://indiabudget.nic.in/es2007-08/esmain.htm. Accessed 30 Jan 2018

Mittal S, Dai H, Fujimori S, Masui T (2016) Bridging greenhouse gas emissions and renewable energy deployment target: comparative assessment of China and India. Appl Energy 166:301–313. Available at: https://www.sciencedirect.com/science/article/pii/S0306261916000118. Accessed 30 Jan 2018

MOEF (2007) India: greenhouse gas emissions 2007. Government of India. Available at: http://www.moef.nic.in/downloads/public-information/Report_INCCA.pdf. Accessed 30 Jan 2018

NCAER (2009). Climate change impact on the Indian economy—a CGE modelling approach. National Council of Applied Economic Research

NIES (2006) Energy snapshot tool (ESS): manual. National Institute of Environmental Studies, Tsukuba

O'Neill B, Kriegler E, Riahi K, Ebi K, Hallegatte S, Carter T, Mathur R, Van Vuuren D (2013) A new scenario framework for climate change research: the concept of shared socioeconomic pathways. Climatic Change 122(3):387–400. Available at: https://link.springer.com/article/10.1007/s10584-013-0905-2. Accessed 30 Jan 2018

Parikh K, Karandikar V, Rana A, Dani P (2009) Projecting India's energy requirements for policy formulation. Energy 34(8):928–941. Available at: https://www.sciencedirect.com/science/article/pii/S036054420800296X. Accessed 30 Jan 2018

Planning Commission (2002) Report of the committee on Indian vision 2020. Government of India, New Delhi. Available at: http://www.teindia.nic.in/files/reports/ccr/pl_vsn2020.pdf. Accessed 30 Jan 2018
Planning Commission (2006) Integrated energy policy. Government of India, New Delhi. Available at: http://planningcommission.gov.in/reports/genrep/rep_intengy.pdf. Accessed 30 Jan 2018
Planning Commission (2014) The final report of the expert group on low carbon strategies for inclusive growth. New Delhi
Riahi K, Kriegler E, Johnson N, Bertram C, den Elzen M, Eom J, Schaeffer M, Edmonds J, Isaac M, Krey V, Longden T, Luderer G, Méjean A, McCollum D, Mima S, Turton H, van Vuuren D, Wada K, Bosetti V, Capros P, Criqui P, Hamdi-Cherif M, Kainuma M, Edenhofer O (2015) Locked into copenhagen pledges—implications of short-term emission targets for the cost and feasibility of long-term climate goals. Technol Forecast Soc Change 90:8–23. Available at: https://www.sciencedirect.com/science/article/pii/S0040162513002539. Accessed 30 Jan 2018
Shukla P (2006) India's GHG emission scenarios: aligning development and stabilization paths. Curr Sci 90(3). Available at: http://www.iisc.ernet.in/~currsci/feb102006/384.pdf. Accessed 30 Jan 2018
Shukla P, Chaturvedi V (2011) Sustainable energy transformations in India under climate policy. Sustain Dev 21(1):48–59. Available at: http://onlinelibrary.wiley.com/doi/10.1002/sd.516/abstract. Accessed 30 Jan 2018
Shukla P, Chaturvedi V (2012) Low carbon and clean energy scenarios for India: analysis of targets approach. Energy Econ 34(3):S487–S495. Available at: https://www.sciencedirect.com/science/article/pii/S0140988312001028. Accessed 30 Jan 2018
Shukla P, Dhar S, Mahapatra D (2008) Low-carbon society scenarios for India. Clim Policy 8(Supp):S156. Available at: http://www.tandfonline.com/doi/abs/10.3763/cpol.2007.0498. Accessed 30 Jan 2018
Shukla P, Dhar S, Fujino J (2010) Renewable energy and low carbon economy transition in India. J Renew Sustain Energy 2(3):031005. Available at: http://aip.scitation.org/doi/abs/10.1063/1.3411001?journalCode=rse
Shukla P, Garg A, Dholakia H (2015) Energy-emissions trends and policy landscape for India. Allied Publishers Pvt. Ltd., Mumbai
Shukla PR, Mittal S, Liu JY, Fujimori S, Dai H, Zhang R (2017) India INDC assessment: emission gap between pledged target and 2°C target. In: Fujimori S, Kainuma M, Masui T (eds) Post-2020 climate action. Springer, Singapore
TERI (2006) National energy map for India: technology vision 2030. GOI, New Delhi. Available at: http://www.teriin.org/div/psa-fullreport.pdf. Accessed 30 Jan 2018
TERI (2013) The energy report—India: 100% renewable energy by 2050. WWF India, New Delhi
UN (2018) United Nations Population Division|Department of Economic and Social Affairs. Available at: http://www.un.org/en/development/desa/population/. Accessed 30 Jan 2018
UNPD (2003) World population prospects the 2002 revision. UN. Available at: http://www.un.org/esa/population/publications/wpp2002/WPP2002-HIGHLIGHTSrev1.PDF. Accessed 30 Jan 2018
Van Sluisveld M, Gernaat D, Ashina S, Calvin K, Garg A, Isaac M, Lucas P, Mouatiadou I, Otto S, Rao S, Shukla P, Van Vliet J, Van Vuuren D (2013) A multi-model analysis of post-2020 mitigation efforts of five major economies. Clim Change Econ 04(04):1340012
World Bank (2011) Energy intensive sectors of the Indian economy: path to low carbon development. Energy sector management assistance program (ESMAP). World Bank, Washington, DC
World Bank (2018) The world bank, world development indicators. Available at: http://www.worldbank.org/. Accessed 30 Jan 2018

Greenhouse Gas Emissions in India's Road Transport Sector

Namita Singh, Trupti Mishra and Rangan Banerjee

Abstract India is one of the fastest growing major economies of the world. The transport sector of India is the third most GHG (Greenhouse gas) emitting sector, where the major contribution comes from the road transport sector. The chapter examines the trend of CO_2 emission from transport sector (1970–2012) and trend of annual vehicle registration in India (2001–2012). The chapter also gives an account of the travel characteristics of India and explains the fleet models applied to obtain the age and technology distribution of vehicles on the road. Different methodologies adopted for GHG estimation can be explained via two independent approaches based on two sets of data such as fuel sold and vehicle kilometers traveled. The study analyzed the Tier 1, Tier 2, and Tier 3 methods for GHG emission estimation given by IPCC and segregates the studies done on India based on these approaches. Uncertainty in CO_2 emission from different studies for Indian transport sector was high initially, but it has reduced in recent studies due to refinement in the input parameters, emission factors, and associated methodologies. The study addresses the gap in the estimation study and discusses the policies in India for reducing emissions from transport sector such as auto-fuel policy, shifting to four-stroke vehicles from two-stroke vehicles, policies for alternative fuels like CNG and biofuels, and incentives provided for the diffusion of electric vehicles in India.

N. Singh (✉) · T. Mishra · R. Banerjee
Interdisciplinary Program (IDP) in Climate Studies, Indian Institute of Technology Bombay, Mumbai 400076, Maharashtra, India
e-mail: namita.msu@gmail.com

T. Mishra
e-mail: truptimishra@iitb.ac.in

R. Banerjee
e-mail: rangan@iitb.ac.in

T. Mishra
Shailesh J. Mehta School of Management, Indian Institute of Technology Bombay, Mumbai 400076, Maharashtra, India

R. Banerjee
Department of Energy Science and Engineering, Indian Institute of Technology Bombay, Mumbai 400076, India

C. Venkataraman et al. (eds.), *Climate Change Signals and Response*,
https://doi.org/10.1007/978-981-13-0280-0_12

Keywords Greenhouse gas emission · Mitigation policy · Road transport

1 Greenhouse Gas Emissions

1.1 Introduction

Greenhouse gases (GHGs) are responsible for causing global warming on earth. These gases trap the heat in the atmosphere and increase the temperature of the earth. Anthropogenic activities such as deforestation, fossil fuel consumption, and atmospheric emissions majorly from industries and transport sectors have led to greenhouse gas emissions. According to Intergovernmental Panel on Climate Change (IPCC) Fifth Assessment Report which was published in 2014, human involvement is evident in increasing GHG emissions. Suitable mitigation and adaptation strategies must be adopted to curtail the threat of climate change (Jeyalakshmi et al. 2015).

Greenhouse gases which are responsible for global warming consist of carbon dioxide (CO_2), nitrous oxide (N_2O), methane (CH_4), hydrofluorocarbons (HFCs), sulfur hexafluoride (SF_6), and perfluorocarbons (PFCs). Here, CO_2 emission accounts for the major GHG emissions from transport sector which is around 95%. Methane (CH_4) and nitrous oxides (N_2O) consist of 1%, and the leakage from air-conditioning systems of hydrofluorocarbons (HFCs) accounts for the rest of the 3% of GHG emissions. Ozone, aerosol, and carbon monoxide (CO) are also transport sector emissions (Atabani et al. 2011).

Fossil fuel CO_2 emission from the Indian transport sector have increased by approximately 5 times from 50 Mt in 1970 to 242 Mt in 2012 (derived from World Bank Indicator data). Whereas, CH_4 and N_2O emissions have increased by about 45%, and fluorinated gases have increased from 0.4% in 1970 to 2% in 2010. In Asia, China and India are the major economies responsible for the increase in GHG emissions (Nejat et al. 2015). Globally transport sector is responsible for 24% of total CO_2 emissions from fuel combustion in 2015 (which is second most CO_2 emitting sector after electricity and heat, i.e., 42%) (IEA 2017). The current study focuses on GHG emission from the transport sector of India and examines the growth trend of vehicle population on road. The study analyzes the transport (travel) characteristics of India and focuses on the methodologies adopted for the CO_2 emission estimation and the variation in input variables resulting in the uncertainty in CO_2 emission from the transport sector. In the end, the study highlight the policies adopted by the government for emission reduction and the gaps needed to be addressed for robust analysis of GHG emission.

1.2 Greenhouse Gas Emissions from the Transport Sector of India

In India along with economic growth, the urban areas have also increased. This has led to increase in transport infrastructure and transport vehicles. This portion of the chapter will focus on trend analyis of vehicular population on road, which is one of the leading causes of anthropogenic CO_2 emission from the Indian transport sector.

Over the years, the growth rate of CO_2 emission from the overall transport sector of India has increased exponentially as shown in Fig. 1. The growth rate of CO_2 emission from transport sector was not linear from the time period 1970 to 2012. The graph is plotted for the different time periods with different growth rates for a log of CO_2 emission from the transport sector. The data for the growth analysis are obtained for the duration of 1970 to 2012 from World Development Indicators reported by the World Bank. The growth rate of CO_2 emission from the time period 1970 to 1980 was 2.7%, from 1980 to 1990 was 2%, from 1990 to 2000 was 4%, and from 2000 to 2012 was 6.9%. Various reforms for reduction in emissions from transport sector were introduced which should have reduced emission with time such as introduction of four-stroke engine vehicles in mid-1990s, adoption of emission norms like auto-fuel policy like Bharat Stages I, II, and III (which are similar to European emission norms) in India, and introduction of fuels like compressed natural gas (CNG) and liquefied petroleum gas (LPG) in vehicles during the period 2000–2012. But with an increase in private motor vehicles on road, along with an increase in the distances traveled and deterioration of public transport (Ramachandra et al. 2015), the emission has increased with respect to previous decades from urban road transport. The major fuels in these motor vehicles are petrol and diesel, and even though stringent emission norms were introduced in India, lack of their implementation and missing infrastructure has lead to uncontrolled emission from the transport sector.

Emission of greenhouse gases like methane (CH_4) and nitrous oxide (N_2O) is very less from road transport sector of India. Methane emission has increased by fourfold from 1980 to 2000 (from 2.9 to 11.6 kt), but its growth rate has dropped from 8% during 1980–1990 to 6% during 1990–2000. Similarly, for N_2O, road transport contribution in its emission is negligible, but their amount has doubled from 1980 to 2000 (from 0.23 to 0.9 kt). The growth rate during the period 1980–1990 was 8% and during the period 1990–2000 was 6%. Emission reduction in both the gases has been attributed to the introduction of four-stroke engine vehicles in two-wheelers and three-wheelers in the market in the mid-1990s (Singh et al. 2008).

Assume the vehicles registered are the vehicles used on road and are responsible for the increase in emissions. Figure 2 shows the trend in vehicle registered data for two-wheelers, cars, jeeps, taxis (all three considered as four-wheelers), buses, and goods vehicles. Here, the registered vehicle population data for the years 2001–2012 are taken from MORTH (2015), and growth trend for two-wheelers, four-wheelers, buses, and goods vehicles is shown.

The graphs examine the growth in annual vehicle registration for motor vehicles in India. The trend analysis shows the exponential growth in vehicles over the years.

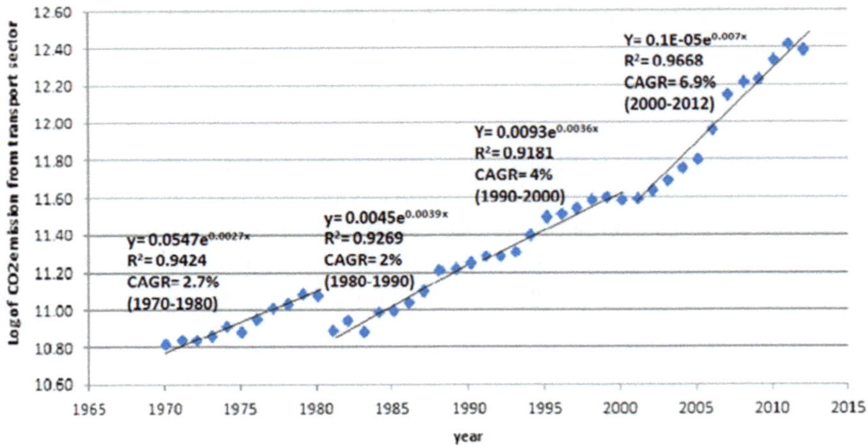

Fig. 1 Growth trend of log of CO_2 emission from transport sector of India for different time period

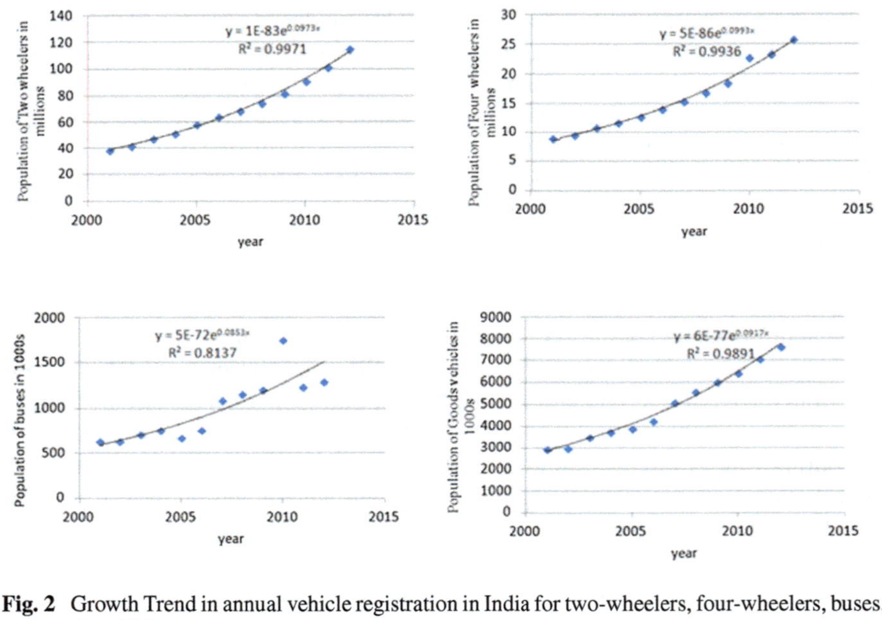

Fig. 2 Growth Trend in annual vehicle registration in India for two-wheelers, four-wheelers, buses, and goods vehicles

For the time period 2001–2012, the growth rates for two-wheelers, four-wheelers, buses, and goods vehicles are 10.5, 10.2, 6.7, and 9.1%, respectively. The registered vehicles are subjected to fleet model which separates the vehicles of different technology based on their age and hence takes care of the phasing out or old vehicles which are no longer on road. The methods applied in the Indian context are Weibull distribution (Baidya and Borken-Kleefeld 2009) and logistic distribution (Pandey and

Venkataraman 2014) to study age distribution of vehicles on road. Understanding the fleet model using logistic function approach, the model required time series vehicle registration data, vehicle sales data, and age distribution of vehicles at a particular year (reference year) to estimate the survival function parameters.

$$V_{c,a}(t) = V_{c,0}(t - a) \times \mathrm{Suf}_{c,a} \tag{1}$$

$$\mathrm{Su}(a) = \frac{1}{1 + \mathrm{e}^{[\propto(1 \frac{a}{L50})]}} \tag{2}$$

Here, in Eq. 1, $V_{c,a}$ (t) is the population of the vehicle in year t, having category c and age a. $V_{c,0}$ $(t - a)$ is the new vehicle registered in the year $t - a$, and $\mathrm{Suf}_{c,a}$ is the survival fraction applied to the vehicles present in the given category of a vehicle having aged a. In Eq. 2, survival fraction is the ratio of survival rate Su for vehicles of age a to the vehicles of age 0. Here, α is the shape factor which indicates the start of significant retirement for the vehicles of given category. And $L50$ represents the age by which half of the vehicles of the particular age have retired. The values for α and $L50$ are as follows—α for two-wheelers, three-wheelers (auto rickshaws), four-wheelers, buses, and goods vehicles are −2.9, −2.9, −5.2, −4.5, and −4.5, respectively. Whereas, $L50$ for two-wheelers, three-wheelers (auto rickshaws), four-wheelers, buses, and goods vehicles are 10.1, 10.1, 19.8, 13, and 13 years respectively (Pandey and Venkataraman 2014).

With the increase in vehicle population, energy consumption by vehicles has also increased. From the overall transport sector in India (road, water, air, and rail), approximately 73% of GHG emissions come from road transport sector in 2011 (Gupta and Singh 2016).

1.3 Travel and Energy Demand for Transport Sector of India

In India, urban road passenger transport consists of road passenger and road freight transport travel demand. Road passenger transport is made up of passenger-carrying vehicles such as motorcycle, scooters, cars, and buses, whereas road freight transport demand consists of trucks and heavy diesel vehicles which carry only goods. Similarly, rail transport also consists of passenger and freight travel demand. The growth rate of road passenger transport demand (which is passenger kilometers traveled) is 11% from 1831.6 billion passenger kilometers in 2000 to 6351.2 billion passenger kilometers in 2012. For road freight transport demand, the growth rate is 8% from 467 billion tonnes kilometer in 2000 to 1212.4 billion tonnes kilometers in 2012. Similarly, for rail passenger kilometers, the growth rate is 8% from 430.7 billion passenger kilometers in 2000 to 1046.5 billion passenger kilometers in 2012, whereas for rail freight transport demand, the growth rate is 7% starting from 305.2 billion tonnes kilometers in 2000 to 667.6 billion tonnes kilometers in 2012 (MORTH 2015). Figure 3 shows the travel demand share for the year 2012 for India. It can be

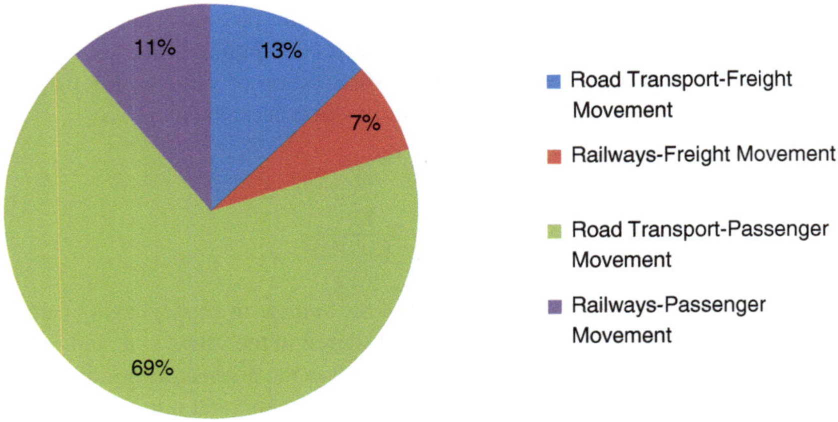

Fig. 3 Share of road and rail transport demand in India for the year 2012

seen that passenger road transport demand (on-road vehicles like two-wheelers, cars, three-wheelers, and buses) dominates the travel demand in India, followed by road freight transport demand (goods carrying vehicles like trucks) which is followed by rail passenger and freight transport demand. Here, travel demand in terms of billion kilometres travelled is indicated by 'movement' of each mode type.

Diesel and petrol are the major transportation fuel for Indian vehicles such as cars, buses, trucks and two-wheelers, and in small percentages, CNG is also used in auto rickshaws, taxis, and buses. The transport sector in India accounts for 70% of diesel consumption (both direct and retail sales) at all India level. Diesel is used in both transport and non-transport sector. Light commercial vehicles (LCHs), heavy commercial vehicles (HCVs), and buses account for about 38%; whereas cars and utility vehicles (UVs) are responsible for about 22% of diesel sales in India. The growth rate of diesel consumption for Indian transport sector is 7% from 1970–71 to 2012–13. For petrol, two-wheelers are reported to have maximum consumption at 61.42% for the duration of 1970–71 to 2012–13. This could be mainly because of growing middle-class income group in the country. Cars in India running on petrol are reported to be around 34%, which is lower than diesel cars. This could be due to the price difference between diesel and petrol, and consumers perceive diesel cars to be economically optimal. Two-wheelers and cars together account for 95% of the total consumption of petrol. The growth rate of petrol consumption in India is 5.7% since 1970–71 to 2012–13 (MOP&NG 2013).

This increase in fuel (petrol and diesel) consumption is responsible for the increase in energy consumption. The concentration of air pollutants and GHG gases emitted from the exhaust of vehicle has increased with increase in vehicle population (Pucher et al. 2005). These air pollutants have an adverse impact on the health of the people. Hence, one can say that urban transportation has also degraded the health of the people along with causing global warming. The ever-increasing demand for conventional fuel (petrol and diesel) has led to research in alternate technology/fuels which can diffuse in the market and replace the petroleum fuels.

Climate change mitigation strategies are actions which limit the rate of long-term climate change. For the transport sector, various policies have been enacted and suggested as mitigation strategies such as auto-fuel policy in India, blending of biofuel in petrol and diesel, phasing out of old and polluting vehicles, diffusion of electric vehicles, and increasing number of buses or improving the public transport sector are few policies or initiatives to mention. In order to see the impact of policies or strategies on transport emission, one need to first calculate the GHG emissions or build an emission inventory for a reference year for a city/country and then calculate the impact of each strategy in terms of emission reduction from the reference year value. The estimation of emission is done by either fuel consumption approach or activity data approach. The methodology adopted is either top-down or bottom-down at the city or country level.

2 Approach Toward Estimating GHG Emissions from Transport Sector

2.1 Emission Estimation Methodology

The GHG emission inventory at the country level for road transport sector consist of methodologies suggested by IPCC (1996). Figure 4 shows the top-down approach and bottom-up approach together for GHG emission estimation from the transport sector. The Tier 1 approach calculates CO_2 emissions by multiplying estimated fuel sold with a default CO_2 emission factor.

The Tier 2 approach is similar to Tier 1 approach, except that country-specific carbon contents of the fuel sold in road transport are used. CH_4 and N_2O emissions are significantly affected by the distribution of emission controls in the fleet. The Tier 3 approach requires detailed, country-specific data to generate activity-based emission factors for vehicle subcategories and may involve national models. Several studies are available for India where Tier 3 approach is used for emission estimation.

In Fig. 4, the input values are represented by dashed arrows and output values are represented by solid arrows. The bottom-up approach consists of information related to vehicle numbers, age distribution of vehicles on the road, fuel used in vehicles, technology subdivision in vehicles, vehicle activity data, fuel efficiency, and emission factors for each vehicle type. The activity data consist of per capita trip rate, vehicle population, average trip length, and relative modal share data for developing travel demand. The occupancy factor is the load capacity of each vehicle type and gives the per capita emission. As mentioned earlier, the fleet model is applied to get the age distribution of vehicles on road applying either logistic distribution or Weibull distribution. In the bottom-up approach, fuel consumption is calculated by data of vehicle population (obtained after applying fleet model) of different vehicle types, fuel share in vehicles, average vehicle kilometer traveled, fuel efficiency of vehicles, and fuel density, whereas in the top-down approach directly fuel consumed data at the aggregate level is used. After obtaining the fuel consumption value, it is multiplied by the emission factor to estimate the GHG emissions. The bottom-up approach is robust and estimates emission at the vehicle/mode level running on different fuel type.

The emission factors applied are either default IPCC emission coefficients or developed at a country level for different vehicle types. Emission of each greenhouse gas is calculated by multiplying fuel consumption (obtained either by bottom-up or top-down approaches) to their respective emission factor. The final outcome varies in the studies depending on the values of input factors and assumptions.

2.2 *Emission Estimation Studies Done for Indian Transport Sector*

Several studies have been done on India to estimate CO_2 emission from the road transport sector. The study done by Baidya and Borken-Kleefeld (2009) has grouped the studies on the basis of emission estimation method. These methods are as follows:

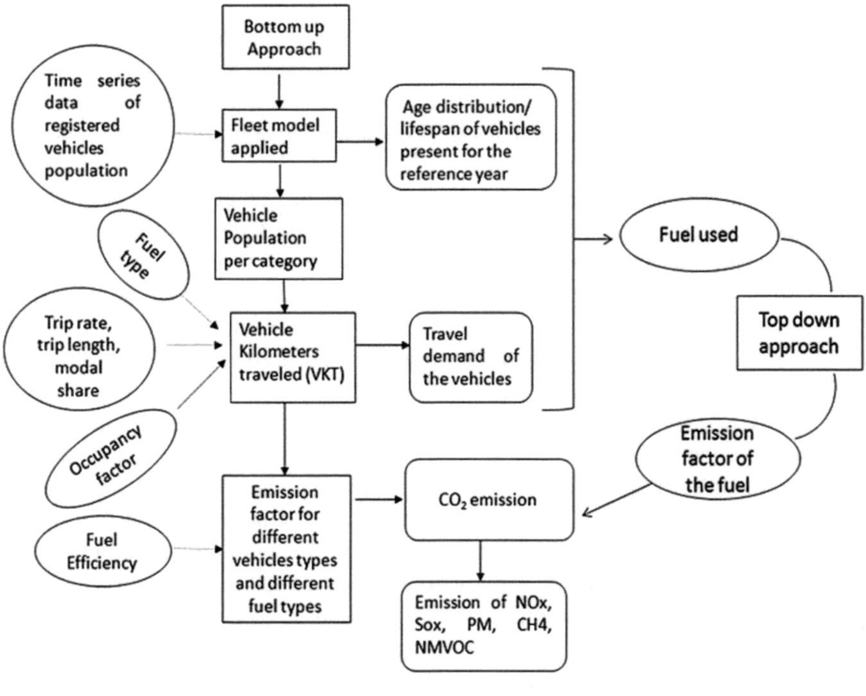

Fig. 4 Bottom-up and top-down approaches for emission estimation

(a) Fuel consumption and total road,
(b) Fuel consumption and vehicle category,
(c) Vehicle kilometer and vehicle category, and
(d) Vehicle kilometer and vehicle subcategory.

In the current study, this is further updated and studies showing emission estimation till the year 2013 are included based on the similar division of groups given by Baidya and Borken-Kleefeld (2009). Table 1 demonstrates the CO_2 emission for the different years from different studies. The studies are listed chronologically where emissions from the year 2013 are listed first and emissions from the year 1980 are listed at the end.

Uncertainty in the final output is dependent on the values of the input parameters, assumptions, type of methodological approaches such as bottom-up or top-down, and emission factors. With the increase in a number of input variables, the uncertainty in the final output increases. Attempts have been made to decrease the uncertainties in the emission estimation by addressing issues such as vehicle activity data like total kilometers traveled and model share and fuel share in different categories of vehicles. The uncertainty in CO_2 emission also results due to variation in emission factors applied in different studies for different vehicle types.

Figure 5 shows the uncertainty in final CO_2 emissions from different studies over different years at 95% confidence interval. For the same year, the difference in input variables results in different CO_2 emission. Uncertainty in CO_2 emission measured from the literature for the year 1990 was 62.36 ($\pm$8) Tg; for the year 1994, it was 73.56 ($\pm$12) Tg; for the year 1995, it was calculated to be 88.85 ($\pm$24); for the year 2000, it came out to be 127.74 ($\pm$49) Tg; and for the year 2005, it was 144 ($\pm$24) Tg. Uncertainty in the final output (CO_2 emission) was high initially, but it has reduced in recent studies due to refinement in the input parameters, emission factors, and associated methodology.

3 Policy Implications in the Indian Transport Sector

3.1 Policies in Transport Sector of India

Effective policy implementation is very important for removing the discrepancies between supply and demand for the transport sector. Policies are the strategies which can be applied to reduce emissions and energy consumption at national or local level. Reducing CO_2 emissions will require some strict actions from the government side. These actions include auto-fuel policy where the objective is to improve the quality of fuel standard and apply strict emission norms for emission reduction by implementing Bharat Stages I, II, III, IV, V, and VI into the motor vehicles. Path of implementation of auto-fuel emission norms for future years has been stated by the Government of India with improvement in vehicle technology. As per India's Intended Nationally Determined Contribution (INDC), it was suggested to increase

Table 1 Studies on Indian transport sector

Study	Tg of CO_2	Year	Group
Singh et al. (2017) (I)	469.07	2013	c
Sadavarte and Venkataraman (2014)	162.00	2010	d
Ramachandra and Shwetmala (2009)	258.10	2003–2004	c
Sadavarte and Venkataraman (2014)	103.00	2005	d
ADB (2006)	220.00	2005	d
Baidya and Borken-Kleefeld (2009) Max	146.85	2005	d
Baidya and Borken-Kleefeld (2009) Ref.	146.33	2005	d
Garg et al. (2006)	143.00	2005	a
Fulton and Eads (2004)	135.38	2005	c
Baidya and Borken-Kleefeld (2009) Min	133.87	2005	d
Baidya and Borkeen-Kleefeld (2009) with CPCB (2000)	133.87	2005	d
IIASA (2008)	133.76	2005	a
Singh et al. (2017)	71.67	2001	c
Sadavarte and Venkataraman (2014)	222.00	2000	d
Van Aardenne et al. (2005)	119.81	2000	a
Garg et al. (2006)	116.00	2000	a
Singh et al (2008) (II)	105.00	2000	a
IIASA (2008)	102.55	2000	a
Borken et al. (2007)	99.59	2000	c
Ohara et al. (2007)	86.37	2000	a
Mittal and Sharma (2003)	42.93	1997	c
Sadavarte and Venkataraman (2014)	82.00	1996	d
Olivier et al. (2002)	98.74	1995	a
Garg et al. (2006)	89.00	1995	a
IIASA (2008)	78.82	1995	a
Ramanathan and Parikh (1999)	79.00	1994	c
MiEF (2004)	71.89	1994	a
Bhattacharya and Mitra (1998)	69.80	1994	b
Garg et al. (2006)	70.00	1990	a
IIASA (2008)	61.67	1990	a
Olivier et al. (2002)	60.46	1990	a
Bhattacharya and Mitra (1998)	57.30	1990	b
Singh et al. (2008) (II)	27.00	1980	a

Singh et al (2017) (I)—(Singh et al 2017); Singh et al (2008) (II)—(Singh et al 2008); Max is the maximum CO_2 emission; Min is the minimum CO_2 emission; Ref. is the CO_2 emission for the reference year; Baidya and Borken-Kleefeld (2009) with CPCB (2000) is CO_2 emission obtained from CPCB emission factors

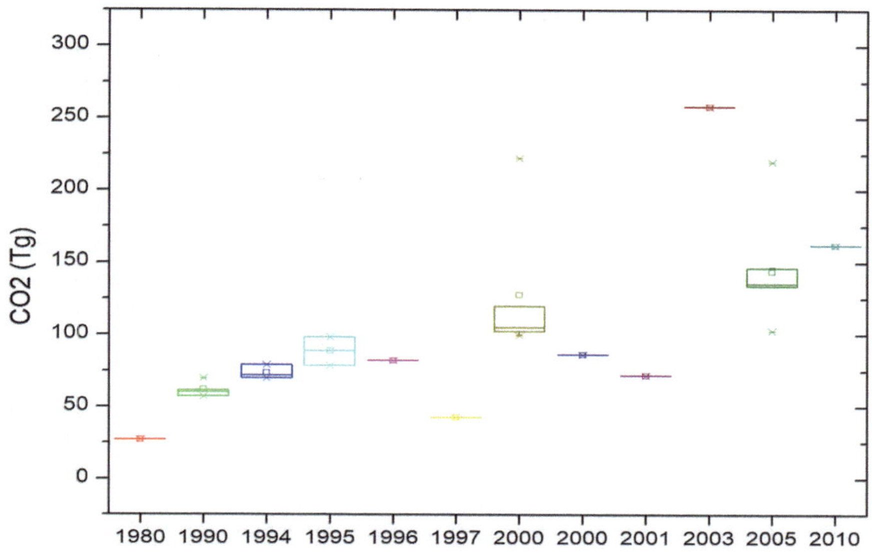

Fig. 5 Uncertainty in CO_2 emissions over the years in the literature

the share of railways (public transport) from 36 to 45% to restrict the emissions. Around 20% blending of biofuels for bioethanol and biodiesel (alternate fuels) by 2017 was suggested by the National Policy on Biofuels. Promotion of electric vehicles (alternate technology) for its diffusion in Indian market is done by providing incentives under Faster Adoption and Manufacturing of Hybrid and Electric Vehicles (FAME) in India. The program includes the development of high-performance electric or hybrid vehicles for Indian scenario, decreasing the capital cost for its purchase, developing charging infrastructure, and creating programs for its promotion and awareness in the consumers (Dhar et al. 2017). Studies also suggest improvement in fuel efficiency for cars and two-wheelers, which will sufficiently reduce the tailpipe emissions from the motor vehicles.

3.2 *Summary of the Study*

The trend analysis of CO_2 emission from Indian transport sector shows exponential growth. The growth rate was 2.7% during the period 1970–1980; it decreased to 2% during the period 1980–1990 and increased to 4% during 1990–2000, which further increased to 6.9% during 2000–2012. This increase is due to increase in motor vehicles on road running on petrol and diesel and also increase in distance traveled by these vehicles. CO_2 is the major GHG emission from the transport sector, whereas the contribution of CH_4 and N_2O is negligible (Singh et al. 2008).

Registered motor vehicles in India have increased during the period 2001–2012. The annual growth rates for two-wheelers, four-wheelers, buses, and goods vehicles were 10.5, 10.2, 6.7, and 9.1%, respectively. The annual growth rate of road passenger and freight traffic demand for the same time period (2001–2011) was 11 and 8%, respectively. Similarly, for rail transport, passenger and freight demand were 8 and 7%, respectively, for the duration 2001–2011. Top-down or bottom-up approaches applied for emission estimation at a national level or local level depend on the availability of the data. Uncertainty in CO_2 emission depends on uncertainty in input variables, the methodological approach applied, and the emission factors used. Uncertainty in the final output (CO_2 emission) was high initially, but it has reduced in recent studies due to refinement in the input parameters, emission factors, and associated methodology.

The gap which can be addressed from the study is the discrepancy in the data for the input variables and emission factors which results in uncertainty in GHG (CO_2) emissions from several studies for the same year. The gap in policy implementation is addressed by acknowledging the lack of awareness in people with respect to incentives and benefits of electric vehicles for the environment and lack of infrastructure which is required for the speedy diffusion of alternate fuels and technologies.

References

ADB (2006) Energy efficiency and climate change considerations for on-road transport in Asia. Asian Development Bank, Manila, Philippines

Atabani AE, Badruddin IA, Mekhilef S, Silitonga AS (2011) A review on global fuel economy standards, labels and technologies in the transportation sector. Renew Sustain Energy Rev 15:4586–4610. https://doi.org/10.1016/j.rser.2011.07.092

Baidya S, Borken-Kleefeld J (2009) Atmospheric emissions from road transportation in India. Energy Policy 37:3812–3822. https://doi.org/10.1016/j.enpol.2009.07.010

Bhattacharya S, Mitra AP (1998) Greenhouse gas emissions in India for year 1990. Centre on global change. National Physical Laboratory, New Delhi, India

Borken J, Steller H, Meretei T, Vanhove F (2007) Global and country inventory of road passenger and freight transportation: fuel consumption and emissions of air pollutants in 2000. Transportation research record, 2011, https://doi.org/10.3141/2011-14

Dhar S, Pathak M, Shukla PR (2017) Electric vehicles and India's low carbon passenger transport: a long-term co-benefits assessment. J Clean Prod 146:139–148. https://doi.org/10.1016/j.jclepro.2016.05.111

Fulton L, Eads G (2004) IEA/SMP model documentation and reference case projection. International Energy Agency/World Business Council for Sustainable Development

Garg A, Shukla PR, Kapshe M (2006) The sectoral trends of multigas emissions inventory of India. Atmos Environ 40:4608–4620

Gupta M, Singh S (2016) Factorizing the changes in CO_2 emissions from Indian road passenger transport: a decomposition analysis. Stud Bus Econ 11:67–83. https://doi.org/10.1515/sbe-2016-0036

IEA (2017) CO_2 emissions from fuel combustion highlights (2017 edition)

IIASA (2008) GAINS for India & South Asia: BL India SNAP1 Categories. Available at: (http://www.iiasa.ac.at/web-apps/apd/gains/)

IPCC (Inter-Governmental Panel on Climate Change) (1996) IPCC good practice guidance and uncertainty management in national greenhouse gas inventories. Cambridge University Press, New York

Jeyalakshmi S, Parameswaran V, Mathew J et al (2015) Statistics related to climate change—India

MiEF(2004) India's initial national communication to the United Nations framework convention on climate change. Ministry of Environment and Forests, Government of India, New Delhi, pp. 38–41

Ministry of Petroleum and Natural Gas (MOP&NG) (2013) All India study on sectoral demand of diesel and petrol, report by petroleum planning and analysis cell

Ministry of Road Transport and Highways (MORTH) (2015) Motor transport statistics, New Delhi. https://data.gov.in/catalog/total-number-registered-motor-vehicles-india

Mittal ML, Sharma C (2003) Anthropogenic emissions from energy activities in India: generation and source characterization (Part-II): emissions from vehicular transport in India

Nejat P, Jomehzadeh F, Taheri MM et al (2015) A global review of energy consumption, CO_2 emissions and policy in the residential sector (with an overview of the top ten CO_2 emitting countries). Renew Sustain Energy Rev 43:843–862. https://doi.org/10.1016/j.rser.2014.11.066

Ohara T, Akimoto H, Kurokawa J, Horii N, Yamaji K, Yan X, Hayasaka T (2007) An Asian emission inventory of anthropogenic emission sources for the period 1980–2020. Atmos Chem Phys 7:6843–6902

Olivier JGJ, Berdowski JJM, Peters JAHW, Bakker J, Visschedijk AJH, Bloos JPJ (2002) Applications of EDGAR emission database for global atmospheric research. ISBN: 90 5851 077 8, Global Change NOP-NRP Report 410200051, RIVM rapport 773301001, 155 p

Pandey A, Venkataraman C (2014) Estimating emissions from the Indian transport sector with on-road fleet composition and traffic volume. Atmos Environ 98:123–133. https://doi.org/10.1016/j.atmosenv.2014.08.039

Pucher J, Korattyswaropam N, Mittal N, Ittyerah N (2005) Urban transport crisis in India. Transp Policy 12:185–198. https://doi.org/10.1016/j.tranpol.2005.02.008

Ramachandra TV, Shwetmala (2009) Emissions from India's transport sector: statewise synthesis. Atmos Environ 43:5510–5517. https://doi.org/10.1016/j.atmosenv.2009.07.015

Ramachandra TV, Aithal BH, Sreejith K (2015) GHG footprint of major cities in India. Renew Sustain Energy Rev 44:473–495. https://doi.org/10.1016/j.rser.2014.12.036

Ramanathan R, Parikh JK (1999) Transport sector in India: an analysis in the context of sustainable development. Transport policy 6:35–45. https://doi.org/10.1016/S0967-070X(98)00030-4

Sadavarte P, Venkataraman C (2014) Trends in multi-pollutant emissions from a technology-linked inventory for India: I. Industry and transport sectors. Atmos Environ 99:353–364. https://doi.org/10.1016/j.atmosenv.2014.09.081

Singh A, Gangopadhyay S, Nanda PK et al (2008) Trends of greenhouse gas emissions from the road transport sector in India. Sci Total Environ 390:124–131. https://doi.org/10.1016/j.scitotenv.2007.09.027

Singh R, Sharma C, Agrawal M (2017) Emission inventory of trace gases from road transport in India. Transp Res Part D Transp Environ 52:64–72. https://doi.org/10.1016/j.trd.2017.02.011

Van Aardenne J, Dentener F, Olivier JGJ, Peters JAHW (2005) The EDGAR 3.2 Fast Track 2000 dataset (32FT2000). Netherlands Environmental Assessment Agency, Rijswijk

A State-Level Methodology for Estimating Present-Day Emissions of Short-Lived Climate Pollutants from Fired Brick Production in India

Kushal Tibrewal, Sameer Maithel and Chandra Venkataraman

Abstract Indian brick sector is dominated by traditional kilns with inefficient combustion technology. However, it has substantial potential in mitigating SLCPs through shifts towards efficient technologies, thus highlighting the need to understand the present-day emission profile from manufacturing of fired-clay bricks. We developed a methodology to estimate the emissions of black carbon (BC), organic carbon (OC), SO_2 and ozone precursors (CO, CH_4, NO_x, NMVOC) at state level. It is used to estimate national brick demand which is distributed among states to estimate state-level production assuming production to follow demand. Fractional contribution of four major kiln technologies in India (Bull's trench kiln, clamp kiln, Zig Zag fired kiln and vertical shaft brick kiln) is estimated for each state to estimate the state-wise share of different kiln technologies in production. Emissions of $PM_{2.5}$, BC, OC and SO_2 are estimated to be 165.9 (142.0–189.7) Gg yr^{-1}, 119.1 (97.6–140.5) Gg yr^{-1}, 9.4 (7.3–11.4) Gg yr^{-1} and 393.6 (314.1–473.1) Gg yr^{-1}. For ozone precursors, the estimates are 2.6 (2.2–3.0) Tg yr^{-1}, 248.4 (137.4–359.4) Gg yr^{-1}, 66.2 (49.2–83.1) Gg yr^{-1} and 64.0 (48.2–79.7) Gg yr^{-1} for CO, CH_4, NO_x and NMVOCs. The states with large share of BTKs contributed most to BC emissions while regions having clamp kilns emitted higher OC, CO and CH_4.

Keywords Brick kilns · Short-lived pollutants · Aerosols · Ozone precursors

---

K. Tibrewal · C. Venkataraman
Interdisciplinary Program (IDP) in Climate Studies, Indian Institute of Technology Bombay, Mumbai 400076, Maharashtra, India
e-mail: kushalt@iitb.ac.in

S. Maithel
Greentech Knowledge Solutions Pvt. Ltd., 197, Indraprastha Apartment, Pocket 3, Sector 12 Dwarka, New Delhi 110078, India
e-mail: sameer@gkspl.in

C. Venkataraman (✉)
Department of Chemical Engineering, Indian Institute of Technology Bombay, Mumbai 400076, Maharashtra, India
e-mail: chandra@iitb.ac.in

C. Venkataraman et al. (eds.), *Climate Change Signals and Response*,
https://doi.org/10.1007/978-981-13-0280-0_13

1 Realizing the Role of Brick Sector in Mitigation of Short-Lived Climate Pollutants

Anthropogenic activities induce changes in composition of atmospheric composition through emissions of long-lived greenhouse gases (LLGHGs) and short-lived climate pollutants (SLCPs) which influence global climate change (Stocker et al. 2013). Although CO_2 is considered to be the most important contributor of climate change (Stocker et al. 2013), other forcing elements, like short-lived climate pollutants (SLCPs), also contribute significantly to climate change (UNEP 2017). SLCPs comprise of black carbon (BC) and organic carbon (OC) (particulate matter constituents) and SO_2 (particulate matter precursor), tropospheric ozone (O_3) and its precursor gases (CH_4, CO, NO_x and NMVOCs) (CCAC 2014). These pollutants are characterized by short atmospheric lifetimes varying from a few days to a few decades contributing to impacts that last for several decades (10–20 years) because of their radiative effects.

Besides influencing global climate change, SLCPs also play a large role in degrading regional air quality. Rising levels of particulate matter have led to increasing mortality rates (Cohen et al. 2017). In addition to this, modelling studies have shown the potential of increasing levels of tropospheric ozone (Ghude et al. 2014) and BC emissions (Burney and Ramanathan 2014) in reducing crop yield. Thus, there is growing consensus on the potential of mitigation strategies targeting SLCPs to alleviate near-term climate change with a potential to reduce warming due to their rapid temperature response (UNEP/WMO 2011) and simultaneously improve air quality (Shindell et al. 2012). Moreover, adoption of policies for improving air quality focuses on reducing NO_x and SO_2 emissions, and reduces the masking of GHG warming by such cooling SLCPs, thus motivating the need for addressing mitigation of warming SLCPs like BC and CH_4 (UNEP 2017).

For India, energy-intensive activities involving traditional combustion practices and extensive use of biomass fuel such as residential cooking, brick manufacturing and open burning of agricultural residue are major emitters of warming SLCPs (Venkataraman et al. 2016). The brick industry in India is the second largest producer of bricks with an annual production of approximately 250 billion bricks and consuming large amount of coal and biomass fuel (Maithel et al. 2012). Use of traditional and inefficient technologies emitting particulate matter along with CO and SO_2 are major causes of degraded air quality around major urban centres of India (Guttikunda and Calori 2013; Kumbhar et al. 2014). Several studies have recognized the potential of brick industry in reducing emissions through shifts towards efficient technology (USEPA 2012; Weyant et al. 2014). Thus, careful accounting of brick production activity, dominant technologies, diversity of fuel mixes and resulting emissions over India is needed to harness the potential of brick industry in mitigating SLCPs.

There are few available emission estimates from Indian brick industries, addressing local air pollution (Guttikunda and Calori 2013; Kumbhar et al. 2014) or made at a gross national level (Pandey et al. 2014) focussing on coal as a single fuel. More recently, the carbon footprint of brick kilns in a region of North India showed an average value of 427.985 kg CO_2/1000 bricks for fixed chimney Bull's trench kilns (FCBTKs) (Maheshwari and Jain 2017). However, there are significant differences in brick-making technology among different parts of the country (Maithel et al. 2012) which include Bull's trench kilns (BTKs) and Zig Zag fired kilns dominating the Northern India while clamp kilns contributing more in the southern states. It is also known that fuel used in brick kilns varies by region and can include mixtures of coal and biomass fuels such as firewood, dry dung, rice husks and other agricultural residues (Maithel et al. 2012). Therefore, it is important to account for regionally varying technologies and the widespread use of biomass in brick kilns.

To address these gaps, the chapter provides a methodology to estimate emissions from brick kilns at a state level in India. The methodology is used to estimate a national brick demand and distribute it among all states. The existing knowledge on regional spread of prevalent technology is utilized to quantify the fractional share of different brick kiln technologies in each state. Average values of emission factors and heating values are calculated from different studies to estimate the final emissions from combustion of a fuel mix (coal and biomass) per kiln type.

2 Estimating Emissions at State Level

This section summarizes the underlying methodology followed to estimate the state-level emissions of SLCP aerosols (PM2.5, BC, OC and SO_2) and ozone precursors (CH_4, CO, NO_x and NMVOCs) from manufacturing of fired-clay bricks in India. The approach for calculating emissions (Eq. 1) begun by estimating the national brick demand 'B' (millions/year). The national demand is distributed to each state 's' using brick utilization fraction 'u_s' (%) to estimate the production at state level assuming production would follow demand. For each state, the production is apportioned into various kiln technologies 'k' using the kiln share '$t_{k,s}$' (%). The amount of brick produced is converted to mass of the bricks produced using a typical weight 'w' (kg) of one block of fired-clay brick. Subsequently for each kiln technology, corresponding specific energy consumption 'SEC_k' (MJ/kg) and fuel characteristics [assumed fuel mix and calorific value of the fuel mix, 'CV' (MJ/kg)] are used to estimate the fuel consumption. Finally, for each pollutant 'p', technology-linked emission factors '$\mathrm{EF}_{p,k}$' (g/kg) are multiplied with the fuel consumption to arrive at the emissions '$E_{p,s}$' (Gg/year). The overall methodology is represented in Fig. 1.

$$E_{p,s} = \sum_k \frac{B \times u_s \times \mathrm{SEC}_k \times t_{k,s} \times w}{\mathrm{CV}} \times \mathrm{EF}_{p,k} \tag{1}$$

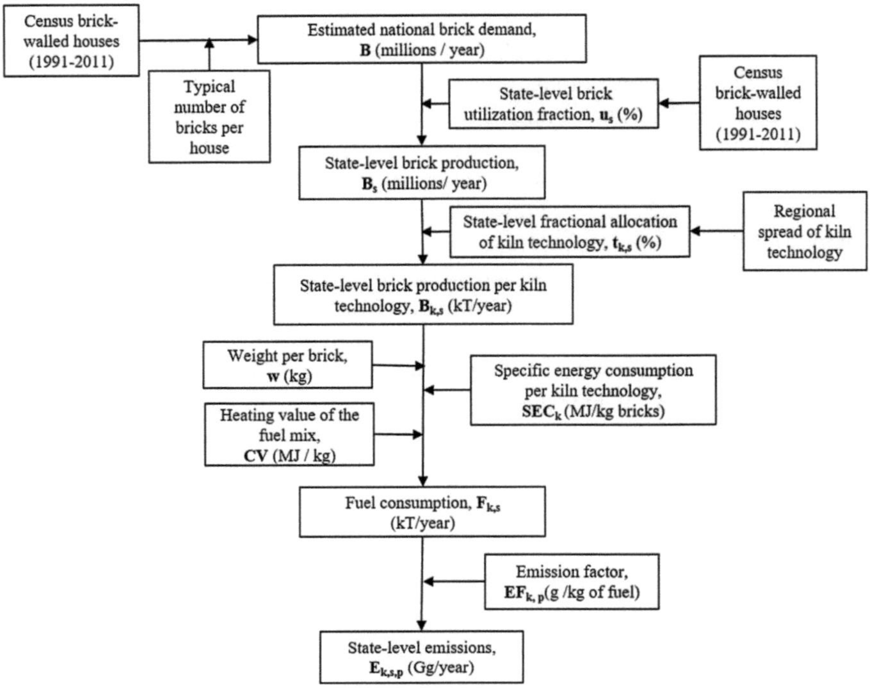

Fig. 1 Methodology to estimate state-level emissions from brick manufacturing

2.1 *Activity Data*

2.1.1 Estimated National Brick Demand

The brick industry in India is an unorganized sector. Unavailability of accurate data on number of kilns and their locations makes it difficult to know the exact brick production and its regional difference. Thus, an attempt is made to estimate annual demand of burnt clay bricks and distribute among states to estimate state-level production, assuming demand is equal to production because of transportation barriers in supply of raw materials and finished products. Demand in brick sector is driven by need for housing due to growth of population and urbanization and strong growth in construction sector (KASPL 2015). Since exact share of bricks for various purposes is not available, the demand in brick is distributed as 50% for new constructions, 10% for bricks used in repair works. As 40% of the brick production is in the form of damaged or low-quality bricks (GKSPL 2014), the total production of bricks to meet the demand is considered as 1.6 times the demand. Activities requiring brick under new constructions are assumed to have similar share as cement. With this assumption, out of total bricks used in new constructions, 67% is used for housing, 13% for infrastructure, 11% for commercial construction and 9% for industrial construction (KASPL 2015).

National demand per year of fired-clay brick for housing is estimated using trend in the number of brick-walled houses as reported in census of India for 1991, 2001 and 2011 (Census of India 2011). The following approach is applied to estimate the probable number of brick-walled houses constructed in 2015. Data for state-wise total number of brick-walled houses is collected from census for the years 1991, 2001 and 2011. An exponential trend-line is fitted for each state to analyse the growth in number of brick-walled houses up to 2015. The slope of the trend-line for any year gives the number of newly constructed brick-walled houses in that year. Besides covering the trend well, the use of exponential trend-line is favoured to fit the data as the slope of the trend-line at each year is representative of the fact that with increasing growth in national population, the growth in newly constructed houses must also increase to cater to the increasing housing demand. The total bricks required for newly constructed houses are calculated assuming an average built-up area of 1000 ft.^{2} per house with specific brick consumption of 13 bricks per sq. ft. of built-up area (Happho 2017).

2.1.2 State-Level Brick Utilization Fraction

The aim here is to distribute the national brick demand among the states. Since majority of demand is driven by housing, therefore a proxy, in form of brick utilization fraction, is developed to distribute the estimated national brick demand among the states proportionally to number the newly constructed brick-walled houses in each state. Brick utilization fraction for each state is the ratio of number of brick-walled houses constructed in each state to the total number of brick-walled houses constructed in the country for that year. For estimating brick utilization fraction 'u_s' (Eq. 2), the number of brick-walled houses constructed in 2015 for each state 'H_s' and for the whole nation 'H' is calculated using trend from census as discussed in previous section.

$$u_s = \frac{H_s}{H} \tag{2}$$

The brick production for each state 'B_s' is calculated by multiplying the brick utilization factor for each state to the national estimated brick demand 'B' (Eq. 3).

$$B_s = u_s \times B \tag{3}$$

2.1.3 Fractional Allocation of Kiln Technology for Each State

There are mainly four types of technologies currently used in manufacturing fired-clay bricks. Maithel et al. (2012) provide an estimate of the contribution of each of the kiln technology in the national production and their regional spread. BTK contributes 66% to national production spread all over the country covering states in the Indo-Gangetic plains (IGP), north, east, south and west. Clamp kilns with its 25% share cover predominantly central, western and southern states. Zig Zag fired kilns contributes 8% to the national production with production happening mostly in West Bengal and northern states. Finally, vertical shaft brick kilns (VSBK) contributing a meagre 1% are in Central India and with few others in Odisha and Kerala. This information is used to apportion the brick production into different kiln technologies for each state.

Using the regional spread as reported, the first step is to identify the states covered by each kiln technology. The respective national contribution of number of bricks by VSBK, Zig Zag fired kilns and clamp kilns is distributed among the states covered by each of them in proportion of the brick utilization fraction for those states. This resulted in the number of brick produced by VSBK, Zig Zag fired and clamp kilns, respectively, in each state. For bricks produced by BTKs in each state, the summation of bricks produced by these three kiln types is subtracted from the total brick amount for the state. Finally, the kiln share '$t_{k,s}$' for each kiln type 'k' is calculated (Eq. 4) as the ratio of number of brick produced by a kiln type in a state '$B_{k,s}$' to the total brick produced in the state 'B_s'.

$$t_{k,s} = \frac{B_{k,s}}{B_s} \tag{4}$$

2.2 Technology-Linked Emission Factor

Emissions of eight pollutants are estimated by using mean emission factors for each kiln type assuming a fixed fuel mix. Emission factors vary for different kiln technologies and depend on the fuel mix used. BTK and clamp kilns are the most polluting technologies (Maithel et al. 2012). Clamp kilns have the highest emission factors for particulate matter followed by BTKs due to intermittent feeding of fuel. Zig Zag fired kilns, although similar to BTKs in structural designs, has lower BC emissions due to better firing practice and continuous feeding of fuel. Zig Zag fired kilns also have the lowest emissions for CO indicating most efficient combustion of the fuel. The estimated emission factors suggest that VSBK are the most eco-friendly technologies with the least emissions of particulate matter (PM2.5) and black carbon (BC) due to steady-state combustion conditions and use of internal fuel. Emissions of SO_2 primarily depend on the sulphur content of the fuel. Thus, kilns using large amounts of coal-based fuels emit larger SO_2 as compared to kilns using biofuels.

Mean emission factors for each kiln type are calculated from measured emission factors for various fuel mixes from several studies for each pollutant (Appendix). Measured emission factors are used for particulate matter (PM2.5, BC and OC) from 11 Indian brick kilns with varying fuel mixes (Weyant et al. 2014). Another study based on same brick kilns is used to obtain emission factors for SO_2 (Rajarathnam et al. 2014). Emission factors for CO are also based from (Weyant et al. 2014) for BTK and VSBK while for Zig Zag fired and clamp kilns, they are obtained from two studies (Weyant et al. 2014; Stockwell et al. 2016). Stockwell et al. (2016) also reported emission factors for CH_4 and NO_x Zig Zag fired kilns and clamp kilns. Since no measured emission factors are found for BTK and VSBK, values for Zig Zag fired kilns are assumed to be same for them. NMVOCs emission factors are based on values reported for biofuels burnt in brick kilns (Christian et al. 2010) which are assumed to be same for all fuel mixes and all kiln type.

2.3 Uncertainty Estimation

Uncertainties in emissions are calculated analytically, assuming the underlying uncertainties in all input quantities to be normally distributed. Key sources of uncertainties include brick production number, specific energy consumption per kiln type and emission factors which vary with kiln technology and fuel mix. Uncertainty in bricks production amount is calculated as the standard deviation in the estimate of annually built brick-walled houses. Standard deviation in the estimates of brick-walled houses is calculated using the errors obtained in the coefficients of regression for the trend in brick-walled houses from 1991 to 2011. Uncertainty in specific energy consumption is taken as reported in (Weyant et al. 2014). Mean and standard deviation in emission factors for each pollutant per kiln technology are calculated from values obtained from multiple studies, as discussed in the previous section. Final uncertainties in emissions are calculated by combining uncertainties in individual parameters using rule of quadrature. For emissions having relative uncertainties (ratio of standard deviation to arithmetic mean) greater than 30%, lognormal distribution is assumed and the emissions are reported as means with upper and lower bounds calculated using geometric standard deviation.

3 Technology-Specific Brick Production Per State

3.1 National Brick Demand

From the slope of the trend of brick-walled houses from 1991 to 2011 (Census of India 2011) for the year 2015, the total number of newly constructed brick-walled houses is estimated to be 6.2 ± 0.5 million. The regression coefficients and the slope of the exponential trend at year 2015 are presented in Appendix. Assuming an average built-up area of 1000 $ft.^2$ per house, with specific brick requirement of 13 bricks per sq. ft. of built-up area, the national brick demand for housing is estimated to be 82.2 ± 23.8 billion. Since it is 67% of the demand under new constructions, total brick demand for new constructions (including housing, infrastructure, industrial and commercial construction) is 122.7 ± 35.6 billion. Finally, with an assumption that 50% of the total national brick demand is for new constructions while the rest is used for repair work and to account for the damaged and low-quality bricks, the total national brick demand is estimated to be 245.4 ± 71.2 billion.

3.2 State-Level Brick Production

Assuming demand to be equal to production, the national brick demand is distributed among the states using the brick utilization fraction (Fig. 2). The states covering the Indo-Gangetic plains (IGP) accounted for the most brick utilization with nearly 44% of the total national demand. Uttar Pradesh, Bihar and West Bengal contribute 15%, 8% and 8%, respectively, to national demand producing approximately 37.9 ± 0.8, 20.4 ± 2.6 and 20.3 ± 3.0 billion bricks annually. Although Delhi as the national capital territory (NCT) is found to contribute only 2%, considering national capital region (NCR—includes NCT and parts of Haryana, Rajasthan and UP) the contribution is 9% highlighting the rapid urbanization that underwent in the recent years in that region. Gujarat, Rajasthan, Madhya Pradesh and Maharashtra are the major contributors in the Western and Central India contributing 6%, 5%, 3% and 8%, respectively. Southern India contributed nearly 14% with AP and Tamil Nadu being the major states producing 17.2 ± 10 and 9.4 ± 8.2 billion bricks, respectively. The brick production at the state level estimated using this methodology is presented in Appendix.

3.3 Fractional Allocation of Kiln Technology for Each State

Even though brick industry is such an energy-intensive sector with significant share in energy consumption and driven by heavy demand, there has been very few initiatives to promote energy efficiency and emission control. It is not until 1990s when the first set of emission standards were proposed for brick industry. The standards laid down the maximum limit for concentration of particulate matter (750–1000 mg/Nm^3) and minimum stack height for optimal dispersion of sulphur dioxide (12–30 m) for different kiln capacities. The introduction of emission standards in 1996 led to the shift from moving chimney BTKs to fixed chimney BTKs and encouraged adoption of newer technologies (Maithel and Uma 2000). Despite introduction of advanced technologies as early as 1970s for Zig Zag fired kilns and 1995 for VSBK, there is no large-scale implementation (Maithel et al. 2012). Recently, due to increasing pollution levels, the Central Pollution Control Board (CPCB) directed all State Pollution Control Boards to provide status on conversion of natural draft to induced draft brick kilns with rectangular shape and Zig Zag setting (CPCB 2017).

The estimated share of different kiln technologies for each state is shown in Fig. 2. The states in the IGP and north-eastern parts of the country are predominant

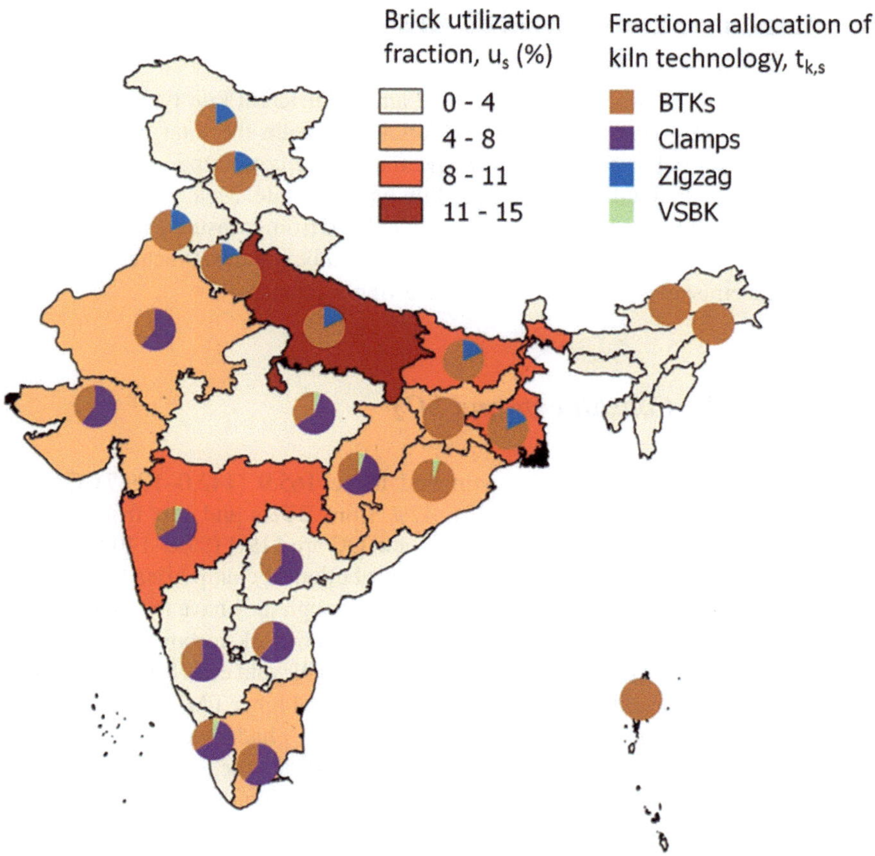

Fig. 2 State-level brick utilization fraction (u_s) and fractional allocation of kiln technology ($t_{k,s}$)

users of BTKs (greater than 80%) and Zig Zag fired kilns, while the central and the peninsular regions are dominated by clamp kilns (greater than 60%). VSBKs contributing to a meagre 1% to the national production are present mostly in Odisha, Madhya Pradesh and Kerala. However, understanding the actual prevalence of brick-making technology is highly important as it influences the energy consumption and emissions of various pollutants the most. Thus, a thorough exploration is required to extract the ground truth on the extent of various technologies under operation through field surveys, government listings or geo-spatial tagging through satellite imagery.

4 State-Level Emissions of SLCPs

The state-wise brick production distributed among different kiln type is multiplied with the mass of a typical brick, 2.9 kg/brick, to obtain the total mass of bricks produced per state. For each kiln type, specific energy consumption (Appendix) is used to estimate the total energy use per kiln type per state. Mean heating values of fuel mix (Appendix) for each kiln type are calculated from measured values (Weyant et al. 2014) to estimate the fuel mix consumed. Finally, the mean factors are used to estimate the emissions discussed in the following sections.

4.1 Aerosols: Constituents and SO_2

National emissions of $PM_{2.5}$ are estimated to be 165.9 (142.0–189.7) Gg yr^{-1} (Table 1), with 50% from clamp kilns, 47% from BTKs and rest from Zig Zag fired kilns and VSBK. The emissions of BC and OC are 119.1 (97.6–140.5) Gg yr^{-1} and 9.4 (7.3–11.4) Gg yr^{-1} (Table 1). The contribution of clamp kilns (44%) to BC emissions is lower than that of BTKs (55%) even though they have the highest emission factors for $PM_{2.5}$ as they are found to have lower black carbon-to-total carbon ratio (BC/TC) than BTKs due to batch firing process (Weyant et al. 2014). Therefore, states such as Maharashtra and Andhra Pradesh have higher $PM_{2.5}$ emissions with lower BC emissions than Uttar Pradesh even though they had lower brick production as they are dominated by clamp kilns (Fig. 3). Emissions for SO_2 are estimated to be 393.6 (314.1–473.1) Gg yr^{-1} (Table 1). Since emissions for SO_2 are primarily dependent on the sulphur content of fuel used, thus states using predominantly coal would result in greater SO_2 emissions as compared to those using biomass fuels. However, in this study emission factors are kiln specific assuming a common fuel mix for each kiln type for all states. Thus, more accurate spatial variation can be generated by collecting information of various fuel mixes used across different regions and using state-specific emission factor for each kiln type based on fuel mix.

The emission estimates are compared well with previously published study (Pandey et al. 2014) for $PM_{2.5}$, BC and OC (164, 114 and 11 Gg yr^{-1}, respectively). The SO_2 estimates (357 Gg yr^{-1}) are lower than this study due to different assumptions in emission factors and fuel used. Comparing the emissions with the total national emissions from India (Pandey et al. 2014; Sadavarte and Venkataraman 2014), brick industry contributed approximately 10% to the total BC emissions. Several studies have asserted replacing traditional brick kilns with newer technologies, particularly VSBK, will help in reducing BC emissions (Rajarathnam et al. 2014; Weyant et al. 2014). There have been limited efforts to motivate use of efficient and eco-friendly technologies in brick industry. The emission standards proposed in 1996 were revised in 2009 to include limitations by type of kiln technology (1200 mg/Nm^3 for the downdraft kiln; 750–1000 mg/Nm^3 for BTKs; and 250 mg/Nm^3 for the induced draft BTK, Hoffmann and vertical shaft kiln) (MoEFCC 2009) but were not

Table 1 State-level emissions of SLCPs from brick kilns in India (Gg yr^{-1})

States	Aerosols and SO_2				Ozone precursors			
	$PM_{2.5}$	BC	OC	SO_2	CO	CH_4	NO_x	NMVOC
Jammu and Kashmir	2.9	2.2	0.1	7.9	33.4	0.1	1.3	1.3
	(1.1–6.3)	(0.9–4.6)	(0.1–0.2)	(4.1–13.7)	(14.2–67.2)	(0.0–0.3)	(0.2–4.3)	(0.2–4.1)
Himachal Pradesh	1.3	1.0	0.1	3.6	15.2	0.0	0.6	0.6
	(0.5–2.9)	(0.4–2.2)	(0.0–0.1)	(1.8–6.6)	(6.1–31.7)	(0.0–0.1)	(0.1–2.0)	(0.1–1.9)
Punjab	2.8	2.1	0.1	7.6	32.1	0.1	1.3	1.2
	(0.8–7.5)	(0.6–5.7)	(0.0–0.3)	(2.4–18.4)	(8.9–83.2)	(0.0–0.3)	(0.2–4.5)	(0.2–4.3)
Chandigarh	0.1	0.1	0.0	0.3	1.1	0.0	0.0	0.0
	(0.0–0.3)	(0.0–0.2)	(0.0–0.0)	(0.0–0.8)	(0.2–3.5)	(0.0–0.0)	(0.0–0.2)	(0.0–0.2)
Uttarakhand	3.1	2.3	0.1	8.2	34.8	0.1	1.4	1.3
	(1.2–6.5)	(0.9–4.7)	(0.1–0.2)	(4.3–14.3)	(14.8–69.9)	(0.0–0.3)	(0.2–4.5)	(0.2–4.3)
Haryana	3.6	2.7	0.1	9.6	40.6	0.1	1.6	1.5
	(1.4–7.7)	(1.1–5.6)	(0.1–0.2)	(4.9–17.0)	(16.9–82.7)	(0.0–0.3)	(0.3–5.3)	(0.3–5.0)
Delhi	2.0	1.6	0.1	5.6	23.5	0.1	0.8	0.8
	(0.0–11.8)	(0.0–9.8)	(0.0–0.4)	(0.1–33.3)	(0.5–139.9)	(0.0–0.3)	(0.0–5.0)	(0.0–4.8)
Uttar Pradesh	17.0	12.8	0.6	45.6	192.5	0.5	7.6	7.3
	(6.6–36.1)	(5.2–26.3)	(0.3–1.2)	(23.9–79.3)	(81.7–387.3)	(0.1–1.6)	(1.3–24.9)	(1.3–23.8)
Bihar	9.1	6.9	0.3	24.5	103.5	0.3	4.1	3.9
	(3.5–19.7)	(2.7–14.4)	(0.2–0.6)	(12.3–44.0)	(42.5–212.8)	(0.0–0.9)	(0.7–13.5)	(0.7–12.9)
West Bengal	9.1	6.8	0.3	24.4	103.1	0.3	4.1	3.9
	(3.4–19.8)	(2.7–14.5)	(0.2–0.6)	(12.0–44.3)	(41.8–213.8)	(0.0–0.9)	(0.7–13.5)	(0.7–12.8)
Jharkhand	6.0	5.0	0.2	17.1	71.5	0.2	2.5	2.4
	(2.1-13.6)	(2.0-10.4)	(0.1-0.4)	(8.5-30.7)	(28.5-150.1)	(0.0-0.6)	(0.3-9.3)	(0.3-8.8)
Odisha	4.8	4.0	0.2	13.5	59.4	0.1	2.1	2.0
	(1.7–10.9)	(1.6–8.3)	(0.1–0.3)	(6.6–24.7)	(24.2–122.7)	(0.0–0.5)	(0.3–7.4)	(0.3–7.1)
Chhattisgarh	9.5	6.4	0.7	20.3	180.9	23.8	3.6	3.5
	(4.3–18.3)	(1.6–17.8)	(0.2–1.8)	(3.4–67.4)	(91.7–321.8)	(2.7–92.6)	(0.6–12.3)	(0.6–11.4)
Madhya Pradesh	8.3	5.6	0.6	17.7	157.8	20.7	3.1	3.0
	(3.8–16.0)	(1.4–15.5)	(0.1–1.6)	(3.0–58.8)	(80.0–280.9)	(2.4–80.8)	(0.5–10.7)	(0.5–10.0)
Rajasthan	12.4	8.5	0.8	26.7	231.7	30.5	4.6	4.5
	(5.5–24.2)	(2.1–23.2)	(0.2–2.4)	(4.6–87.6)	(113.7–422.1)	(3.5–119.2)	(0.7–15.9)	(0.8–14.8)
Gujarat	14.9	10.2	1.0	32.0	277.8	36.6	5.5	5.4
	(6.8–28.5)	(2.6–27.7)	(0.3–2.8)	(5.6–104.6)	(140.6–495.2)	(4.2–142.5)	(0.8–19.0)	(0.9–17.6)
Maharashtra	19.7	13.3	1.4	41.8	373.3	49.1	7.4	7.2
	(6.8–45.0)	(2.9–39.6)	(0.3–4.1)	(6.4–145.3)	(136.3–827.4)	(5.0–198.8)	(1.0–26.4)	(1.1–24.7)
Andhra Pradesh	17.2	11.8	1.2	37.0	321.1	42.3	6.3	6.2
	(4.9–44.4)	(2.3–36.7)	(0.2–3.8)	(5.3–131.7)	(91.0–825.4)	(3.9–177.5)	(0.8–23.7)	(0.9–22.3)
Karnataka	7.3	5.0	0.5	15.7	136.8	18.0	2.7	2.6
	(1.5–22.4)	(0.8–17.3)	(0.1–1.8)	(1.9–60.1)	(26.7–425.5)	(1.4–80.2)	(0.3–10.7)	(0.3–10.2)
Kerala	1.0	0.7	0.1	2.1	19.0	2.5	0.4	0.4
	(0.0–6.8)	(0.0–4.6)	(0.0–0.5)	(0.0–14.5)	(0.1–129.5)	(0.0–17.0)	(0.0–2.5)	(0.0–2.5)
Tamil Nadu	9.4	6.4	0.6	20.2	175.2	23.1	3.5	3.4
	(1.8–29.6)	(0.9–22.7)	(0.1–2.4)	(2.3–78.1)	(31.8–563.6)	(1.7–104.1)	(0.4–14.0)	(0.4–13.2)
North-east	3.7	3.1	0.1	10.5	44.1	0.1	1.5	1.5

(continued)

Table 1 (continued)

States	Aerosols and SO_2				Ozone precursors			
	$PM_{2.5}$	BC	OC	SO_2	CO	CH_4	NO_x	NMVOC
	(1.7–7.0)	(1.6–5.5)	(0.1–0.2)	(6.3–16.6)	(22.3–78.5)	(0.0–0.3)	(0.3–4.6)	(0.3–4.4)
Union territories	0.6	0.5	0.0	1.8	7.5	0.0	0.3	0.3
	(0.1–2.1)	(0.1–1.8)	(0.0–0.2)	(0.7–3.8)	(4.5–11.8)	(0.0–0.1)	(0.0–1.4)	(0.0–1.4)
Total	165.9	119.1	9.4	393.6	2635.7	248.4	66.2	64.0
	(142.0–189.7)	(97.6–140.5)	(7.3–11.4)	(314.1–473.1)	(2239.7–3031.6)	(137.4–359.4)	(49.2–83.1)	(48.2–79.7)

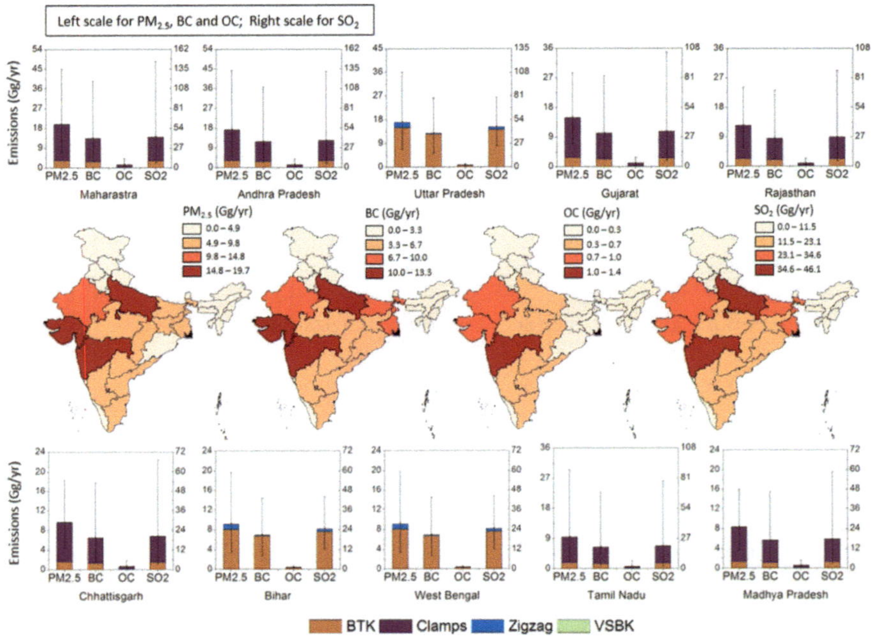

Fig. 3 Regional distribution and major emitters of aerosols ($PM_{2.5}$, BC, OC) and SO_2 emissions from brick kilns (Gg/yr)

stringent enough to induce much change (Maithel et al. 2012). Draft of new emission standards has been proposed in 2015 (MoEFCC 2015) with stricter limits for BTKs. Other than regulating stack emissions, there were initiatives to promote resource-efficient bricks, such as hollow and perforated bricks, which consume less energy and resources for their productions (UNDP 2009), fly ash bricks and non-fired bricks such as autoclaved aerated concrete (AAC) blocks.

4.2 Ozone Precursors

National emissions of CO are estimated as 2.6 (2.2–3.0) Tg yr^{-1} with 63% from clamp kilns and 35% from BTKs (Table 1). Emissions of CO are an indicator of incomplete combustion of fuel as the carbon is not allowed to fully convert into CO_2. Among the kilns, Zig Zag fired kilns are found to have the lowest CO emission factors as the firing pattern allows large amount of time for the fuel to get burnt and clamps have the highest. Even the highly efficient technology such as VSBK has greater CO emission factors than Zig Zag fired kilns due to use of internal fuel as limited supply of air is available at the surface of the fuel for complete combustion. CH_4 is primarily emitted due to use of biomass fuel. Clamp kilns are assumed to be operated mostly using larger share of biomass in the fuel mix. Thus, the national emissions of 248.4 (137.4–359.4) Gg yr^{-1} are contributed 99% by clamp kilns (Table 1). Since clamp kilns dominated emissions of both CO and CH_4, western and southern states such as Maharashtra, Andhra Pradesh, Gujarat, Rajasthan and Tamil Nadu are among the top emitters (Fig. 4). Emissions of NO_x and NMVOCs are estimated to be 66.2 (49.2–83.1) Gg yr^{-1} and 64.0 (48.2–79.7) Gg yr^{-1} (Table 1).

When compared with the emissions from previous study (Pandey et al. 2014), emissions of CO (2.6 Tg yr^{-1}) are comparable to this study while CH_4 emissions (5 Gg yr^{-1}) are 50 times lower than this study. The reason for such a significant difference can be attributed to assumption in the fuel mix. Coal was assumed to be

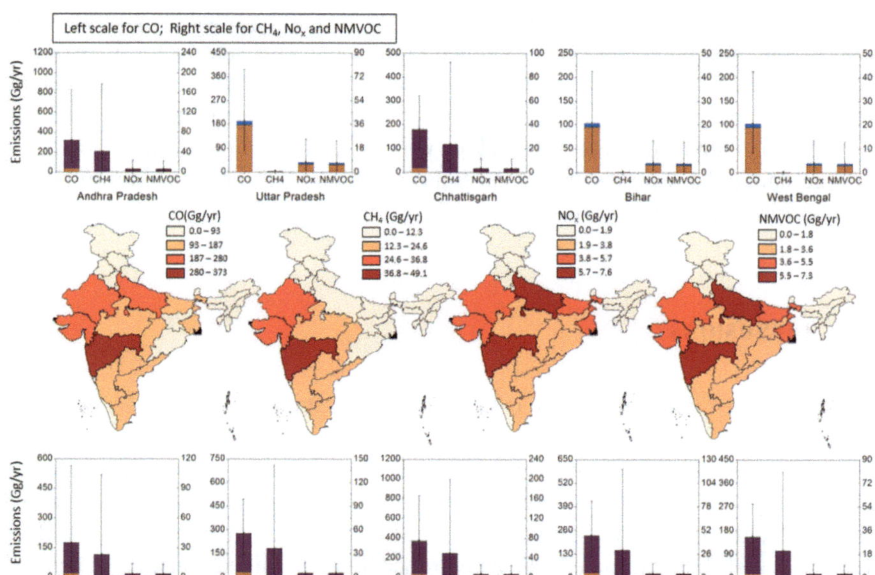

Fig. 4 Regional distribution and major emitters of CO, CH_4, NO_x and NMVOC emissions from brick kilns (Gg/yr)

the only fuel with very low emission factors in all the kilns (Pandey et al. 2014), whereas this study assumed a mean emission factor from a combination of different fuel mixes, particularly use of biomass fuel from clamp kilns. Emissions of NO_x (139 Gg yr^{-1}) are higher while of NMVOCs (8 Gg yr^{-1}) are lower than this study, which again can be attributed to different values of emission factors used.

4.3 Uncertainty in Emissions

Confidence bounds in emissions are calculated analytically from combination of uncertainties in brick production, specific energy consumption of kiln type and the respective emission factors. Uncertainties in brick production are represented as plus or minus one standard deviation. A lognormal distribution is assumed to represent uncertainties in final emissions for each state since the combined relative uncertainty (standard deviation/mean) is greater than 30% for almost all states. The standard deviation in the sum of emissions of a given pollutant from different technologies and for nation from different states is estimated as the sum of the individual uncertainties in quadrature.

Although the relative uncertainties for individual states are greater than 30%, the values for national total after combing all states are found to be 14% for $PM_{2.5}$, 18% for BC, 20% for SO_2, 15% for CO, 45% of CH_4 and 25% for NO_x and NMVOCs. Uncertainties are greater for CH_4, NO_x and NMVOCs due to large uncertainties in emission factors. Since reported emission factors had no reported uncertainty or are unavailable, 100% uncertainty is assumed in the mean emission factor. Reasons for uncertainties in other pollutants primarily arise due to lack of information on fuel mix used across different states. A single emission factor is considered for a particular kiln type averaged from measured values with different fuel mixes, thus giving rise to large standard deviations in the final emission factors. The upper and lower bounds in emissions for each state are presented in (Table 1).

5 Conclusions

A state-level emission inventory is developed to estimate the emissions of SLCPs from manufacturing of fired-clay bricks in India. A methodology is developed to estimate national brick demand, using trends in numbers of brick-walled houses, and assumptions in share of brick demand for other constructions (commercial, industrial and infrastructure), repair works and to account for damaged bricks. Brick demand is distributed to state level using trends in brick-walled houses at state level. Assuming that production would follow demand, it is found that Uttar Pradesh, Bihar, West Bengal and Maharashtra are the largest production centres for fired bricks. Using state-wise kiln technology data (with BTKs dominating in north and eastern states and clamp kilns dominating in western and southern states) and technology-linked emission factors averaged across fuel mixes (coal and biomass), emissions of different SLCPs are calculated.

Emissions of PM2.5, BC, OC and SO_2 are estimated to be 165.9 (142.0–189.7) Gg yr^{-1}, 119.1 (97.6–140.5) Gg yr^{-1}, 9.4 (7.3–11.4) Gg yr^{-1} and 393.6 (314.1–473.1) Gg yr^{-1}. For ozone precursors, the estimates are 2.6 (2.2–3.0) Tg yr^{-1}, 248.4 (137.4–359.4) Gg yr^{-1}, 66.2 (49.2–83.1) Gg yr^{-1} and 64.0 (48.2–79.7) Gg yr^{-1} for CO, CH_4, NO_x and NMVOCs. States with large share of BTKs (such as Uttar Pradesh, Bihar and West Bengal) contributed most to the BC emissions while regions having clamp kilns (Maharashtra, Gujarat and Rajasthan) emitted higher amounts of OC, CO and CH_4. Further improvements in inventory methodology will require survey-based data allowing estimation of brick production, along with better understanding of regional spread of various kiln technologies and fuel types, as well as field measurements of emission factors.

Appendix

Mean emission factors across different fuel mixes (g/kg of fuel mix)

	$PM_{2.5}$	BC	OC	SO_2	CO	CH_4	NO_x	NMVOC
BTK	3.3	2.7	0.1	9.3	38.9	0.1	1.4	1.3
Clamp kilns	3.7	2.4	0.3	7.4	74.7	11.2	1.3	1.3
Zig-Zag fired	2.0	0.3	0.1	3.1	14.3	0.1	1.4	1.3
VSBK	1.3	0.4	0.1	1.7	44.5	0.1	1.4	1.3

The values presented here are mean emission factors averaged over different fuel mixes from a particular kiln type. These are compiled from different studies Christian et al. 2010; Weyant et al. 2014; Rajarathnam et al. 2014; Stockwell et al. 2016

Regression equation and coefficients of the trend in brick-walled houses (1991–11)

$y = a \times \exp(b \times x)$

States	a	b	Slope of trend in the year 2015
Andaman and Nicobar	227.63	0.16	2124
Andhra Pradesh	5904848.59	0.03	435,651
Arunachal Pradesh	4299.61	0.08	2677
Assam	520339.8	0.06	139,315
Bihar	4550067.07	0.04	516,356
Chandigarh	166726.18	0.02	5345
Chhattisgarh	362278.92	0.08	245,034
Dadra and Nagar	10256.72	0.08	6632
Daman and Diu	12169.58	0.07	5421
Delhi	2135088.94	0.03	103,920
Goa	14795.87	−0.01	0

$y = a \times \exp(b \times x)$			
States	a	b	Slope of trend in the year 2015
Gujarat	4601356.95	0.03	376,987
Haryana	3011436.06	0.03	202,514
Himachal Pradesh	298337.37	0.06	75,914
Jammu and Kashmir	390776.35	0.07	166,838
Jharkhand	587839.55	0.08	316,568
Karnataka	3114619.37	0.03	185,559
Kerala	2261188.15	0.01	25,741
Lakshadweep	68.11	0.23	4592
Madhya Pradesh	3817165.79	0.03	213,803
Maharashtra	7148854.43	0.03	505,731
Manipur	19439.43	0.07	7176
Meghalaya	11911.74	0.09	9137
Mizoram	4496.32	0.08	2483
Nagaland	26132.27	0.05	5050
Odisha	1596779.42	0.05	261,614
Puducherry	91670.27	0.05	14,598
Punjab	3925453.5	0.02	159,985
Rajasthan	1724557.89	0.05	314,330
Sikkim	1577.46	0.16	11,867
Tamil Nadu	6408504.31	0.02	237,771
Tripura	32094.21	0.08	17,429
Uttar Pradesh	12603818.11	0.03	960,686
Uttarakhand	286329.81	0.08	173,434
West Bengal	4895899.75	0.04	514,708

State-level distribution of total brick production across different kiln types

States	BTKs	Clamp kilns	Zig-Zag fired	VSBK	State-level total production, B_s (million/year)
Andaman and Nicobar	84				84
Andhra Pradesh	6648	10,520			17,168
Arunachal Pradesh	105				105
Assam	5490				5490

States	BTKs	Clamp kilns	Zig-Zag fired	VSBK	State-level total production, B_s (million/year)
Bihar	16,696		3652		20,348
Chandigarh	173		38		211
Chhattisgarh	3259	5917		480	9656
Dadra and Nagar	261				261
Daman and Diu	214				214
Delhi	4095				4095
Goa					0
Gujarat	5753	9103			14,856
Haryana	6548		1432		7981
Himachal Pradesh	2455		537		2992
Jammu and Kashmir	5395		1180		6575
Jharkhand	12,475				12,475
Karnataka	2832	4481			7312
Kerala	342	622		50	1014
Lakshadweep	181				181
Madhya Pradesh	2844	5163		419	8425
Maharashtra	6727	12,212		991	19,930
Manipur	283				283
Meghalaya	360				360
Mizoram	98				98
Nagaland	199				199
Odisha	9797			513	10,310
Puducherry	575				575
Punjab	5173		1131		6305
Rajasthan	4797	7590			12,387
Sikkim	468				468
Tamil Nadu	3629	5741			9370
Tripura	687				687
Uttar Pradesh	31,064		6794		37,858
Uttarakhand	5608		1227		6835
West Bengal	16,643		3640		20,283
India	**161,958**	**61,348**	**19,631**	**2454**	**245,390**

Specific energy consumption per kiln type

Sp. energy consumption, SEC_k (MJ/kg brick)			
Kiln, k	Mean	SD	SD/mean
BTK	1.2	0.2	0.2
Clamp kilns	2.9	0.0	0.0
ZigZag fired	1.1	0.1	0.1
VSBK	0.7	0.3	0.4

Source Weyant (2014)

Heating value of fuel mix (MJ/kg)

Heating value of fuel mixes, CV (MJ/kg)			
Kiln	Mean	SD	SD/mean
BTK	23.8	4.9	0.2
Clamp kilns	23.3	8.4	0.4
ZigZag fired	20.4	0.3	0.0
VSBK	15.1	10.8	0.7

Source Weyant et al. (2014)

References

Burney J, Ramanathan V (2014) Recent climate and air pollution impacts on Indian agriculture. Proc Natl Acad Sci 111:16319–16324. https://doi.org/10.1073/pnas.1317275111

CCAC (2014) Climate and clean air coalition—annual report. Climate and Clean Air Coalition (CCAC), Paris

Census of India (2011) Census of India. http://www.censusindia.gov.in/. Accessed 15 Feb 2018

Christian TJ, Yokelson RJ, Cárdenas B et al (2010) Trace gas and particle emissions from domestic and industrial biofuel use and garbage burning in central Mexico. Atmos Chem Phys Discuss 9:10101–10152. https://doi.org/10.5194/acpd-9-10101-2009

Cohen AJ, Brauer M, Burnett R et al (2017) Estimates and 25-year trends of the global burden of disease attributable to ambient air pollution: an analysis of data from the Global Burden of Diseases Study 2015. Lancet 389:1907–1918. https://doi.org/10.1016/S0140-6736(17)30505-6

CPCB (2017) CPCB direction to State Pollution control boards. http://www.indiaenvironmentportal.org.in/files/file/direction_brick-kiln_27.06.17.pdf. Accessed 15 Feb 2018

Ghude SD, Jena C, Chate DM et al (2014) Reductions in India's crop yield due to ozone. Geophys Res Lett 41:5685–5691. https://doi.org/10.1002/2014GL060930

GKSPL (2014) Factsheets about brick kilns in South and South-East Asia. Greentech Knowledge Solutions Pvt Ltd., New Delhi, India

Guttikunda SK, Calori G (2013) A GIS based emissions inventory at 1 km × 1 km spatial resolution for air pollution analysis in Delhi, India. Atmos Environ 67:101–111. https://doi.org/10.1016/j.atmosenv.2012.10.040

Happho (2017) Quantities of building materials for 1000 Sqft house construction. http://www.happho.com/choose-building-materials-estimate-cost-quantities-house-construction/. Accessed 15 Feb 2018

KASPL (2015) Human resource and skills requirement in the construction material and building hardware sector (2013–17, 2017–22)

Kumbhar S, Kulkarni N, Rao AB, Rao B (2014) Environmental life cycle assessment of traditional bricks in western Maharashtra, India. Energy Procedia 54:260–269. https://doi.org/10.1016/j.egypro.2014.07.269
Maheshwari H, Jain K (2017) Carbon footprint of bricks production in fixed chimney bull's trench kilns in India. Indian J Sci Technol 10:1–11. https://doi.org/10.17485/ijst/2017/v10i16/112396
Maithel S, Uma R (2000) Environmental regulations and the Indian brick industry. Environ Pract 2:230–231
Maithel S, Lalchandani D, Malhotra G et al (2012) Brick kilns performance assessment, emissions measurements, & a roadmap for cleaner brick production in India. New Delhi
MoEFCC (2009) Press Release. http://www.moef.nic.in/legis/ep/543_E.pdf. Accessed 15 Feb 2018
MoEFCC (2015) Press Release. http://www.moef.nic.in/sites/default/files/BrickKilns.PDF. Accessed 15 Feb 2018
Pandey A, Sadavarte P, Rao AB, Venkataraman C (2014) Trends in multi-pollutant emissions from a technology-linked inventory for India: II. Residential, agricultural and informal industry sectors. Atmos Environ 99:341–352. https://doi.org/10.1016/j.atmosenv.2014.09.080
Rajarathnam U, Athalye V, Ragavan S et al (2014) Assessment of air pollutant emissions from brick kilns. Atmos Environ 98:549–553. https://doi.org/10.1016/j.atmosenv.2014.08.075
Sadavarte P, Venkataraman C (2014) Trends in multi-pollutant emissions from a technology-linked inventory for India: I. Industry and transport sectors. Atmos Environ 99:353–364. https://doi.org/10.1016/j.atmosenv.2014.09.081
Shindell D, Kuylenstierna JCI, Vignati E et al (2012) Simultaneously mitigating near-term climate change and improving human health and food security
Stocker TF, Qin D, Plattner GK et al (2013) IPCC, 2013: climate change 2013: the physical science basis. Contribution of Working Group I to the Fifth Assessment Report of the Intergovernmental Panel on Climate Change, 1535 pp
Stockwell CE, Christian TJ, Goetz JD et al (2016) Nepal ambient monitoring and source testing experiment (NAMaSTE): emissions of trace gases and light-absorbing carbon from wood and dung cooking fires, garbage and crop residue burning, brick kilns, and other sources. Atmos Chem Phys 16:11043–11081. https://doi.org/10.5194/acp-16-11043-2016
UNDP (2009) Energy efficiency improvements in the Indian Brick Industry. United Nations Development Program. www.resourceefficientbricks.org/
UNEP (2017) The emissions gap report 2017. United Nations Environment Programme (UNEP), Nairobi
UNEP/WMO (2011) Integrated assessment of black carbon and tropospheric ozone
USEPA (2012) Reducing black carbon emissions in South Asia: low cost opportunities
Venkataraman C, Ghosh S, Kandlikar M (2016) Breaking out of the box: India and climate action on short-lived climate pollutants. Environ Sci Technol 50:12527–12529. https://doi.org/10.1021/acs.est.6b05246
Weyant C (2014) Emissions from combustion technologies with a focus on brick making and in-home cooking. Thesis, Master of Science. University of Illinois, Urbana-Champaign
Weyant C, Athalye V, Ragavan S et al (2014) Emissions from South Asian Brick Production

Adaptation and its Socioeconomic Facilitators in the Marine Fishing Community of Maharashtra, India

Krishna Malakar and Trupti Mishra

Abstract Adaptation is crucial for countering the impacts of climatic and environmental changes, and sustaining livelihood. Communities adapt to various changes by intensifying their efforts to improve yields from their livelihood or diversifying into other forms of profession. Various socioeconomic factors facilitate adaptation in communities. However, facilitators of adaptation in fishing communities have seldom been studied. This study investigates adaptation and its facilitators in the marine fishing community of Maharashtra (in India) at two different time periods. The study assesses three important adaptation strategies in the community: motorization, mechanization, and diversification. It is found that mechanization and diversification among the community have increased over time. The association between adaptation and the proposed socioeconomic facilitators is examined through regression analysis. It is found that education, cooperative membership, and lower poverty influence different adaptation strategies. The regression estimates broadly lead to the acceptance of the proposed hypotheses, although different conclusions could be made for each of the adaptation strategies in both the time periods. Understanding these facilitators can assist in planning adaptation and developmental policies. The findings of the study can help design state-level adaptation programs for the marine fishing community.

Keywords Adaptation · Marine fishing · Facilitators · Motorization
Mechanization · Diversification · Driver · Maharashtra · India

K. Malakar · T. Mishra
Interdisciplinary Program (IDP) in Climate Studies, Indian Institute of Technology Bombay, Mumbai 400076, Maharashtra, India
e-mail: krishnamalakar26@gmail.com

T. Mishra (✉)
Shailesh J. Mehta School of Management, Indian Institute of Technology Bombay, Mumbai 400076, Maharashtra, India
e-mail: truptimishra@iitb.ac.in

C. Venkataraman et al. (eds.), *Climate Change Signals and Response*,
https://doi.org/10.1007/978-981-13-0280-0_14

1 Vulnerability and Need of Adaptation in Marine Fishing

Ensuring sustainability of natural resource-dependent livelihoods requires adaptation to various environmental changes. Currently, the oceans and its biodiversity are strained by climate change, pollution, destruction of natural marine habitats and others (Daw et al. 2009). Among these stresses, changes in climate can impact marine life tremendously by altering some of the physical and chemical properties of the sea, leading to changes in distribution and population of marine species (Roessig et al. 2004). This can substantially impact yields and livelihoods of the marine fishing sector. In South Asia, climate change is likely to decrease marine fishing yields by up to 20% (approximately) by 2050 (Barange et al. 2014). Studies (Vivekanandan 2011; Zacharia et al. 2016) have suggested that the marine fishing sector of India is also vulnerable to climate change, and hence needs to adapt to a changing fish catch at the earliest to sustain livelihoods.

Adaptation to a declining fish catch involves strategies which can increase yields. Using improved fishing gear and technology, that increase individual fish catch, can be said to be forms of private adaptation in the community. Various socioeconomic factors capacitate and facilitate communities to adapt. For instance, being educated might lead to better access to information and technical knowledge, which might help to make adaptation choices (Below et al. 2012). Similarly, having assets and finances can enable to acquire improved technologies and change livelihood strategies (Nhemachena et al. 2014). Examining the diverse socioeconomic facilitators of adaptation can contribute immensely to designing adaptation and developmental policies. Such assessments can help identify the important socioeconomic factors that need to be focused upon for developing community livelihoods and adaptation. Although marine fishing is in acute need of adaptation, the existing literature lacks studies on understanding private adaptation strategies of fishermen and their socioeconomic facilitators. Hence, this study aims to assess adaptation and its facilitators in marine fishing livelihoods in the state of Maharashtra in India.

Maharashtra is an important marine fishing state in India. Among the eleven coastal states in the Indian mainland, Maharashtra stands fifth in its fish landings (CMFRI 2014). But the state experienced 5.2 and 23% decline in its fish catch in 2014 and 2015, respectively compared to the past year (CMFRI 2013, 2014, 2015). It has one of the longest coastlines and comprises of 456 fishing villages (CMFRI 2010). Adaptation to reduced yields is crucial for sustaining the livelihoods, of 81,492 families in Maharashtra, dependent on marine fishing (CMFRI 2010). Thus, this study focuses on understanding adaptation in the marine fishing community of Maharashtra in India. The study examines adaptation and its facilitators at two different time periods 2005 and 2010 to analyze any changes undergone by the community over the years. Identifying the important facilitators of adaptation may help policy makers to initiate state-level actions to develop the sector. The approach of the study may also be useful to understand adaptation and its facilitators in the other marine fishing states of the country.

The following section describes some of the relevant literature, approach, and hypotheses of the study. Section 3 presents the data and methods used. Sections 4 and 5 discuss the results and conclusion, respectively.

2 Adaptation Strategies and Their Facilitators

Adaptation to climate change and other environmental impacts is crucial in marine fishing. There have been studies investigating the vulnerability (Islam et al. 2014; Metcalf et al. 2015) and adaptive capacity (Cinner et al. 2015) of coastal fishing communities, but private adaptation and their facilitators in these communities have not been examined extensively in the literature. Thus, this study aims to identify some of the facilitators of private adaptation in the Maharashtrian marine fishing community. The available literature on marine fishing and other natural resource-dependent sectors, such as agriculture, widely suggests that various socioeconomic factors can determine adaptation (Deressa et al. 2009; Islam et al. 2014; Metcalf et al. 2015). The IPCC (2001) [Intergovernmental Panel on Climate Change] also recognizes various socioeconomic characteristics such as finances, technology, skill, and infrastructure as important determinants of adaptive capacity. This study is an attempt to understand the association between some of these socioeconomic variables and implementation of adaptation strategies in marine fishing.

Education can be an important determinant of adaptive capacity and adaptation, as it might assist in taking informed adaptation decisions (Deressa et al. 2009). Education can improve access to information and technology (Tambo and Abdoulaye 2012). Thus, education might help intensify fishing efforts as well as diversify into other sources of livelihoods (Rahut et al. 2014). It has been considered as an important indicator of adaptive capacity and vulnerability in many studies (Deressa et al. 2009; Blythe et al. 2014; Islam et al. 2014; Metcalf et al. 2015). Blythe et al. (2014) considers education as an asset that facilitates adaptation, and the assessment further segregates education into primary and secondary levels. This study, too, includes different levels of education, namely primary, secondary, and above secondary to understand their differential influence on adaptation.

Finances and other forms of economic capital help invest in adaptation technologies and strategies. It is an intrinsic part of adaptive capacity and forms an important contributor to various composite indices representing vulnerability and adaptation such as the Social Vulnerability Index (Cutter et al. 2003) and Livelihood Vulnerability Index (Hahn et al. 2009). This study has also considered economics of the community to be a facilitator of adaptation. Studies on analyzing the determinants of adaptation in agricultural communities have found economic variables such as income (Deressa et al. 2009) and assets (Jain et al. 2015) to be significant for undertaking adaptation. Poverty has also been identified as an economic limit to adaptation in the literature (Brooks et al. 2005; Bryan et al. 2013). Thus, the economic health of the community is represented through the percentage of families below the poverty line in the present study. Other economic indicators, such as household income, could not be included in the study as data on these indicators is not available for the community at the village level.[1] Poverty can hamper intensification adaptation strategies that require financial investments. But it is often seen that poverty, an undesirable circumstance, can drive households to diversify their livelihoods to supplement their income (Agyeman et al. 2014).

Cooperative societies are important sources of information and credit among communities. They also help to enhance social networks among community members (Frank et al. 2011). Hence, they can play an important role in building adaptive capacity (Frank et al. 2011) by improving access to information about adaptive technologies/strategies as well as finances required for the same. In Maharashtra, cooperative societies provide credit, insurance of life and gear, and other assistance required for fishing activities, such as supply of ice, diesel, transport, and marketing (Nair et al. 2010). Membership in these fisheries cooperatives may help fishermen to improve their gear and fishing strategies and adapt to a lowering fish catch. These cooperatives can aid intensification of fishing strategies. But being cutoff from such cooperatives can prevent access to intensification opportunities and induce diversification. Thus, this study assesses the relation between the adoption of various adaptation strategies in the marine fishing community and the percentage of fisherfolk population having membership in fishing cooperative.

Adapting to a declining fish catch requires improved gear to increase yield. Motorized and mechanized boats can substantially improve fish catch (Reddy 2016). Hence, fishermen generally intensify their fishing activities through motorization and mechanization. Diversification is also an important strategy to supplement one's livelihood (Hussein and Nelson 1998). Hence, this study considers the percentage of motorized and mechanized boats (of all boats including non-motorized) and percentage of fishermen who have diversified as indicators of adaptation. The study attempts to assess the facilitators of implementation of these adaptation strategies.

From the above discussion, the present study hypothesizes the following:

H1: Education is positively associated with intensification, namely motorization and mechanization, as well as diversification.
H2: Poverty is negatively associated with intensification, namely motorization and mechanization, but positively with diversification.
H3: Membership in cooperative is positively associated with intensification, namely motorization and mechanization, but negatively with diversification.

3 Empirically Examining the Facilitators of Adaptation

The present study uses village-level data for all the variables on adaptation and socioeconomic facilitators. The required data for the study has been taken from the Marine Fisheries Census of Maharashtra 2005 and 2010 (CMFRI 2005, 2010). These

[1] This is a village-level study as described in 'Data and Method' section.

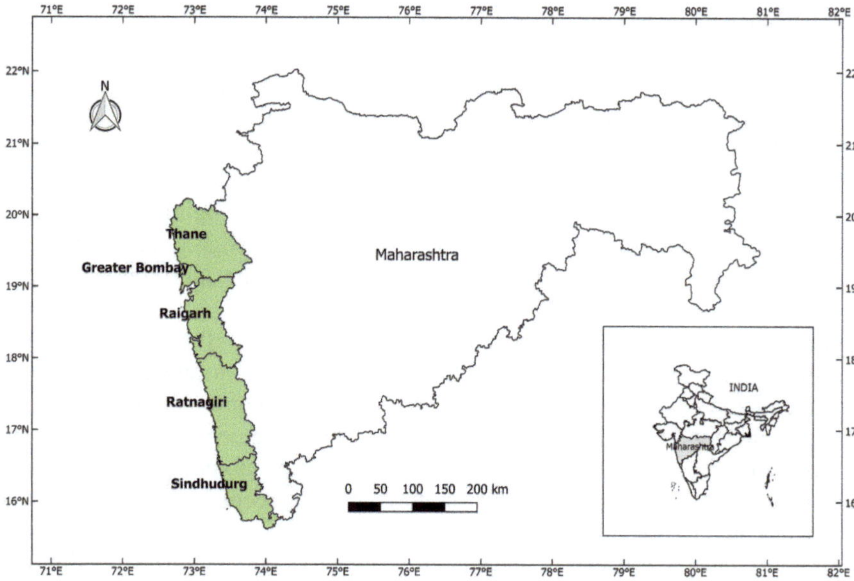

Fig. 1 Map of Maharashtra (coastal districts are highlighted)

are the latest censuses. The earlier census was conducted in 1980 (CMFRI 1981) and is not inclusive of many parameters covered in the 2005 and 2010 censuses. It also does not provide village-level information. The 2005 and 2010 Marine Fisheries Censuses provide district- and village-level demographic and fishing-related information of the community in the state. Maharashtra had five coastal districts in 2005 and 2010: Thane, Greater Mumbai, Raigad, Ratnagiri, and Sindhudurg.[2] This study aims to utilize the available micro-level regional data on the community, as it can help capture the variability in the state. Hence, the unit of analysis in the present study is village. In 2005, there were 406 fishing villages. The number increased to 456 in 2010. But because of inconsistency in the data, 71 and 97 villages have been deleted from the analysis for 2005 and 2010, respectively. Thus, the estimates of 2005 are based on 335 fishing villages, and findings of 2010 are from data on 359 villages. Figure 1 shows a map of Maharashtra and its coastal districts.

An overview of the data is obtained through its descriptive statistics, namely mean, minimum, maximum, and standard deviation. Further, an independent samples t-test is used to understand the differences in means of the variables in 2005 and 2010. This is done to make comparisons between the variables in 2005 and 2010. The test could not be done for assessing differences in higher education levels and poverty. This is because the 2005 census provides data on population having secondary and above secondary education, while the 2010 census reports population

[2]At present, Maharashtra has six coastal districts including Palghar, which was earlier a part of Thane district. It was formed in 2014.

having higher secondary and above higher secondary education. The 2005 census does not provide data on families below poverty line. Because of such inconsistencies, comparisons could not be made for these variables. Also, independent samples t-test is preferred over paired samples t-test as some of the villages included in the 2005 analysis are not the same for 2010. Additionally, Levene's test for equality of variances is conducted before the t-test. If the Levene's test (F) statistic is significant (p value < 0.05), equality in variances is not assumed. A nonsignificant statistic (p value > 0.05) indicates equality in variance. The t-statistic slightly differs with the assumption of equality and inequality in variance. Hence, the appropriate t-statistic is examined in accordance with results of Levene's test.

Fractional logit regression and corresponding marginal effects are calculated to estimate the relationship between the various facilitators and the percentage implementation of adaptation strategies. Fractional logit regression allows proportions/percentages as dependent variables. In this study, the dependent variables are the different adaptation strategies, that is, percentage of motorized boats, mechanized boats (of all boats including non-motorized), or percentage of fishermen who have diversified in the villages. The independent variables are the various possible facilitators of adaptation, that is, percentage of fisherfolk population having different levels of education (primary education in 2005 and 2010, till and above secondary education in case of 2005, and till and above higher secondary education in case of 2010), percentage of families below poverty line (not applicable for 2005 as data is not available) and percentage of fisherfolk population having membership in fishing cooperative. Fractional logit regression depicts the relationship between log of odds ratio and the independent variables. Hence, the marginal effects are also subsequently estimated.

4 Estimates of Adaptation and Its Facilitators

4.1 Adaptation in the Community

The percentages of implementation of different adaptation strategies have changed between the five years 2005–10 (Table 1). In 2005, a very low proportion (0.21%) of the population had diversified. In 2010, levels of diversification increased by almost hundred folds to 21.95%. In 2010, average percentage of motorized boats decreased, whereas percentage of mechanization increased. The t-test (Table 2) also shows that mechanization and diversification have significantly increased in 2010, whereas motorization has decreased. This indicates that mechanization and diversification are the preferred adaptation strategies in the recent time. Also, in both 2005 and 2010, mechanization is more than motorization.

Similar observations can be made from Fig. 2, which shows the scatter plot between percentages of motorized and mechanized boats in the villages of Maharashtra in 2005 and 2010. Motorized boats appear to have decreased massively in 2010,

Table 1 Summary statistics of variables

	Variable	Mean		Minimum		Maximum		Standard deviation	
		2005	2010	2005	2010	2005	2010	2005	2010
1.	Percentage of motorized boats	17.87	6.47	0.00	0.00	100.00	100.00	24.98	17.86
2.	Percentage of mechanization boats	33.30	43.75	0.00	0.00	100.00	100.00	36.64	38.68
3.	Percentage of fishermen who have diversified	0.21	21.95	0.00	0.00	1.00	100.00	0.25	29.49
4.	Percentage of fisherfolk population having primary education	27.59	28.37	0.00	0.00	73.55	83.98	13.89	14.49
5.	Percentage of fisherfolk population having education till secondary (2005) and higher secondary (2010)	31.68	34.84	2.22	0.00	63.04	81.35	14.62	15.47
6.	Percentage of fisherfolk population having education above secondary (2005) and higher secondary (2010)	7.50	4.07	0.00	0.00	35.71	46.43	6.25	5.01
7.	Percentage of families below poverty line[a]	NA	28.27	NA	0.00	NA	100.00	NA	26.48
8.	Percentage of fisherfolk population having membership in fishing cooperative	11.09	16.69	0.00	0.00	53.90	70.03	12.31	18.27

[a]Data not available for 2005; *NA* not available

indicated by the large cluster of points lying on the *y*-axis. These points represent villages having 0% motorized boats and various percentages of mechanized boats. Also, the 2010 plot has points mostly aligned toward the *y*-axis, that is, indicating greater mechanization, whereas plot of 2005 is more spread out, representing a mix of distribution between motorized and mechanized boats in the villages.

Table 2 Estimates of *t*-test

	Variable	Mean in 2005	Mean in 2010	*t*-statistic	*p* value	Levene's test for equality of variances
1.	Percentage of motorized boats	17.87	6.47	6.87***	0.000	Equal variances not assumed as $F = 70.276$, $p = 0.000$
2.	Percentage of mecha-nization boats	33.30	43.75	−3.65***	0.000	Equal variances assumed as $F = 3.034$, $p = 0.082$
3.	Percentage of fishermen who have diversified	0.21	21.95	−13.97***	0.000	Equal variances not assumed as $F = 551.777$, $p = 0.000$
4.	Percentage of fisherfolk population having primary education	27.59	28.37	−0.72	0.473	Equal variances assumed as $F = 0.055$, $p = 0.814$
5.	Percentage of fisherfolk population having membership in fishing cooperative	11.09	16.69	−4.76***	0.000	Equal variances not assumed as $F = 57.590$, $p = 0.000$

***Significant at 99% confidence level

4.2 Education, Poverty and Cooperative Membership

The average percentage of fisherfolk population who has obtained primary education increased in 2010 (Table 1), but it is not a statistically significant change (Table 2). Comparison of the higher levels of education is difficult for the two years. This is because the Marine Fisheries Census 2005 and 2010 provide data on different levels of education. The 2005 census presents data on the population having education till and above secondary level. On the other hand, the 2010 census documents data on population having education till and above higher secondary level. The 2005 census also does not provide data on the number of families below poverty line. In 2010, about 28.2% of the fisherfolk families were poverty stricken. Average membership in fishing cooperative increased significantly by 5.6% in 2010 (Tables 1 and 2).

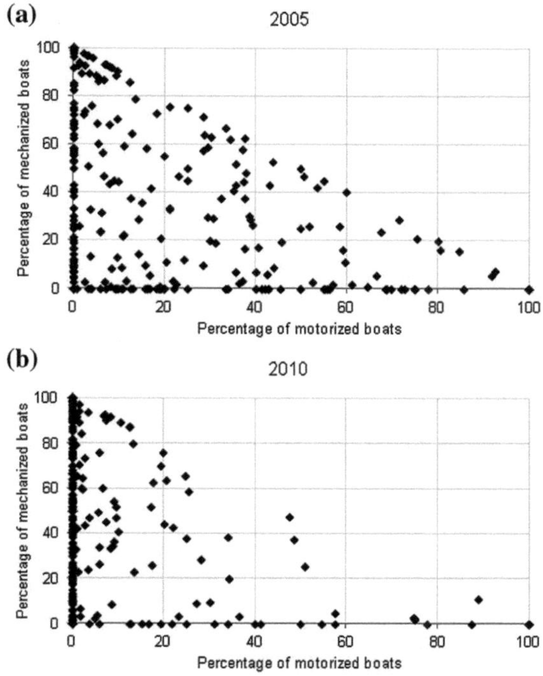

Fig. 2 Scatter plot showing percentage of motorized and mechanized boats in **a** 2005 and **b** 2010 [Authors' representation based on data from CMFRI (2005, 2010)]

Thus, the data implies that only one facilitator of adaptation, namely cooperative membership, improved over the five years in 2010.

4.3 Assessment of the Facilitators of Adaptation

The regression estimates for motorization indicate very different results for 2005 and 2010 (Table 3). The regression is significant at 99% confidence level and has a pseudo-R^2 value of 0.021 for 2005. On the other hand, the regression does not produce any significant result for 2010. This indicates that none of the independent variables can predict motorization, and the model is unable to explain the variability in the dependent variable. This might be a result of low percentage as well as randomness in adoption of motorization as a strategy in 2010. However, the estimates for 2005 indicate that secondary education having a marginal effect of 0.241 (Table 4) is important for adopting motorization.

In case of mechanization, the regression estimates for both the years are significant (Table 3). But, the model fit is slightly better for 2010 as compared to 2005. The pseudo-R^2 values are 0.032 and 0.067 in 2005 and 2010, respectively. The 2010 estimates reveal primary education to be a significant parameter but it has a negative coefficient signifying that it does not promote mechanization. Again, the

Table 3 Coefficient estimates of regression models in 2005 and 2010

	Motorized		Mechanized		Diversified	
	2005	2010	2005	2010	2005	2010
Percentage of fisherfolk population having primary education	−1.221	−1.334	0.401	−1.568**	0.151	−0.569
Percentage of fisherfolk population having education till secondary (2005) and higher secondary (2010)	1.677**	1.422	−1.212*	−1.115*	1.825***	2.217***
Percentage of fisherfolk population having education above secondary (2005) and higher secondary (2010)	1.833	−1.277	−0.119	0.955	−0.489	−1.076
Percentage of families below poverty line[a]	NA	0.131	NA	−1.878***	NA	0.729**
Percentage of fisherfolk population having membership in fishing cooperative	0.964	0.549	3.315***	1.918***	−0.396	−2.089***
Constant	−2.004***	−2.919***	−0.805***	0.721**	−6.732***	−1.771***
Pseudo-R^2	0.021***	0.015	0.032***	0.067***	0.005**	0.043***

[a]Data not available for 2005; *NA* not applicable
*Significant at 90% confidence level
**Significant at 95% confidence level
***Significant at 99% confidence level

estimates of both years show that secondary education, a significant variable, negatively influences mechanization. This indicates that education is not driving adoption of mechanization. Many fishermen, who are not educated, have possibly been able to mechanize their boats. This might have been possible through some other facilitators of adaptation, such as availability of subsidies, loans, which have not been captured in the study. However, being a member in cooperatives, which can be a source of information and finances, is highly significant to have a mechanized boat in both the years. Membership in fishing cooperative has a positive marginal effect of 0.706 and 0.430 in 2005 and 2010, respectively (Table 4). It is also found that mechanization is more likely to be in villages having low poverty, indicated by a negative coefficient for the same.

Table 4 Marginal effects of the regression models in 2005 and 2010

	Motorized		Mechanized		Diversified	
	2005	2010	2005	2010	2005	2010
Percentage of fisherfolk population having primary education	−0.176	−0.080	0.085	−0.352***	0.0003	−0.093
Percentage of fisherfolk population having education till secondary (2005) and higher secondary (2010)	0.241**	0.085	−0.258*	−0.250*	0.004***	0.362***
Percentage of fisherfolk population having education above secondary (2005) and higher secondary (2010)	0.264	−0.077	−0.025	0.214	−0.001	−0.176
Percentage of families below poverty line[a]	NA	0.008	NA	−0.421***	NA	0.119**
Percentage of fisherfolk population having membership in fishing cooperative	0.139	0.033	0.706***	0.430***	−0.0008	−0.341***

[a]Data not available for 2005; *NA* not applicable
*Significant at 90% confidence level
**Significant at 95% confidence level
***Significant at 99% confidence level

Diversification, as a strategy, is possibly undertaken by population having secondary education (Table 3). Results for both the years show it to be a relevant parameter. In 2010, the marginal effects show that having higher secondary education has the maximum influence (0.362) on diversification among all the significant drivers (Table 4). This indicates that having mid-level education and qualification enhance the probability to diversify. Fishermen having lower education might not be having opportunities to diversify. Further, it can be observed from the estimates of 2010 that membership in fishing cooperative has a negative association with diversification. This indicates that not being a member in cooperatives is probably related to lower information dissemination among fishermen about various opportunities and credit schemes that can help intensify their fishing efforts. Thus, lower membership results in higher levels of diversification in the villages. It is also seen that poverty and diversification have a positive coefficient, revealing that poverty increases the probability of fishermen to diversify their livelihood.

4.4 Comparing the Facilitators of Different Adaptation Strategies

The hypothesis (H1) that education positively influences adaptation is rejected in case of mechanization. In case of motorization, the hypothesis is accepted with respect to secondary education in 2005. H1 is also accepted for diversification, indicated by its positive association with secondary and higher secondary education in 2005 and 2010, respectively. Thus, not all levels of education have significant positive relationship with adaptation in the community. The hypothesis H1 is accepted in only a few cases. The highest level of education, that is, percentage of fisherfolk population having education above secondary (2005) and higher secondary (2010) is significant for none of the cases. Thus, different levels of education can have different extent of association with adaptation. It appears that secondary and higher secondary education majorly increases the likelihood of diversification. This can be observed from highly significant (at 99%) positive coefficients and marginal effects for the same in both the years. Overall, it might be said that, apart from diversification, education may not be much relevant for enabling adaptation through motorization and mechanization in the community in Maharashtra.

The hypothesis (H2) regarding association between poverty and adaptation is accepted in most cases. H2 is tested for the year 2010 only as data of 2005 is not available. The hypothesis that poverty is negatively and positively associated with mechanization and diversification, respectively, is accepted. The proposed hypothesis defining a positive relationship between poverty and diversification does not mean to signify that poverty is desirable for diversification. Poverty is a desperate condition which might drive communities and individuals to multiply their sources of income. No significant coefficient is obtained for the relationship between poverty and motorization.

Lastly, the hypothesis (H3) about cooperative membership and adaptation is also mostly accepted. Membership in cooperative is positively associated with mechanization at 99% confidence level. But, cooperative membership is not significantly associated with motorization leading to rejection of the hypothesis. This indicates that mechanization, being capital intensive, might be largely supported by cooperatives (through provision of credits, diesel subsidies as well as financial networks) compared to motorization. Membership in cooperative is negatively related with diversification, again at 99% confidence level, in 2010. It is not associated significantly with diversification in 2005.

5 Insights and Way Forward

Identifying facilitators of adaptation can help plan programs and policies to enhance adaptive capacity. Although marine fishing is likely to be adversely impacted by climate and various environmental changes, such studies are very limited for the sector. This study aims to understand adaptation and its socioeconomic facilitators in the marine fishing sector of Maharashtra, India. Marine fishing in Maharashtra is experiencing a declining catch, and adaptation is crucial. This study has made a comparative assessment of adaptation in the sector at two time periods, 2005 and 2010. This has assisted in understanding changes in patterns of adaptation over time. It is observed that mechanization of boats and diversification increased in 2010. The percentage of motorized boats decreased in 2010. Fishermen have preferred to adapt by mechanization and through diversification, in 2010 as compared to 2005.

Fractional logit regression is utilized to assess the association between the adaptation strategies and their socioeconomic facilitators. Three different adaptation strategies are considered: motorization, mechanization, and diversification. Socioeconomic facilitators, namely education, poverty, and cooperative membership, are hypothesized to be related to adaptation. The results slightly vary for the two years. The proposed hypotheses regarding the relationship between the strategies and facilitators are mostly accepted. It is found that secondary education is positively related to motorization (only in 2005). Secondary education (in 2005 and 2010) can also influence diversification. Overall, the results indicate that education might not be a relevant facilitator of intensification, namely motorization and especially mechanization, but is important for diversification. Cooperative membership is an important facilitator of mechanization as it is positively associated with it in both the years 2005 and 2010. Poverty is negatively and positively associated with mechanization and diversification, respectively, leading to acceptance of hypothesis H2. Further, low cooperative membership is related to diversification (in 2010).

The findings of the study can serve as a guide for designing state-level action plans for improving the livelihoods of the community. Special policies and schemes focused on the community need to be designed. Programs directed toward poverty alleviation, improving education, and cooperative management are important for capacitating the community to adapt and sustain their livelihoods. The study is limited as it is based on secondary data and could assess a limited number of facilitators. Future work can involve understanding a larger set of adaptation strategies, such as change in fishing habits, duration and a greater number of social, cognitive, and economic facilitators. The study also could not include any variable indicating adverse impacts of pollution, climate change, and others on declining fish catch, which might initiate the community to adapt. This is majorly because of data as well as methodological constraints to estimate village-level biophysical vulnerability of fisherfolk to multiple stresses impacting fish catch and their livelihoods.

This study provides a general understanding of adaptation in Maharashtra using a representative village-level dataset. The findings and approach of the study can add to the limited literature on assessing private adaptation in marine fishing communities, particularly those located in India. Further, the approach of the study can be possibly used for investigating facilitators of adaptation in other states of the country and other regions.

References

Agyeman BAS, Asuming-Brempong S, Onumah EE (2014) Determinants of income diversification of farm households in the western region of Ghana. Q J Int Agric 53:55–72

Barange M, Merino G, Blanchard JL et al (2014) Impacts of climate change on marine ecosystem production in societies dependent on fisheries. Nat Clim Chang 4:211–216. https://doi.org/10.1038/NCLIMATE2119

Below TB, Mutabazi KD, Kirschke D et al (2012) Can farmers' adaptation to climate change be explained by socio-economic household-level variables? Glob Environ Chang 22:223–235. https://doi.org/10.1016/j.gloenvcha.2011.11.012

Blythe JL, Murray G, Flaherty M (2014) Strengthening threatened communities through adaptation: insights from coastal Mozambique. Ecol Soc 19. https://doi.org/10.5751/es-06408-190206

Brooks N, Adger WN, Kelly PM (2005) The determinants of vulnerability and adaptive capacity at the national level and the implications for adaptation. Glob Environ Chang 15:151–163. https://doi.org/10.1016/j.gloenvcha.2004.12.006

Bryan E, Ringler C, Okoba B et al (2013) Adapting agriculture to climate change in Kenya: household strategies and determinants. J Environ Manag 114:26–35. https://doi.org/10.1016/j.jenvman.2012.10.036

Cinner JE, Huchery C, Hicks CC et al (2015) Changes in adaptive capacity of Kenyan fishing communities. Nat Clim Chang 5:872–876. https://doi.org/10.1038/nclimate2690

CMFRI (1981) Marine fisheries information service. Cochin, India

CMFRI (2005) Maharashtra marine fisheries census 2005. Kochi, India

CMFRI (2010) Maharashtra marine fisheries census 2010. Kochi, India

CMFRI (2013) Marine fish landings in Maharashtra for the year 2013 (in tonnes). http://www.cmfri.org.in/Maharashtra.html. Accessed 14 Oct 2016

CMFRI (2014) Marine fish landings in Maharashtra for the year 2014 (in tonnes). http://www.cmfri.org.in/mh2014.html. Accessed 14 Aug 2015

CMFRI (2015) Marine fish landings in Maharashtra for the year 2015 (in tonnes). http://www.cmfri.org.in/mh2015. Accessed 15 Sep 2017

Cutter SL, Boruff BJ, Shirley WL (2003) Social vulnerability to environmental hazards. Soc Sci Q 84:242–261. https://doi.org/10.1111/1540-6237.8402002

Daw T, Adger WN, Brown K (2009) Climate change and capture fisheries: potential impacts, adaptation and mitigation. In: Cochrane K, Dee Yong C, Soto D, Bahri T (eds) Climate change implications for fisheries and aquaculture: overview of current scientific knowledge, FAO fisheries and aquaculture technical paper No. 530. FAO, Rome, pp 107–153

Deressa TT, Hassan RM, Ringler C et al (2009) Determinants of farmers' choice of adaptation methods to climate change in the Nile Basin of Ethiopia. Glob Environ Chang 19:248–255. https://doi.org/10.1016/j.gloenvcha.2009.01.002

Frank E, Eakin H, López-Carr D (2011) Social identity, perception and motivation in adaptation to climate risk in the coffee sector of Chiapas, Mexico. Glob Environ Chang 21:66–76. https://doi.org/10.1016/j.gloenvcha.2010.11.001

Hahn MB, Riederer AM, Foster SO (2009) The livelihood vulnerability index: a pragmatic approach to assessing risks from climate variability and change—a case study in Mozambique. Glob Environ Chang 19:74–88. https://doi.org/10.1016/j.gloenvcha.2008.11.002

Hussein K, Nelson J (1998) Sustainable livelihoods and livelihood diversification. IDS working paper 69

IPCC (2001) Climate change 2001: impacts, adaptation, and vulnerability: contribution of working group II to the third assessment report of the intergovernmental panel on climate change. Cambridge University Press, Cambridge

Islam MM, Sallu S, Hubacek K, Paavola J (2014) Vulnerability of fishery-based livelihoods to the impacts of climate variability and change: insights from coastal Bangladesh. Reg Environ Chang 14:281–294. https://doi.org/10.1007/s10113-013-0487-6

Jain M, Naeem S, Orlove B et al (2015) Understanding the causes and consequences of differential decision-making in adaptation research: adapting to a delayed monsoon onset in Gujarat, India. Glob Environ Chang 31:98–109. https://doi.org/10.1016/j.gloenvcha.2014.12.008

Metcalf SJ, van Putten EI, Frusher S et al (2015) Measuring the vulnerability of marine social-ecological systems: a prerequisite for the identification of climate change adaptations. Ecol Soc 20. https://doi.org/10.5751/es-07509-200235

Nair SR, Pandey SK, Sharma A, Salim SS (2010) Performance of fisheries cooperatives in India: an evaluation of primary societies in Thane district of Maharashtra. J Fish Econ Dev 11:16–27

Nhemachena C, Hassan R, Chakwizira J (2014) Analysis of determinants of farm-level adaptation measures to climate change in Southern Africa. J Dev Agric Econ 6:232–241. https://doi.org/10.5897/JDAE12.0441

Rahut DB, Ali A, Kassie M et al (2014) Rural livelihood diversification strategies in Nepal. Poverty Public Policy 6:259–281. https://doi.org/10.1002/pop4.75

Reddy M (2016) An economic analysis of fisheries development programmes in Coastal Karnataka-a case study of Dakshina Kannada district. University of Mysore

Roessig JM, Woodley CM, Cech JJ, Hansen LJ (2004) Effects of global climate change on marine and estuarine fishes and fisheries. Rev Fish Biol Fish 14:251–275. https://doi.org/10.1007/s11160-004-6749-0

Tambo JA, Abdoulaye T (2012) Climate change and agricultural technology adoption: the case of drought tolerant maize in rural Nigeria. Mitig Adapt Strateg Glob Chang 17:277–292. https://doi.org/10.1007/s11027-011-9325-7

Vivekanandan E (2011) Climate change and Indian marine fisheries. Kochi, India

Zacharia PU, Dineshbabu AP, Thomas S et al (2016) Relative vulnerability assessment of Indian marine fishes to climate change using impact and adaptation attribute. Kochi, India

Biodiesel Blending in India—Analysis of National Biofuel Policy

Sreenivas Chigullapalli, Anand B. Rao and Anand Patwardhan

Abstract Biofuels are receiving increased attention due to their potential to enhance the energy independence in the transportation sector with simultaneous climate change mitigation by reducing GHG emissions. Prevailing market prices for the fossil-based diesel do not favour the production of biodiesel in India. To be able to make biodiesel production in India a commercial success, we may need to have strong technological base supported by policy support mechanisms. If produced sustainably, biodiesel may offer a part of the solution for problems such as energy security, import dependence for energy, rural employment generation, and climate change mitigation.

1 Introduction

In energy and climate policies, CO_2 emissions resulting from the consumption of transportation fuels are getting increasing attention. Emissions from the transport sector in India are rising continuously due to increases in passenger and freight transport demand. The mitigation strategies reduce emissions

(i) By reducing distance travelled;
(ii) By shifting to more efficient modes of transportation;
(iii) By increase in fuel efficiency;
(iv) By reducing the carbon intensity of transportation fuels (g CO_2/MJ).

S. Chigullapalli · A. B. Rao
Interdisciplinary Program (IDP) in Climate Studies, Indian Institute of Technology Bombay, Mumbai, India

A. B. Rao (✉)
Centre for Technology Alternatives for Rural Areas (CTARA), Indian Institute of Technology Bombay, Mumbai, India
e-mail: a.b.rao@iitb.ac.in

A. Patwardhan
School of Public Policy, University of Maryland, College Park, USA

C. Venkataraman et al. (eds.), *Climate Change Signals and Response*,
https://doi.org/10.1007/978-981-13-0280-0_15

In global scenarios, options (iii) and (iv) form the main mitigation options (Kejun 2014).

India is heavily dependent on international markets to meet its crude oil requirement. Currently, India meets 82% of its crude oil demand through imports, leaving the country susceptible to global price fluctuations (British Petroleum 2016). India's energy policy envisages moving towards more sustainable sources of energy. In order to achieve the twin objectives of energy security and sustainable energy, there has been increased focus on the potential of biofuels in recent times. Biofuels are produced from renewable biological sources and are considered viable alternatives or supplements to fossil fuels and their usage has the potential to reduce vehicular emissions. Increased use of biofuels (which are produced within the country) also helps in ensuring significant foreign exchange savings, besides revitalizing the rural economy through economic opportunities across the value chain.

Transportation biofuels production has received wider attention across the globe to make transportation energy mix less carbon intensive. But the carbon neutrality of the biofuel depends on the inputs and processes involved in the production and consumption of the biofuels and the fossil fuel it is substituting. Biofuels are the traditional and versatile sources of transportation energy with renewed interest due to its climate mitigation potential assuming CO_2 neutrality and the renewable nature of feedstocks. The rationale behind biofuel policies have been energy security, climate change mitigation. Developing economies have also focused on the social dimensions of biofuel value chains, indicating the inclusion of local poor for poverty alleviation and rural development. However, ecological and social dimensions of biofuel value chain are often overlooked or understudied.

1.1 Policies and Initiatives for Transportation Biofuels in India

In response to combat the problem of climate change, India launched a multi-pronged mission known as "National Action Plan on Climate Change (NAPCC)" in 2008 (Government of India 2008). During the 12th Five-Year Plan, Government of India (GoI) added another mission "*National Bio-energy Mission*" to NAPCC as ninth mission to promote and encourage a favourable environment to attract large-scale private sector investments in bioenergy.

The Government of India approved the "*National Policy on Biofuels*" in December 2009. The biofuel policy extends its support for the use of biofuels to supplement fossil-based transport fuels (bioethanol for blending with petrol and biodiesel for bending with diesel) with an indicative target of 20% blending for petrol and diesel by 2020. The policy expects that biodiesel and bioethanol should be included under the "*Declared Goods*" to ensure inter-state and intra-state unrestricted movement. Biodiesel has no taxes and duties as per the national policy on biofuels (MNRE).

As part of promotion of clean energy and safe, smart and sustainable green transportation network initiatives, India's "*Intended Nationally Determined Contributions (INDC)*" to UN Framework Convention on Climate Change (UNFCCC) also adopts aspirational target of 20% of biofuel blending in fossil-based transportation fuels (UNFCCC 2015).

To avoid the conflict of fuel versus food security, the biofuel policy proposes to raise the feedstocks on degraded or wastelands that are not suited for agriculture. In this connection, GoI launched "*National Biodiesel Mission*" identifying *Jatropha curcas* as suitable tree-borne oilseed for biodiesel production. An ambitious target covering 11.2–13.4 million hectares of degraded or wasteland under *Jatropha* cultivation by the end of the 11th Five-Year Plan (2007–2012) was expected but the competing uses for the land did not allow for the successful implementation of the programme (Bandyopadhyay 2015).

1.2 Transportation Biofuels in India

India provides a good market for biofuels since 82% of its crude oil is imported from around the world in 2016 (British Petroleum 2016). It is anticipated that over 94% of its crude oil will be imported by 2030 if energy trends continue on their current trajectory (Bandyopadhyay 2015). GoI proposes to reduce its crude oil import dependence by 10% by 2022 in several ways:

(i) Increasing domestic production of crude oil and crude oil-based fuels;
(ii) Promoting energy efficiency in the use of fossil fuels;
(iii) Encouraging greater use of alternative fuels.

Biofuels have been an option in the basket of alternative fuels that substitute fossil fuels. Use of biofuels in the transport sector will partly reduce the import dependence on crude oil.

India achieved bioethanol blend rate of 3.3% in gasoline (1.1 billion litres of ethanol blending in gasoline) in the year 2016. This higher gasoline blend rate can be attributed to fixed fuel ethanol prices, excise duty exemptions, and surplus ethanol availability. India imported 400 million litres of ethanol (classified as undenatured, fuel use ethanol at a price of US$0.54 per litre) out of which 80% was from USA (Sindelar 2017).

Ethanol blending programme (EBP) in India is likely to expand in near future but at a slower pace because of the fuel ethanol supply deficit. The industrial ethanol demand is partly met by international imports. The supply deficit is expected to grow as the demand for ethanol blending is continued to rise with demand for gasoline. Due to this effect, the gasoline blending rate in petrol has fallen to 2% (0.7 billion litres of ethanol blending in gasoline) in the year 2017.

Although the policy environment for the promotion of biofuels in India appears to be positive, the expected target of 20% biofuel blending in transportation fuels seems a remote possibility. The observed blending rate for the biodiesel in India is

Table 1 Statistics about biodiesel in India (Aradhey and Jonn 2016)

Calendar year	2010	2011	2012	2013	2014	2015	2016	2017
Production (in million litres)	90	102	115	120	130	135	140	150
Production capacity								
Name plate capacity (in million litres)	450	450	460	465	480	480	500	500
Capacity Use (%)	20.00	22.70	25.00	25.80	27.10	28.10	28.00	30.00
Feedstock use(in 1000 tonnes)								
Used cooking oil	38	42	48	49	50	50	52	56
Animal fats and tallow	6	6	7	7	6	5	6	6
Other oils	50	58	65	70	75	85	85	90
Market penetration								
Biodiesel, on-road use (in million litres)	26	30	35	38	40	45	50	60
Diesel, on-road use (in million litres)	42,625	45,520	49,343	49,354	49,605	52,239	56,111	58,422
Blend rate (%)	0.06	0.07	0.07	0.08	0.08	0.08	0.09	0.1
Diesel, total use (in million litres)	71,041	75,866	82,23	82,256	82,674	87,064	93,518	97,371

far below the pre-decided target of 20%. The blending rate achieved during 2017 is 0.1% (Sindelar 2017). Several agronomical and economic constraints resulted in the underachievement of the proposed targets. India is having limited production capacity for commercial biodiesel manufacture since biodiesel has to compete with diesel for its commercial success. Administered prices for crude oil products were heavily subsidized, whereas biodiesel purchase price is far below the cost of production in most of the cases.

"The demand for diesel is five times higher than the demand for petrol in India" (Aradhey and Jonn 2016). But while the ethanol industry is well established, the biodiesel industry is still in its initial stages of development. Since the demand for edible vegetable oil exceeds supply, the government has decided to use non-edible oil from *J. curcas* seeds as biodiesel feedstock. The growth in the biodiesel industry has been driven by tax incentives for blending biodiesel with diesel for energy security and climate change mitigation reasons. Table 1 shows the trend in biodiesel production and consumption patterns in India.

1.3 Issue of Economic Viability

Before looking at the climate change mitigation benefits that biodiesel can offer, one has to realistically assess the chances that a market for biodiesel will emerge in India. So far, few private farmers and corporate investors have engaged in non-edible oil crops (mainly *J. curcas*). The market for biodiesel in India has not yet emerged, because biodiesel (minimum purchase price by oil marketing companies is Rs. 26.5/litre, but the cost of production is 20–50% higher than MPP according to the feedstock used) is not cost competitive with conventional diesel at current market prices (Rs. 66.5/litre, in Mumbai as on 21st Jan 2018). This is due to a number of reasons:

i. Government of India subsidizes the price of conventional diesel, keeping it artificially low.
   a. Negative externalities of using conventional diesel are not reflected in its price.
   b. Crude petroleum is still outside the ambit of GST.
ii. Biodiesel production needs to be more economically viable.
   a. Most of the oil-bearing crops are wild plants.
   b. Untapped resources are not given priority.
   c. Expected yields of non-edible oil-bearing crops have not been materialized without inputs on dry and marginal lands.
   d. Yields are higher on fertile farmlands, but non-edible oil-bearing crops yield low RoI compared to alternative cash crops (due to lack of clear policy support on minimum support price, i.e. MSP for feedstocks).
   e. Competing uses for the non-edible oil feedstocks making the prices of the feedstocks higher than expected with the time.
iii. Moreover, the biodiesel blending is just a recommended measure according to the national biofuel policy, making progress in the biodiesel industry much slower.

2 Experience and Learning from Biofuel Blending Programmes from Other Countries

Brazil has placed both regulatory and economic incentives to develop its biofuel industry to secure captive market. It has developed its biofuel policy on a well-established sugarcane–bioethanol industry and on an emerging biodiesel industry. The incentives include tax breaks, accessibility to credit through financial institutions and blending mandates. Despite its success from an economic and emission reductions point of view, the social benefits of sugarcane–ethanol industry have not been materialized. This is due to large size landholdings and improper allocation of burdens and benefits leaving the rural poor as seasonal migrant labour working under infeasible working environments (Bastos Lima 2013).

To balance and compensate for the imbalance created by the sugarcane–ethanol industry, Brazil started its socially oriented biodiesel programme that promoted castor bean cultivation among smallholders and to create a synergy among contract farming systems and biodiesel industry. This led to smallholders' dependence on single crop coupled with poor implementation process led to broken contracts translating into a situation where farmers left with a non-marketable and non-edible crop. With the creation of a subsidiary biofuel industry to engage with marginal farmers, the distribution of high-quality seeds, changes in the contract terms to allow more leverage to contract farmers have resulted in increasing success of biofuel policy in Brazil (Bastos Lima 2012).

Indonesia promoted feedstock cultivation for biofuel industry on "marginal lands". Subsidies and tax breaks were the incentives to sugarcane and palm oil industries to produce ethanol and biodiesel, respectively. It also created a favourable legal environment to attract long-term private sector investments and blending mandates to guarantee market penetration of biofuels. Despite the efforts put in, the fuel ethanol has not been a success due to more attractive end markets for sugar industry. The cultivation of "marginal lands" with Jatropha faced problems due to lack of viable and established biodiesel value chain. Although palm oil-based biodiesel seems economically viable, it has shown mixed social implications.

India's national biofuel policy has also based upon the well-established sugarcane agro-industry and smallholder integration through biodiesel value chains. Indian national biofuel policy makes use of similar policy instruments such as tax breaks and blending mandates. But unlike Brazil and Indonesia, it utilizes only sugarcane molasses (not sugarcane juice) as a feedstock for ethanol production, due to tight sugar supplies. In India, the bioethanol value chain has much larger participation of sugarcane farmers. All the value addition is captured by the bioethanol industry while the conditions of sugarcane farmers remain basically unchanged. Therefore, there is hardly a socially transformative element to the Indian ethanol blending policy.

The Indian national biofuel policy, on the other hand, has promoted Jatropha plantations on "marginal and/or wastelands" for promoting biodiesel blending. Realized lower seed yields in comparison to reported yields, lack of policy support on minimum support price for the seeds, lack of committed buyers led to failure of Jatropha cultivation in India.

3 Discussion

Emphasizing on social aspects of sustainability has been a priority of most of the developing economies. India's national biofuel policy aimed at domestic renewable energy production by creating employment generation. However, a careful analysis reveals that in practice policy seem to lose its focus to rural development needs. Biofuel policy in India has been marked by ambitious blending targets but lack clear vision about the complexity of the value chain and grass root realities.

Governments should avoid top-down grand plans such as the Indian National Biodiesel Mission and instead develop strategies in more participatory ways. The careful observation of biofuel policy indicates a gap between the implementation of the biofuel value chains and the biofuel policy instruments usually adopted. Although the biofuel policy framework to promote the biodiesel in India is very encouraging, experience has shown that the efforts have not translated into what was expected on the production and commercialization fronts to meet the country's biofuel blending requirements. This calls for a re-examination of the biofuel policy from various stages of the biodiesel supply chain.

References

Aradhey A, Jonn S (2016) India biofuels annual. https://gain.fas.usda.gov/Recent%20GAIN%20Publications/Biofuels%20Annual_New%20Delhi_India_6-24-2016.pdf

Bandyopadhyay KR (2015) Biofuel promotion in India for transport: exploring the grey areas. TERI Policy Br

Bastos Lima MG (2012) An institutional analysis of biofuel policies and their social implications lessons from Brazil, India and Indonesia. 1–2

Bastos Lima MG (2013) Adjusting biofuel policies to meet social and rural development needs: analysing the. Policy Brief Int Policy Cent Incl Growth 40:1–5

British Petroleum (2016) BP statistical review of world energy. https://www.bp.com/content/dam/bp/pdf/energy-economics/statistical-review-2016/bp-statistical-review-of-world-energy-2016-full-report.pdf

Government of India M of E and F (2008) National Action Plan on Climate Change (NAPCC). http://www.moef.nic.in/downloads/home/Pg01-52.pdf

Kejun (2014) Climate change 2014—mitigation of climate change—summary for policymaker-s—IPCC

MNRE National Policy on Biofuels. https://mnre.gov.in/file-manager/UserFiles/biofuel_policy.pdf

Sindelar S (2017) India biofuels annual. GIAN Rep

UNFCCC (2015) India INDC to UNFCCC

Cross-Scale Institutional Linkages in Climate Change Responses: An Indian Perspective

D. Parthasarathy, Yash Shethia and N. C. Narayanan

Abstract India's National Action Plan on Climate Change (NAPCC) and its constituent eight National Missions stress on devising plans for climate change action at regional and local scales; these cater to various agro-ecological environments and address the different socio-economic situations prevalent in these. These as well as the various State Action Plans on climate change envisage active involvement of local institutions such as Panchayati Raj Institutions, Water Users' Associations, JFM/ CFM/ FRA Committees, Van Panchayats, Village Councils, and Urban Local Bodies. The importance of sub-national and local activities is stressed in Sect. 15.2.1.3 of IPCC AR 5. At the other end of the spectrum are the various high-powered committees headed in many instances by central and state ministers and advisory committees comprised of senior bureaucrats and from the scientific establishment. As India moves ahead in implementing the NAPCC and its component mission activities, it is likely to face a number of challenges relating to how issues of institutional linkages are handled. Our study based on critical review of the literature, assessment of case studies, and field research analyzes promising institutional designs that could facilitate cross-scale and horizontal linkages, focusing on the idea of polycentric governance system as a possible solution. Effects of higher institutions on local institutions are identified; in the process, instances of multilevel processes in the governance of climate change issues are studied, where the boundaries and divisions between local, national and global scales and between state and non-state actors have been blurred and disrupted. The need to redefine the scale and scope of state action going from a single to multiple modes of governing that reinforce and negate each other is pointed out. The chapter will outline policy suggestions for enhancing state

D. Parthasarathy (✉) · N. C. Narayanan
Interdisciplinary Programme in Climate Studies, Indian Institute of Technology Bombay, Mumbai, India
e-mail: dp@hss.iitb.ac.in

N. C. Narayanan
e-mail: ncn@iitb.ac.in

Y. Shethia
World Wildlife Fund India, New Delhi, India
e-mail: yash.shethia@gmail.com

C. Venkataraman et al. (eds.), *Climate Change Signals and Response*,
https://doi.org/10.1007/978-981-13-0280-0_16

and institutional capabilities to deal with climate change impacts and identifying new research needs.

Keywords NAPCC · National missions · Institutional linkages · Polycentric governance

1 Climate Change and India

India is among the countries expected to be greatly and mostly negatively affected by climate change. In an economy closely linked to its natural resource base and a very large section of its population directly dependent on the environment for their livelihoods, projected changes in climate that impact on water availability, forest quality, wetlands, agriculture, and large and sensitive ecosystems such as the Himalayas are bound to adversely affect the livelihoods of a very large section of the people (NAPCC 2009).

1.1 Observed Changes, Trends and Impacts in India

The Asia IPCC AR4 divides the world's most populous continent into seven subregions; South Asia is characterized as being physiographically diverse and ecologically rich in both its natural and agricultural domains and having five of the top twenty megacities of the world (Cruz et al. 2007: 472). The report has several subsections and begins by highlighting current sensitivity and vulnerability. The key climate trends, variability and extreme events in the case of India include temperature rise, increase in frequency of hot days and multiday heat waves, extreme rain events during the summer monsoons, reduced rainy days in some parts of country, severe and recurrent flooding, frequent droughts, and changes in frequency and intensity of cyclones and storms (Cruz et al. 2007).

The Asia IPCC AR4 goes on to give a sector-wise depiction of the observed impacts looking at agriculture and food production, water resources, human health and natural ecosystems. While the AR4 paints a general picture of a decline in the production of rice, maize and wheat across Asia over the past few decades mainly due to increased water stress attributing it partly to an increase in temperatures, increasing frequency of El Nino and a decline in the number of rainy days (Cruz et al. 2007), the NMSA claims that the yields of wheat and paddy have been negatively affected across parts of India due to the aforementioned factors (NMSA 2010).

Water shortages in India are attributed to rapid urbanization and industrialization, the growth of population and inefficient use of water, and the changing climate is only further aggravating the situation. The heat waves at the beginning of the twenty-first century, i.e. from 2000 to 2006, are reported to have claimed many lives mainly among the poor, labourers and elderly in the states of Andhra Pradesh and Orissa (Cruz et al. 2007).

Poverty and poor hygiene in addition to bacterial proliferation due to higher temperatures in South Asia are held responsible for endemic morbidity and mortality from diarrhoea disease. Climate-related factors such as severe floods, El Nino-related droughts, changes in sea surface temperatures and rainfall together with non-climatic factors such as poverty, lack of access to clean drinking water and the poor or absent sewerage systems are reported to have led to the outbreak of infectious diseases (Cruz et al. 2007).

Fragile ecosystems such as wetlands and coral reefs that are biodiversity rich and economically important have been severely affected, the former affected by the decline in precipitation and droughts and the latter due to coral bleaching triggered by the 1997–98 El Nino event. The coral reefs of South Asia in particular are reported to have been severely affected by the event, and a substantial area of the coral reefs was lost. Wetland ecosystems are severely degraded and threatened due to the decline in precipitation and droughts along with encroachment-related degradation (Cruz et al. 2007).

1.2 Projected Impacts and Vulnerabilities to Climate Change for India

The Asia IPCC AR4 suggests a significant acceleration in warming for Asia over what was recorded in the last century; while this varies from one sub-region to another, South Asia is predicted to witness a warming greater than the global mean warming. Most of the modelling done for South Asia under the AR4 suggests a decrease in precipitation during the winter months of December, January and February. Extreme weather events such as heat wave and excessive precipitation are projected to increase across South Asia. There is also an enhanced risk of tropical storms and cyclones along the coastal regions of South Asia (Cruz et al. 2007). What follows is a sector-wise look at the projected impacts and vulnerabilities from climate change in India.

i. Agriculture, Irrigation and Food Security

Modelling exercises to assess impact of climate change on crop yields indicate a decline of up to 30% in South Asia. Some studies indicate a decline in yield potential of wheat and maize from between 2 to 5% for a temperature rise of 0.5–1.5 °C in India. Rice production in South Asia could decline by 3.8% by the end of the twenty-first century and net cereal production by 4 to 10% even under the most conservative climate change scenario (Cruz et al. 2007).

The NMSA singles out parts of western Rajasthan, southern Gujarat, Madhya Pradesh, Maharashtra, northern Karnataka, northern Andhra Pradesh and southern Bihar as being particularly vulnerable. The mission document quotes information from a research project by ICAR saying there would be a reduction in agriculture yields of between 4.5 and 9% in the medium term leading to a fall of up to 2%

per annum in GDP growth, and a reduction in agricultural yields by more than 25% in the long term if no adaptation measures are undertaken (NMSA 2010). Changes in climate will not only affect crop production per unit area but also the area of production. The Asia IPCC AR4 quoting a 2003 FAO report states that more than 28 million hectares of farmlands in South and South East Asia will need a substantial increase in irrigation in order to sustain productivity. For a 1 °C increase in temperature, the demand for water for irrigation in the arid and semi-arid regions of Asia will rise by at least 10% (Cruz et al. 2007).

ii. The Livestock Sector

The livestock sector including milk, meat and poultry is also likely to be adversely affected as higher temperatures limit production of palatable grasses and herbs, increase heat stress among the animals and affect the availability of water leading to increased incidences of disease in animals and reduced milk yields (Cruz et al. 2007). Reduction in grazing land would also indirectly contribute to a shrinking of the livestock sector (NMSA 2010).

iii. The Himalayan Ecosystem

Glacier-fed rivers such as the Indus, Ganga and the Brahmaputra are a lifeline for a very large section of the population of South Asia. Around 15,000 Himalayan glaciers together form a large reservoir 12,000 km^3 of freshwater. The Himalayan glaciers are thinning and receding rapidly, and at current estimates of melting, it is predicted that the perennial Himalayan Rivers may become seasonal in the near future severely affecting South Asian economies creating havoc particularly in the north-Indian plains (Cruz et al. 2007).

In addition to glacier recession, the NMSHE also lists variations in the volume of the flow of water in the Himalayan Rivers, unsustainable ecological changes, deforestation and degradation, biodiversity loss, creation of conditions leading to natural disasters, and the dislocation of mountain societies closely dependent on the Himalayan ecosystems among the possible impacts of global warming on the Indian Himalayas.

iv. Water

India is expected to face a serious water crisis in the future; it is forecast that the gross per capita availability of water in India will decline from around 1820 m^3 per annum in 2001 to 1140 m^3 per annum by 2050. The projected decrease in precipitation during the winter months is likely to result in lesser storage and increased water stress. Fewer rainy days with increased intensity of rainfall are predicted; this is expected to adversely affect groundwater recharge potential as the run-off would be high and would also increase the likelihood of floods (Cruz et al. 2007). Overall, however, a decline in the run-off is projected for all river basins in India except Narmada and Tapti, and a decline by more than two-thirds is predicted in the Sabarmati and Luni river basins (NAPCC 2009). India's ground and surface water sources particularly along the coasts will be increasingly threatened by salinity ingress due to rising sea levels as a direct outcome of climate change (Cruz et al. 2007).

A listing of the most vulnerable areas in India due to the impact of climate change on water resources reads as follows—areas that are prone to drought and flood, coastal areas, deficit rainfall areas, areas where the groundwater is over-exploited or in a critical or semi-critical stage, areas with water quality issues, and river basins that are snow-fed (National Water Mission 2011).

v. Coastal Areas

Conservative estimates of sea level increases by the end of the twenty-first century are pegged at 40 cm higher than what they are present. This is expected to swell the number of people in coastal areas flooded each year from thirteen million to ninety-four million. Almost 60% of the increase will happen in South Asia, and a large part of this in India. An increase in inundation of lowlands, loss of coastal marshes and wetlands, erosion of beaches and increase in salinity in coastal areas is likely due to the projected rise in sea level. Moreover with an increase in the intensity of cyclones predicted coastal areas are likely to witness a rise in loss and damage to life and property (Cruz et al. 2007).

vi. Population Increase, Pressure on Natural Resources and Energy Consumption

India's population is expected to reach close to 1.7 billion by 2050; in a country with an already high population density, this is likely to result in a rise in pressure on natural resources as demands for goods and services go up. Urbanization and industrialization together with intensification in agriculture would lead to intensified land and water use. While these phenomena are expected to substantially increase India's energy consumption, changes in climate will also influence energy consumption and in turn the CO_2 emissions from the region (Cruz et al. 2007).

Forestlands already affected and threatened by the demand for provisioning services such as fuel wood, fodder and change in land use to agriculture and from the "forces of de-greening operating across the country" as the Green India Mission document puts it will be rendered further vulnerable due to the change in climate (GIM 2010; Cruz et al. 2007).

vii. Migration

India would also be affected by both internal migration (Cruz et al. 2007) and migration from neighbouring countries particularly Bangladesh and Nepal as both of these countries are hotspots of climate change with large vulnerable populations and India for several reasons is an attractive destination (Khaladkar et al. 2009). Internal migration could be in the nature of sudden spikes in rural to urban migration due to a combination of factors such as low precipitation together with depressed economic conditions or when an extreme weather event force those caught in it to migrate (Cruz et al. 2007). Migration from the outside could be a result of a sudden and large influx again due to an extreme weather event or a slow and sometimes steady influx due to gradual changes in climate in the neighbouring country/countries (Khaladkar et al. 2009).

2 India's Adaptation Strategy in the Face of Climate Change

In response to rising concerns about climate change impacts and the need to formulate a robust strategy to address vulnerabilities arising from actual and projected impacts, in June 2008 India released the National Action Plan on climate change (NAPCC). The NAPCC is both an acknowledgement and response from India to the scientific conclusions of IPCC AR4 that clearly posits climate change as a global problem and one that holds potentially serious threats for India (EPW 2008). The opening paragraph of the NAPCC document is in many ways a succinct expression of India's perspective on climate change. It acknowledges that climate change is a threat to India's rapid economic growth that it is not really a problem of India's making and yet the country must engage with the international community in order to deal with the threat in a collective and cooperative manner, to do this India first needs a national adaptation strategy and second the country needs to seize the opportunity to bring greater ecological sustainability in its quest for development (NAPCC 2009).

India's National Action Plan on climate change (NAPCC) is now almost a decade old. Recognizing state jurisdiction over several aspects of the NAPCC particularly in adaptive action the central government in 2010 asked state governments to develop SAPCC.[1] The objective behind this was also to bring some coherence across the different states in climate change responses (Dubash 2012). Critical reviews (Dubash 2009, 2013; Dubash et al. 2013; Dubash and Joseph 2016; Dubash and Jogesh 2014; Byravan and Rajan 2013) have shown that both the national- and state-level regional plans are characterized by unevenness and tardiness in strategizing, institution formation and execution. The NAPCC envisaged the implementation of Action Plans through a set of eight National Missions covering the areas of water, energy, habitats and agriculture. More missions are on the anvil—focusing on coastal zones, health and waste management. A key problem in the Action Plans and Missions pointed out by critical studies is the lack of adequate attention to problems of institutions, namely the need for appropriate institutional design, and the key imperative to ensure institutional collaborations that are both cross-scale and cross-agency.

While recognizing the complexity of developing an action plan for a country as large and diverse as India, a country with multiple social, economic and political problems, inequality, poverty and development challenges, this paper argues that the NAPCC as currently designed (together with the National Missions) reflects severe flaws, loopholes and gaps that need to be addressed, so that the document can become a living one—used by government agencies, NGOs, researchers, and the market for strategizing their actions in the realm of climate change. The specific focus of this chapter is on institutional problems, in particular focusing on the neglect of cross-scale and horizontal linkages in the action plan and the missions at local, regional and national levels. This neglect is also seen in critical assessments of the NAPCC and National Missions, in part due to the excessive focus on mitigation, as compared

[1] A 12 point list of guiding principles was issued by MoEF for preparation of the SAPCC.

to adaptive strategies that address climate risks. Where institutions are identified in some of the national missions, these are ad hoc, not geared to co-benefits, and a failure to coordinate across institutions and sectors is evident—e.g. across water, energy, forestry, habitat planning, industry and health. Disaster management institutions are largely ignored. It is argued that a large country like India with a federal structure, and the need for polycentric governance to address both top-down and bottom-up imperatives, the design of the action plan and of missions, should take into account the substantial theoretical and empirical work on institutions carried out by social scientists in the broad area of environmental governance. This is especially important in the context of state governments, district administrations and regional development bodies increasingly coming up with their own action plans and strategies to address climate change-induced vulnerabilities.

3 A Critique of the NAPCC

The NAPCC would not have been an easy document to write and put together (EPW 2008). It is even contended that the document was hastily written and launched when it was, so as to ease the pressure on India from developed countries prior to an important international summit that the Prime Minister was scheduled to attend (Bidwai 2012). It might be more fair however to treat it as a plan that despite its several drawbacks and limitations is a significant development coming as it did at a time just prior to which Indian policy makers were not particularly engaging with climate change issues seeing it as a problem not of India's creation, and therefore, the mitigation of which was not the country's problem. It must also be pointed out that its own framers admit it is an evolving document (EPW 2008).

Some of the proposed actions are new and bold, there is a renewed thrust on some activities, and an attempt to deepen existing approaches; however, one also gets an impression that many existing programmes and activities are being billed as actions aimed at dealing with climate change (EPW 2008: 5). What follows is a critique of the various missions, which is followed by a critical analysis of the action plan from the perspective of the need to address cross-scale and horizontal institutional linkages.

1. The NMSA—The domain of the NMSA is vast, and agricultural activity is prevalent on around 46% of India's land area (NMSA 2010) and concerns three-fifths of all livelihoods in India (Bidwai 2012). It is both comprehensive and ambitious and has a welcome emphasis on the need for convergence with the other NAPCC missions. The Mission's focus on promoting dry land agriculture too is timely and a change from the dominant thinking among policy makers that until quite recently viewed dry land agriculture as a lost cause. Even at their low levels of productivity, the dry lands contribute significantly to the country's output of coarse cereal, maize, chickpea, pigeon pea, groundnut and soya bean (Vittal et al. 2006).

The role of markets figures prominently in the NMSA and is in fact mentioned as one of the mission interventions, wherein focus is on building strong farmer–institution–industry linkages as also on the setting up of food processing industries and exports, while this may be okay as far as medium and large farmers in irrigated tracts of the country are concerned, Lele cautions that in order to build resilience into dry land agricultural systems, these may need to be not fully plugged into the global market as highly inter-connected systems are known to be extremely sensitive to even minor climatic changes leading to massive fluctuations (Lele 2006).

Finally, there is the view that there is excessive focus on biotechnology-driven interventions and given the complex nature of the effects of climate change on agriculture it is not too certain if the choice of this path over pursuing a path of low energy low-input agriculture based on traditional farming systems and practices in combination with good watershed management is a sound one (Bidwai 2012).

2. The National Water Mission—The National Water Mission has a complex task at hand, while there is no doubt that climate change will only exacerbate an already stressed water situation in India (Cruz et al. 2007; NAPCC 2009) some researchers point out that climate change policies themselves may adversely impact freshwater resources and ecosystems resulting in maladaptation (Pittock 2011).

   Pittock examined the national climate change policies of nine governments, including India to see if these address the interplay between the various sectors including the inevitable increase in demands for water for activities such as energy production, carbon sequestration and climate change adaptation as also to discern the presence of institutional mechanisms to mediate between the various demands and facilitate policy integration. An examination of the NAPCC together with India's Energy Policy of 2007, National Water Policy 2002, and the 2006 Environmental Policy revealed no such attempt at integration (Pittock 2011).

3. The NMSHE—There is little doubt that the NMSHE has a herculean task at hand; the Himalayas are physically vast, ecologically complex, occupy a geopolitically delicate space and their fate is closely linked to the lives and livelihoods of a large section of the Asian populace, the Mission does well to acknowledge that a regional approach will have to be taken to protect the Himalayan ecosystem from the impact of climate change (NMSHE 2012).

   Within the country itself, the domain of the NMSHE is very large, and there are twelve Himalayan states spanning the Western and Eastern Himalayas, there is a plethora of institutions working on multitude activities ranging from governance, to research and monitoring of ecosystems, flora and fauna, communities and occupations and the NMSHE through an elaborate institutional design seeks to bring coherence and synergy to all of these (NMSHE 2010).

   The mission document states that 41.5% of the geographical area of the Himalayas in India is under forests, and it represents 47% of the very good forest cover of the country (NMSHE 2010). The document also in different places acknowledges that the Himalayan forests and biodiversity are vital to the livelihoods of the mountain communities (NMSHE 2010) and that the traditional ecological knowledge and local knowledge of forest management practices together with the participation of the local communities is vital to meeting the Mission's goal and objective (NMSHE 2010). However, the Mission does not seem to acknowledge that land ownership and management in several Indian Himalayan states is communal or commons based (Chopra and Dasgupta 2008).

4. The GIM—Unlike most of the other mission documents, the GIM was prepared after a series of public consultations (Bidwai 2012). Also important is the GIM's recognition that JFM is one of the many forest-based institutions and that also included in these are CFM groups in Orissa, *Van Panchayat* in Uttarakhand, traditional village institutions and Village Councils in the Schedule VI areas, and what might eventually emerge as the largest community-based forest institution in the country—the Forest Committees set up under the FRA (GIM 2010). This is a break from past approaches where JFM was sought to be promoted even where older and resilient forest management institutions exist and which has been discussed in some detail in the context of the Himalayan Mission above. The Mission's call for revamping JFM Committees and bringing these under the ambit of the *Gram Sabha* to promote decentralized forest governance and community empowerment (GIM 2010) is implicit if grudging admission that the outcomes of the JFM programme have been far from satisfactory. However, the Mission's attempt to break away from past and straitjacketed approaches to forest governance may, in fact, face the greatest resistance from the forest bureaucracy itself (Bidwai 2012).
5. The National Solar Mission—The Mission has been described as being more in the nature of a statement of intent of India's commitment to harnessing renewable energy sources (Bidwai 2012) but weak on detailing how the mission would be accomplished (EPW 2008). Bidwai goes on to write that some of the mission's ideas are worthy but implementing these would require substantive experimentation, going through a continuous and rigorous process of trial and error, and innovation. Unpacking these processes and the fact that indigenous R&D would need to be coordinated and led such that there is a significant breakthrough in reduction of cost of materials would require "real leadership—political and technological" (Bidwai 2012).
6. The NMEEE—Some commentators hail the measures in the NMEEE as being among the most concrete and forward looking in the NAPCC (EPW 2008) and yet others have criticized the NMEEE as having restricted its scope by not including more energy-intensive industries to the list of the existing nine, and for being complacent about India's growing dependence on coal and as also for contending that India's efforts in reducing emissions are not going to significantly alter the global stock of GHG till 2050 (Bidwai 2012).
7. The National Mission on Sustainable Habitat—Bidwai's observations begin by pointing out the mission primarily deals with urban issues excluding the habitat of the rural people. He also says that the work of many architects and urban planners deploying low-carbon technologies and creating less energy-intensive structures is ignored, this when globally concrete production is estimated to contribute 5%

of GHG emissions. He attributes these deficiencies to the preoccupation of the Ministry of Urban Development that has authored the mission document with "prime realty in the big cities" and accuses the authors of completely bypassing issues related to the ecological footprint of modern urban construction (Bidwai 2012).

On the issue of promotion of urban public transport, the experience has been that there has been more of lip service to options such as the BRTS, whereas state action on the ground has actually been to the detriment of mass public transport, to bolster his contention Bidwai quotes the massive investment in construction of flyovers, the continued incentives to automobile manufacturers, and the presence of a tax regime that taxes buses at rates twenty times higher than cars per vehicle kilometre. Bidwai concludes by saying that while some of the recommendations of the mission are worthy, these do not add up to together and lack a coherent structure (Bidwai 2012).

Despite the criticism and obvious limitations, however, the NAPCC and the Eight Missions are an important first step (EPW 2008) and the country would only have been worse off without these (Bidwai 2012).

4 Cross-Scale Institutional Linkages in Climate Change Responses—The Need and Challenges

The NAPCC and its constituent Mission's stress on devising plans for climate change action at regional and local scales catering to the various agro-ecological environments and addressing the different socio-economic situations prevalent in these. These as well as the SAPCC envisage active involvement on the one hand of local institutions such as *Panchayati Raj* Institutions including the *Gram Sabha* at the village and at hamlet level, Water Users' Associations, JFM/CFM/FRA Committees, *Van Panchayat*, Village Councils and Urban Local Bodies, and also speak of leveraging Traditional Ecological Knowledge and values systems (sic) of local communities in climate change action. At the other end of the spectrum are the various high-powered committees headed in many instances by central and state ministers and advisory committees comprised of senior bureaucrats and from the scientific establishment; and in the case of the NMSHE also envisage some form of regional institutional arrangement of the Himalayan Range States.

India has adopted a co-benefit-based approach to climate action, i.e. implementation of measures that address development objectives and at the same time yield benefits for effectively addressing climate change. In keeping with this approach, the GoI has launched a number of plan and policy development processes (Dubash 2012). There are initial indications that there will be considerable conceptual and implementation challenges to the national- and state-level action plans. A foremost challenge would be to address the linkage between development actions and their climate aspects. In addition to working around issues of financing actions, there are likely to difficulties in implementation of several actions where jurisdictions of the central and state governments and local authorities overlap. As Dubash aptly put it "climate change will have to be integrated as an objective within an already littered institutional landscape." (Dubash 2012).

The emergence of domestic climate policy will in all likelihood pose significant governance and coordination challenges, and the country has already begun to grapple with mainstreaming climate concerns into sectoral policies; this integration which is still in its nascent stages will inevitably give rise to new institutional complexities and politics which nonetheless is welcome as it heralds a deeper engagement with climate change concerns within the country (Dubash 2012).

Centralised authority is in recent decades giving way to authority being dispersed across multiple centres. A basic premise of the move away from unitary government is that there is greater flexibility in the dispersion of governance across multiple jurisdictions than in the concentration of governance in one jurisdiction (Hooghe and Marks 2001).

Coming to the issue of shaping global environmental governance Betsill and Bulkeley (2006?) posit that the local too is an important site that influences the former and that the multilevel governance approach characterized by the links between vertical and horizontal jurisdictions is better placed to grasp the social, political and economic processes that shape global environmental governance (Betsill and Bullkeley 2006).

Decades of work on common property has increasingly proven that the traditional "command and control" mechanisms relying either on state control or private holding of resources are unable to deal appropriately with the complexities of linked social–ecological systems (Bisaro et al. 2010). Thus for the management of resources such as fisheries, forests, grazing lands, watersheds, wildlife, protected areas among others to be effective requires the involvement of multiple parties (Berkes 2009).

There are now increasing instances of multilevel processes particularly in the governance of issues such as climate change where the boundaries and divisions between local, national and global scales, and between state and non-state actors have been blurred and disrupted. It is also true however that a multi-governance perspective does not necessarily mean a weakened state, it is nonetheless about redefining the scale and scope of state action, going from a single to multiple modes of governing that reinforce and negate each other (Betsill and Bulkeley 2006).

It is unlikely that nation states would be able to meet the international commitments they have made on addressing environmental problems including that of climate change without explicitly engaging actors at the sub-national levels (Betsill and Bulkeley 2006). Knowledge and information of the local situation are best available with institutions at that level, whereas the state has the vantage of a broader jurisdictional and spatial scale and access to a variety of tools and techniques that are not available to local institutions (Berkes 2009). Ostrom, in the context of community-governed natural resources wrote that local- and small-scale institutions may be very effective in delivering on many aspects of sustainable development but would in the absence of support from large-scale and higher-level institutions be faced with a threat to their long-term sustenance (Bulkeley and Betsill 2005).

The management of ecosystem services and human well-being is an information-intensive process, in fact, the knowledge of ecosystem dynamics and state of resources—availability, abundance, trends, etc.—is dispersed across local-, regional- and national-level actors. Social–ecological systems may respond and behave differently at different scales, and it is unlikely that any single agency harbours all the knowledge and resources required to effectively manage these (Berkes 2009).

When the density of institutions operating in a social space increases so would the interactions between them, in fact, these interactions would inevitably affect the performance and robustness of these institutions. Institutions interact with each other both horizontally, i.e. at the same level of social organization and vertically for instance between local systems and national systems. Efforts to build linkages and connections between institutions may be pursued to meet collective goals, enhance institutional effectiveness and/or integrate/nest smaller institutional systems into larger comprehensive arrangements[2] (Young 2002).

Berkes lists six effects of higher institutions on local institutions, these being (Berkes 2002):

1. Centralization of decision making, meaning local norms of management are set aside in favour of centralized decisions on management,
2. Shifts in systems of knowledge, i.e. when scientific management systems replace traditional and intuitive knowledge systems,
3. Colonization when local institutions are dismantled to create revenues for the state, an example being when India's colonial government took control over forests and grazing lands,
4. Nationalization of resources, meaning disempowering or reducing the power and influence of local institutions with the ostensible purpose of managing or protecting resource/s better,
5. Increased participation in markets, enabling entry of bigger and more resourceful players into domains that were hitherto restricted to local and smaller players/institutions in order to cater to increased and remunerative market demands,
6. Development policies, i.e. when state policies aimed at development and often supported by multilateral institutions or international development aid agencies disrupt local institutions and their land-use practices.

Certain levels of interventions by higher institutions may also serve the purpose of reviving and strengthening local institutions, among these being state recognition of local-level institutions through enabling legislation. Higher institutions may even assist local institutions by assisting in their cultural and political revitalization, through capacity building and by creating a favourable environment for institution building (Berkes 2002).

Field studies on climate change undertaken in Nepal and Bangladesh corroborate a lot of what Young (2003) and Berkes (2008) have to say about interactions between

[2]Ostrom (2005) describes this as a whole system at one level being a part of a system at another level. Such nested sub-assemblies of part-whole units in complex adaptive systems have also been referred to as holons by Arthur Koestler.

local- and higher-level institutions and hold out significant lessons for India. The field study of communities living in high altitude areas, the middle hills and the Terai region in Nepal revealed that on the whole the changes in climate and the impact of these observed by them did not vary significantly from those that were revealed in the IPCC AR4 report. The authors express the concern that even as the higher institutions in Nepal take time to build knowledge on meaningful action for adaptation and mitigation and develop the means to effectively communicate this knowledge, the peoples' knowledge built on continuous observation- and area-specific interpretation may be ignored which could lead to highly adverse outcomes (Chapagain et al. 2009).

The study in Bangladesh, on the other hand, concluded that in the absence of information or inadequate information and capacities, poor rural farmers remained stoic when faced with climate hazards, the study also expressed that the traditional knowledge in the face of rapid climate change may soon prove inadequate and would be lost if not supplemented with scientific training by higher institutions. The author, however, added that the training by the higher institutions would have to be done using simpler conceptual structures as even communities trained by NGOs about climate change-related hazards were not able to fully comprehend and therefore utilize the knowledge they had been exposed to (Wachinger 2010).

Scientists, decision and policy makers in the face of growing natural disasters, extreme weather events, economic and political crises and long-term changes such as those in demographic and climate are rethinking the traditional predict and control regime approach to consider alternative approaches. Pointing out that there is no panacea or single blueprint nearly all of them call for applying a diagnostic approach and undertaking institutional experimentation to enable learning and arrive at an institutional design suited to a specific context. Both Berkes and Ostrom lead on to some promising institutional designs that could facilitate interactions across scales; while the former calls these cross-scale linkages, the latter refers to the polycentric governance system.

1. Co-management Arrangements between Communities and Governments—A cross-scale institutional linkage that connects local-level management/users with government level through a formal power-sharing arrangement. Examples cited include those of JFM, to the implementation of the rights to resources of indigenous peoples in the United States, Canada, New Zealand and Australia. One reason for putting in place such institutional linkages could be legal but another reason for the growing interest in co-management initiatives is because these enable the leveraging of the strengths of both the local institution and the higher-level institution and doing away with the weakness of each.
2. Multistakeholder bodies—A related form of cross-scale linkage, multistakeholder bodies are not always easily distinguishable from co-management institutions but what does set them apart is that these link multiple user groups and interests, local and regional with the government, and provide a platform for conflict resolution and negotiations among users. The criticism is that multi-stakeholder bodies can become highly diffused with all kinds of stakeholders

becoming a part and reducing its efficacy allowing governments' to use these as a mere discussion forum without conceding any real shared management responsibility (Berkes 2002).

3. Development, Empowerment, Co-management Arrangements—What distinguishes this form of cross-scale linkages is the emphasis on community development and empowerment where co-management is an incidental outcome. Often such an arrangement also includes the involvement of NGOs or other institutions with a capacity building role. There are likely to be both horizontal and vertical cross-linkages. The potential for transmitting the development-empowerment experience at a horizontal level from one local institution to another makes this arrangement an attractive one (Berkes 2002).
4. Citizen Science—A few examples of this kind of cross-linkages are the environmental stewardship groups of Canada, regional associations for watersheds and lake quality in Sweden; many of these initiatives are from industrially developed countries that have robust civil society traditions and well-developed environmental movements but can also count among them the people's science movements of Kerala. These are characterized by citizen action for environmental management, and through the involvement of environmental NGOs, as also the tendency to use a mix of scientific knowledge with local observations (Berkes 2002).
5. Policy Communities and Social Movement Networks—Cross-scale linkages connecting local-level issues with the regional and international agendas. These could be in the nature of epistemic communities comprising scientists, government experts, and NGO representatives who share principled beliefs, notions of validity and policy goals that cut across political and administrative boundaries. An example cited is that of the Mediterranean Action Plan that was arrived at through the work of such an epistemic community that included representatives from nations that were otherwise in conflict. Some therefore also consider epistemic communities as a subset of policy communities. The role of NGOs once again comes to the fore as these institutions take on functions that states are unwilling or are unable to take on.
6. Polycentric Governance System—Ostrom refers to the polycentric governance system whereby citizens organize not just one but several governing mechanisms at differing scales. Each institution has considerable independence to devise and apply norms and rules within a well-defined geographical domain. While the mandate of some may be of general purpose governance others may have niche areas of involvement and engagement. All of these are nested in several levels of general purpose governance. In a polycentric system while local institutions would exercise some authority to determine how a particular resource should be governed and used, they would also benefit from the learning and experiences of peer institutions through horizontal linkages, as also be checked by the larger general purpose governance structures should they (local institutions) be affected by problems such as local tyrannies and inappropriate discrimination. Also, a polycentric system would enable a desirable blend of scientific inputs with local

knowledge as larger knowledge-generating institutions are able to effectively link with multiple local and smaller institutions (Ostrom 2005).

A policy brief prepared by Capri suggests that in order to identify appropriate institutional arrangements for effective responses to climate change, it may be useful to keep in mind both the spatial as well as the timescales. Through the use of examples, the policy brief illustrates the domain of the various institutions and how these may optimally interact, for instance, decisions at the level of an individual farmer to plant a drought-resistant variety of seed would require action at the level of higher institutions to produce the new varieties of seeds and to set in place mechanisms for their distribution. At the hamlet or village level, higher-level institutions could provide local institutions with technical, financial and/or material resources for effective responses and action, for instance assisting a group of farmers in building, operating and maintaining a source of irrigation. The building of horizontal linkages becomes more important further up in the national scale as higher-level institutions engage with each other in national and global interactions and negotiations, though local institutions through what Berkes calls Social Movement Networks may seek relevance and clout at both these levels (Meinzen-Dick et al. 2010).

The policy brief also recommends that institutional arrangements should be designed keeping in mind the timescale. For instance, some institutional arrangements may need to come into play only at the time of an extreme weather event and till its immediate aftermath is dealt with and not all the time. On the other hand, some climate impacts as also the response mechanisms may unfold over a few years and in some cases over an even longer time horizon. The authors point to the significance of property rights especially when there is a time lag between an action and its outcomes for instance efforts in ecological restoration (Meinzen-Dick et al 2010).

In a report prepared for The World Bank, Agarwal et al. (2012) examined closely issues of institutional capacity and cross-scale institutional coordination. The study found that cross-scale relationships are vital to local institutions, particularly for effective local adaptation. It was found that just as in India some national-level effort had and was being made to improve the country's response to climate change through current development strategies but missing in many cases was the link between the local-level adaptation processes set in motion by the local communities and the national-level plans. The study once again highlighted what has been stated earlier in the section that local responses are likely to be limited and inadequate in the face of risks posed by climate change unless these are supported by larger institutional responses in the form of information, capacity building and material resources (Agarwal et al. 2012).

The study also found significant gaps between higher institutions at the horizontal level which in turn adversely affected the response and coping capacities of the local institutions. The failure to share information, coordinate and closely interact at the horizontal level both at the level of local and higher institutions results in available knowledge and resources being underutilized and worse still result in advise and directives to local institutions that may be at cross-purposes. Despite the important role that local institutions could play, they were not well connected to higher-level institutions. The authors thus conclude that given this scenario all of this would result in an institutional response to climate change that falls short of what is theoretically possible (Agarwal et al. 2012).

As it moves ahead implementing the NAPCC and its component mission activities, India is likely to face similar situations and challenges; how issues of institutional linkages are handled will greatly influence India's ability to address and deal with its climate change challenges.

NAPCC and SAPCCs are action plans that are conceptualized in this paper as policies, as documents which lay the ground and offer a framework for conceptualizing climate change, thinking about problems of mitigation and adaptation, and identifying the key sectors which need to be transformed to work towards mitigation and adaptation goals. Thereby, the action plans also, directly and indirectly, come up with an institutional and governance framework for addressing climate change. Any critical assessment of climate governance in India has to adopt a 'policy process' perspective; the attempt here is to regard policy process as strongly related to broader discourses on a problem or issue and hence subjected to the influences of multiple or dominant public discourses on a theme. From a strategic knowledge viewpoint, linking the NAPCC to broader discussions of institutional linkages can yield significant benefits for the effectiveness of knowledge utilization in enhancing the efficiency of implementation of the action plans on climate change.

References

Agarwal A, et al (2012) Climate policy processes, local institutions, and adaptation actions: mechanisms of translation and influence. Wiley Interdisciplinary Reviews: Climate Change 3.6, pp 565–579

Berkes F (2002) Cross-scale institutional linkages: perspectives from the bottom up. In Ostrom E et al. (eds) The drama of the commons. National Academy Press, pp 293–321

Berkes F (2008) Commons in a multi-level world. International journal of the commons 2.1, pp 1–6

Berkes F (2009) Evolution of co-management: role of knowledge generation, bridging organizations and social learning. J Environ Manag 90(5):1692–1702

Betsill MM, Bulkeley Harriet (2006) Cities and the multilevel governance of global climate change. Glob Gov Rev Multilateralism Int Organ 12(2):141–159

Bidwai P (2012) The politics of climate change and the global crisis. Mortgaging our future. Orient Black Swan, New Delhi

Bisaro A, Hinkel J, Kranz N (2010) Multilevel water, biodiversity and climate adaptation governance: evaluating adaptive management in Lesotho. Environ Sci Policy 13(7):637–647

Bulkeley H, Betsill M (2005) Rethinking sustainable cities: multilevel governance and 4. The 'urban' politics of climate change. Environmental politics 14.1, pp 42–63

Byravan S, Rajan SC (2013) An evaluation of India's national action plan on climate change

Chapagain N, Banjade MR (2009) Community forestry and local development: experiences from the Koshi Hills of Nepal. J For Livelihood 8(2):78–92

Chopra, K, Dasgupta P (2008) Nature of household dependence on common pool resources: an empirical study. Econ Polit Wkly: 58–66

Cruz RV (2007) "Asia." Climate Change (2007): Impact, adaptation and vulnerability. Contribution of working group II to the fourth assessment report of the intergovernmental panel on climate change, pp 469–506

Dubash NK (2009) Toward a progressive Indian and global climate politics. Centre for policy research climate initiative. *Working paper* 1

Dubash N (ed) (2012) Handbook of climate change and India: development, politics and governance. Routledge

Dubash NK (2013) The politics of climate change in India: narratives of equity and cobenefits. Wiley Interdisc Rev Clim Change 4(3):191–201

Dubash, NK, Jogesh A (2014) From margins to mainstream? State climate change planning in India as a ‘door opener’ to a sustainable future

Dubash NK, Joseph NB (2016) Evolution of institutions for climate policy in India. Econ Polit Wkly 51(3):45

Dubash N et al (2013) Indian climate change policy. Econ Polit Wkly 48(22):47–62

EPW Editorial (2008) Climate change: not vision, not plan. Econ Polit Wkly 43(28):5–6

Hooghe L, Marks G (2001) Multi-level governance and European integration. Rowman & Littlefield

Khaladkar RM, Mahajan PN, Kulkarni JR (2009) Alarming rise in the number and intensity of extreme point rainfall events over the Indian region under climate change scenario. Indian Institute of Tropical Meteorology

Lélé S (2006) Thinking about ecological sustainability. SEMINAR-NEW DELHI, vol 564. MALYIKA SINGH

Meinzen-Dick R, Markelova H, Moore K (2010) The role of collective action and property rights in climate change strategies. CAPRi Policy Brief 7:1–4

National Action Plan on Climate Change, Ministry of Environment and Forests, Government of India (2007)

Ostrom E (2005) Self-governance and forest resources. Terracotta reader: a market approach to the environment, 12

Pittock J (2011) National climate change policies and sustainable water management: conflicts and synergies. Ecol Soc 16:2

Vittal KPR et al (2006) Crop diversification in rainfed regions of India. In: Sharma KD, Soni B (eds) Land use diversification for sustainable rainfed agriculture, pp 24–70

Wachinger G et al (2010) Risk perception and natural hazards. WP3-report of the CapHaz-Net Projekt. URL: http://www.caphaz-net.org. Synergien zwischen Naturschutz und Klimaschutz–Wasser/Gewässer (management)

Young OR (2002) Institutional interplay: the environmental consequences of cross-scale interactions. The drama of the commons, pp 263–291

Young OR (2003) Environmental governance: the role of institutions in causing and confronting environmental problems. International Environmental Agreements 3.4, pp 377–393

Economic Assessment of Climate Mitigation Pathways (2015–2050) for the Brick Sector in India

Priyanka Jajal, Kushal Tibrewal, Trupti Mishra and Chandra Venkataraman

Abstract India is the second largest producer globally, of fired brick in traditional kiln technologies with 250 million bricks produced yearly. Amongst different sectors, brick production is an important contributor to several SLCPs like black carbon, and ozone precursors like non-methane volatile organic compound and carbon monoxide. This report presented develops and evaluates the evolution of SLCP emissions from India, during 2015–2050, from the brick sector, under two different scenarios of diffusion of cleaner technologies and practices, compared to that under a reference scenario. Net emissions of SLCPs from India in 2015 from the brick sector are estimated at 422.89 GT CO_2 eq. (using GWP-20). Total achievable mitigation of SLCPs in 2050 is 25% under a promulgated policies scenario, while it is 80% under a prospective policies scenario. Mitigation strategies of brick industry require shift from traditional burnt bricks to introducing higher use of unburnt bricks. A cost–benefit analysis shows that SLCP mitigation cost is as low as 59 Rs./tonne of CO_2 eq., which is significantly lower than those estimated for GHG mitigation in India. Actions reducing emissions from brick industry require shift towards manufacturing of unburnt bricks which also reduce warming of the atmosphere and benefit public health.

P. Jajal · K. Tibrewal · T. Mishra · C. Venkataraman
Interdisciplinary Program (IDP) in Climate Studies, Indian Institute of Technology Bombay, Mumbai 400076, Maharashtra, India
e-mail: Jajal.priyanka@gmail.com; 154406005@iitb.ac.in

K. Tibrewal
e-mail: kushalarts@gmail.com

C. Venkataraman
e-mail: chandra@iitb.ac.in

T. Mishra (✉)
Shailesh J. Mehta School of Management, Indian Institute of Technology Bombay, Mumbai 400076, Maharashtra, India
e-mail: truptimishra@iitb.ac.in

C. Venkataraman
Department of Chemical Engineering, Indian Institute of Technology Bombay, Mumbai 400076, Maharashtra, India

C. Venkataraman et al. (eds.), *Climate Change Signals and Response*,
https://doi.org/10.1007/978-981-13-0280-0_17

Keywords SLCP · Mitigation policy · Scenario analysis · Mitigation cost Bottom-up approach

List of Abbreviation

AIM	Asia-Pacific integrated model
BTK	Bull's trench kiln
CBA	Cost–benefit analysis
CCAC	Climate and clean air coalition
CEA	Cost-effective analysis
CGE	Compound general equilibrium
CH_4	Methane
CO	Carbon monoxide
CO_2	Carbon dioxide
GHGs	Greenhouse gases
GWP	Global warming potential
HFCs	Hydrofluorocarbons
LPG	Liquefied petroleum gas
NMVOC	Non-methane volatile organic compound
NOx	Nitrogen oxides
PM	Particulate matters
SLCPs	Short-lived climate pollutants
SO_2	Sulphur dioxide
VSBK	Vertical shaft brick kilns

1 SLCPs—An Underlying Opportunity for India

Global warming is a pressing issue in the current world, where the source of the problem is the greenhouse gases (GHGs) to be blamed. However, the recent research has found that there are two types of pollutants which cause warming of the atmosphere: one the GHGs and the other the short-lived climate pollutants (SLCPs) (Victor et al. 2015). As the name suggests, SLCPs have a shorter lifespan in the atmosphere compared to the GHGs. GHGs can be present in the atmosphere for as long as 100 s of years, whereas SLCPs have a lifetime in the range of days to weeks and maximum of 14 years for methane (CH_4). SLCPs are unique as they exert warming and cooling to the atmosphere, along with which they also cause health problems to humans.

SLCPs are a number of pollutants including, methane (CH_4), hydrofluorocarbons (HFCs), particulate matter (PM), sulphur dioxide (SO_2) and stratospheric ozone (UNEP 2014; Mass and Grennfelt 2016). Specific contents of particulate matter such as organic carbon and black carbon are of utmost importance as they exert negative and positive warming to the atmosphere, respectively. On the other hand, stratospheric ozone cannot be found directly in the atmosphere, and a combination of various other pollutants result in ozone formation. Such pollutants include, carbon monoxide (CO), volatile organic compounds (VOCs) and nitrous oxides (NOx) (Victor et al. 2015). The SCLPs exert higher warming or cooling than CO_2 for a shorter period, making them important to curb the temperature changes in the short run. Moreover, SO_2, PM and CO are the local air pollutants, impacting human health in various manners. Hence, SLCPs have double-fold effect and curbing them would not only clean the air, but also reduce warming of the atmosphere.

India is one of the prominent emitters of SLCP in the world, with emissions coming from residential cooking, agriculture burning, brick manufacturing and transport sector (CCAC 2015; Venkataraman et al. 2016). The mitigation strategies are being laid out by CCAC for SLCP mitigation; however, India is not an active partner in the coalition making it difficult to channelize the resources. Residential cooking stoves are inefficient and use fuelwood, instead of liquefied petroleum gas (LPG) and other energy-efficient fuels. However, the issue of residential cooking is diverse and concerns individual in every household, which makes it difficult to reach the extent of people to sensitize towards the same. Moreover, the economic cost for such a change may not be huge for a country, but an individual may not have the financial capacity to make the shift as India is a developing country with many living under poverty line.

Similarly, SLCPs from transport sector are being taken care of by advancements in vehicles efficiency improvements; however, they require monetary investment at the individual level. The personal choices are difficult to be restricted, especially in a country like India where the population is huge and the economic status of individuals are diverse. Moreover, initiating public transport and encouraging the use of the same has already been promoted by the government. Another scope for SLCP reduction lies in the agriculture waste burning, where the waste from current harvest is burned to prepare the soil for the next harvest. Similar to cooking practices, agricultural practices are part of the culture for individuals, requiring change which demands financial investments along with the habitual change. Various small-scale studies have been undertaken regarding the above-mentioned issues (Ramanathan and Parikh 1999; Reddy and Venkataraman 2002; Venkataraman et al. 2010; Sadavarte and Venkataraman 2014; Busby and Shidore 2017); however, a country-wide success story is difficult to find under the literature.

Brick manufacturing is one of the major SLCP emitting sectors considered for mitigation under SLCP literature (USEPA 2012; CCAC 2015; Ministry of Environment Forest and Climate Change Government of India 2015). The literature does not estimate the emissions or potential mitigation opportunities available in brick production; however, it is widely stated that the emissions from firing the fuel to bake bricks emit SLCPs. Hence, the study looks into the present emissions and mitigation potentials of the same in India. Firstly, brick manufacturing is an industry which is currently unregulated in India, where the industrial regulations do not apply except for a few with emission standards and technology upgradation (Ministry of Environment Forest and Climate Change Government of India 1986). The regulations have set emission standards restricting emissions of sulphur dioxide and particulate matter. Moreover, national brick mission has been launched to guide the progress of the industry (Bhushan et al. 2016). Such control over the industry gives room for policy guidelines and progress of the brick industry.

This section highlighted the importance of studying the brick industry, especially as a mitigation measure. Section 2 introduces brick industry in India, production technologies and current trends in the same. Section 3 is the tools of analysis, where the methodology adopted to develop the chapter is discussed along with the scenario description of each one of them. After which, the prospective futures of the brick industry are discussed in Sect. 4 along with the cost analysis. Finally, the implications of the chapter and its implications for the policy development are described under Sect. 5.

2 Brick Industry in India

The bricks have been one of the prominent building materials since the history of buildings. Along with the increasing building structures, brick production has also been growing in India, making India the second largest producers of bricks after China. With the production coming from about 100,000 kilns across the country, total production of bricks was 250 billion in 2015 (Lalchandani and Mithel 2013; Rajarathnam et al. 2014). The kilns are located in the large cities such as Delhi, Kolkata and Mumbai, with the spread across adjacent states. Moreover, the states of Assam, Karnataka and Tamil Nadu also have spread of kilns around the cities (Maithel 2003).

Bricks have traditionally been produced using sand moulds fired at a higher temperature to develop into a hard cake with the resistance power. However, the advancement in construction material has led to cement blocks which are produced from hardening cement into desired shape and size. The modern bricks do not require any firing and hence less energy consuming option for production of bricks. Fired and non-fired bricks such as cement blocks are the two market-available categories of bricks in the world. However, in India, non-fired bricks have not picked up yet due to unavailability of the bricks and lack of awareness about the option. In the megacities, the non-fired bricks are seen to have made an appearance in recent times; nevertheless, the extent of the same is negligible.

Looking at the fired bricks, they are produced in two types of kilns: one which runs intermittently and the others which are continuous. Intermittent kilns prearrange the bricks and fuel mix to be fired, which runs in a batch (Kumar and Maithel 2016). Once one batch of bricks is fired, the next one is arranged and fired and so on. Clamps are the prominent kilns which use this technique in India, with the share of 25% of total production. However, such arrangements are low in efficiency and lead to higher emissions. Moreover, as there is no structure required for this arrangement, control of emissions is also not possible, making it one of the dirtiest technologies.

The continuous kilns are the ones where the firing of bricks and staking of wet bricks is done simultaneously at different ends of the kiln. This type of kilns has a boundary wall in an oval shape that guides the flow of the currently firing, loading and unloading bricks. Such a kiln allows for a guided flow of flue gases, making it possible to provide a chimney and possible emission reduction measures. Bull's Trench Kiln (BTK), zigzag firing and Vertical Shaft Brick Kiln (VSBK) are the continuous type of kilns with the share of 66%, 8% and 1% of total production, respectively. The continuous kilns have higher fuel efficiency than clamps with varying emissions. BTK has the lowest fuel efficiency and higher emissions compared to other continuous kilns. Zigzag firing is a modified version of BTK, where the flow of firing and flue gases are guided. It is evident that zigzag firing is more efficient than BTK as well as less polluting. VSBK, on the other hand, is most efficient and lowest polluting technology amongst the continuous firing category (Kumar and Maithel 2016). Continuous firing kilns can install an air pollution reduction device in the chimney to reduce the pollution; however, for the GHG emission, no such measures are possible.

As mentioned earlier, the construction activities have been increasingly demanding more brick production year after year. The current trend shows an increase of 6.6% per annum of brick production since 2012, which is expected to continue in the near future (Bhushan et al. 2016). Such a growth rate is evident as the construction requirement in present and the future is going to increase with increasing population and income. Considering the fuel consumption, due to fired brick use, coal is used around 35 million tons (MT), which is 15% of the total industrial coal use. The emissions from coal have not been estimated; however, it is expected to be huge. CCAC is one of the coalitions working on SLCP emission mitigation opportunities, which identified brick production in South Asia as one of the key sectors (Valdés 2016). Specifically, in India, the brick manufacturing is expected to grow in the future at a similar rate. Hence, there is potential for SLCP mitigation by adopting various standards of emissions and technology shifts.

This chapter delves into brick production in the future, where similar growth rate of production is assumed. However, the technology distribution in the current state is not energy efficient and leads to emissions, which can be reduced by shifting to the better-fired technologies or non-fired technologies. Such trends are difficult to determine especially when the current government rules do not show encouragement towards non-fired bricks. A variety of future scenarios are possible in such cases, from which the best economically and environmentally friendly measure should be adopted while forming a policy. Hence, the next section explains a set of possible scenarios for brick manufacturing.

The mitigation of emissions requires investments in monetary terms from the producers. Hence, an economic analysis has also been proposed. The economic analysis has been performed in two ways under the existing literature, namely cost-effective analysis (CEA) and cost–benefit analysis (CBA). CEA values costs in monetary terms with benefits in 'physical' units, which is used to analyse a policy outcome in an economy. On the other hand, CBA calculates costs in monetary terms and benefits in economic terms where various market approaches with the broader definition is adopted. In CBA, benefits are called ecological benefits which incorporate environmental, health and livelihood improvements into account. However, CBA requires an empirical study on the quantification of the benefits achieved by the mitigation, which could be an extension of the current study. CEA analysis gives a cost which can also be called mitigation cost in the climate change analysis. The mitigation cost comparisons between the literatures determine the priority of an action/policy measure based on the cost minimization principle. Hence, in the present study, only the CEA approach is used for the economic analysis where costs are derived in monetary units and the benefits are derived in physical units of CO_2 equivalent.

3 Tools of Analysis

A sectoral analysis has been studied using various approaches such as top-down and bottom-up. Top-down approaches equate demand and supply of the product in the given economy. Such models base their assumptions on an economy where consumers and producers play a key role in achieving the future equilibrium. Computed general equilibrium (CGE) is one such model, which has been used widely for issues related to climate change with the aim to assess policy changes (Farmer and Steininger 1999; Xu and Masui 2009; Antimiani et al. 2015; Li and Jia 2016). On the other hand, bottom-up approach aggregates information at the plant level to project future emissions with the information such as the addition of new plants and capacity expansion in mind. Various application of the approach has been seen in the literature; however, the most popular models include MARKAL (Sulukan et al. 2010) and Asia-Pacific integrated model (AIM) (Wen et al. 2014).

In the current study, various information has been drawn at the national level such as total production and the share of each technology. Current emissions and future emissions are derived using technology-wise energy requirement and emission factors. Such an approach uses plant-level information of technology, energy used along with emissions to build it up to a national aggregate level, which signifies it as a bottom-up approach. The flow chart shown in Fig. 1 explains the methodology adopted for each scenario generation where firstly the yearly production of bricks is divided into the technology-wise production using a share of each one. From the number of bricks produced, the energy required for production is calculated using specific energy and calorific value of the fuel. The fuel used in India is assumed to be a mixture of coal and biofuels from the waste, as the primary survey of brick kilns signifies use of mixed fuels across the country (Weyant et al. 2014). Emissions of each type of pollutant are then calculated using emission factor of that pollutant for a particular technology. Total emission from brick industry has been derived by summing up the emissions from each individual technology for that year.

For the purpose of this study, 2015 is taken as a base year with the analysis period from 2015 to 2030 and then from 2030 to 2050. 2030 has been chosen as a mid-point of the analysis, where various policies would be revised and modified according to the requirement. Three scenarios have been developed: the first one being the reference scenario based on which the other two are assessed. Second is Promulgated Policies scenario where the existing policies are applied rigorously in the sector. Finally, the third is prospective policies scenario in which the targets of the current policies are

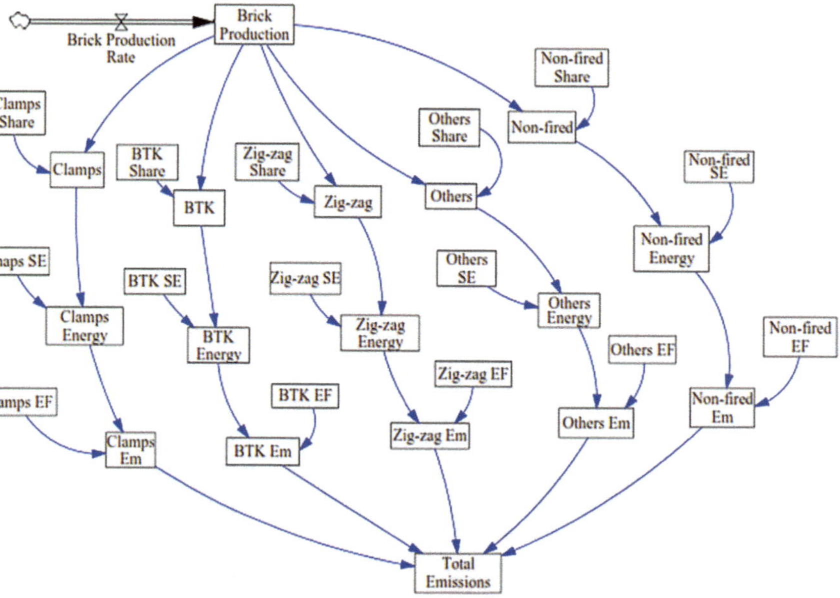

Fig. 1 Emissions calculations for brick manufacturing using specific energy (SE) and emission factor (EF)

achieved earlier and new policies are formed with even more commitment towards the emission reductions. Under each scenario, production growth rate is assumed to stay constant at 6.6% per annum following the current trend.

However, the technology share is assumed to change in each scenario, which is illustrated in Fig. 2. BTK and clamps contribute highest to the overall brick emissions in 2015, as the share of these technologies is highest. The cleaner technologies, such as zigzag, hollow bricks, and non-fired bricks only account for 3% of the total bricks production. The government of India has approached the issue of brick emissions by incorporating the shift from moving chimney BTK to fixed chimney BTK (CPCB 2017). Along with this, the producers are encouraged to convert from BTKs to zigzag, which are less polluting due to high energy efficiency acquired in the design. Hence, the reference scenario assumes pursuance of the similar trend with the highest share of production coming from clamps and BTK combined at 70% in 2030 and slowly reducing to 40% in 2050.

The other two scenarios have been inspired from the literature along with changes deemed required (Venkataraman et al. 2017). The Promulgated Policies have vouched towards high-efficiency fired kilns for the short term (2030) with effectively growing the share of non-fired bricks in the long run. The share of BTK and clamps reduce to 25% by 2050, whereas non-fired bricks would have 45% of share in total production. The high-efficiency fired bricks would have a moderate share of 30% by 2050. Lastly, prospective policies are assumed to promote non-fired bricks in the short term at a

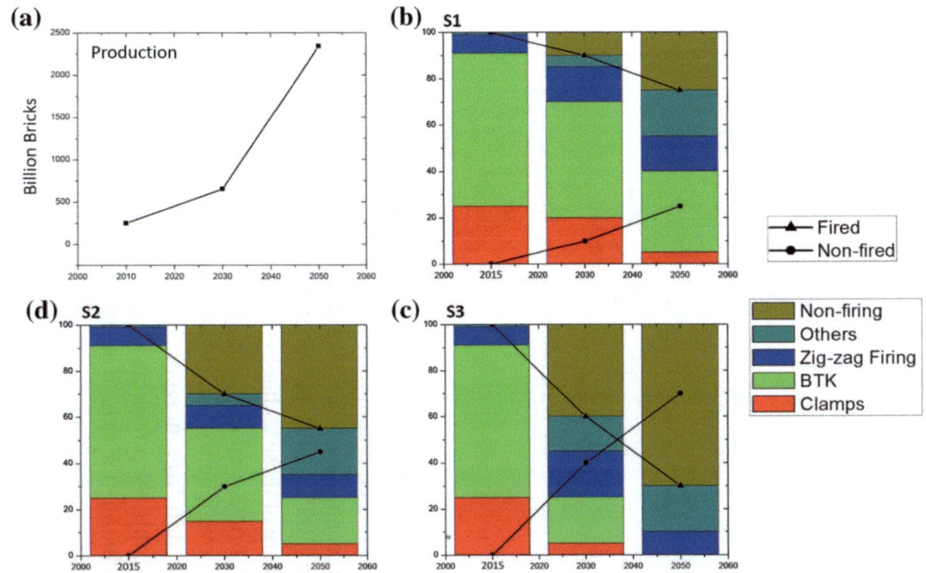

Fig. 2 Assumptions related to scenario formation **a** projected brick production in billion bricks; **b**, **c**, **d** technology shift and proportion of fired and non-fired brick production over 2015, 2030 and 2050 for S1, S2 and S3, respectively

Table 1 Operational and capital costs of brick manufacturing technology-wise, converted into Rs./1000 bricks

Operational cost (per 1000 bricks)	BTK	Clamps	Zig-zag firing	VSBK	AAC (non-fired)
Fuel	416	850	240	182	0
Electricity	0	0	12	9	133
Raw material cost	0	0	0	1855	980
Frieght in	115	370	137	137	90
Freight out	115	370	13.	137	90
Maintenance cost	37	8	54	52	9
Labour cost	20	48	24	1	3
Total Operational Cost	**705**	**1646**	**604**	**2374**	**1306**
		Capital Cost			
Capital Cost	3250000	250000	3900000	20000000	10000000
Years of operation	25	25	25	25	25
Annualized capital cost	502773	38675	6d03328	3095975	1547988
Total Annualized Capital Cost (per 1000 bricks)	**84**	**19**	**121**	**619**	**103**
Total cost (per 1000 bricks)	**789**	**1665**	**725**	**2993**	**1409**

higher rate compared to the fired bricks. Awareness regarding the air pollution and emission standards is assumed to play a major role in the shifts, along with the climate change awareness. By 2030, the share of fired bricks is assumed to reduce drastically to 60% compared to 100% in 2015. Moreover, the trend is assumed to intensify with the share of non-fired bricks at 70% by 2050, where the dirty technologies such as clamps and BTK would have retired from the market.

A cost–benefit analysis of the scenario is undertaken where costs are availed from market place as well as from the information collected under primary survey. To manufacture bricks, the producer has to bear two types of costs, one which are upfront, capital costs required to establish the setting up of the kiln. The other cost is the operational cost, through which daily production is materialized. As capital cost is a one-time cost, it is converted into a daily cost using 15% interest rate, 25 years of operation period and average production capacity of each kiln type. The operational costs include fuel, electricity, raw material, freight in and freight out, maintenance and labour cost. For each type of kiln, total production cost per 1000 bricks is presented in Table 1.

CEA analysis is performed where the costs per scenario in the given year is calculated. The mitigation potential for the same year is acquired from the scenario analysis, wherein the global warming potential (GWP) 20 values are used to make the emissions comparable in CO_2 equivalent. GWP of 20 years is used as the maximum

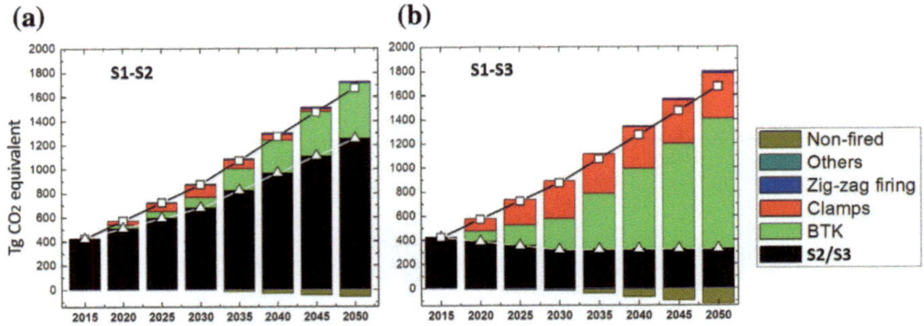

Fig. 3 SLCP emissions change in terms of Tg CO_2 equivalent **a** from promulgated policies (S2) to reference scenario (S1) **b** from prospective policies (S3) to reference scenario (S1)

residence time of SLCP is 10 years, making it impossible for the emissions to exert its warming capacity by 100 years. Moreover, to develop a policy, a reasonable timeframe of 20 years is considered, as in the rapidly changing world, a century becomes too long. Hence, the benefits are accounted in Tg CO_2 eq. for the given scenario in the given year. Finally, a mitigation cost is calculated by dividing cost with the emissions in the given year for the shift in the scenario from reference scenario towards promulgated policies (S2) or prospective policies (S3).

4 Future of Bricks

As production of bricks is expected to increase at 6.6% per annum growth rate throughout the study period, the total production of bricks increases up to 2.3 trillion bricks per year by the end of 2050. As per the technology distribution, reference scenario is the least developed with highest energy consumption leading to highest emissions. SLCPs are the focus of the study, specifically the ones which show positive GWP20 value, such as BC, CO, N_2O, CH_4 and NMVOC. The GWP20 values have been used from the IPCC fifth assessment report (Myhre et al. 2013). Promulgate policies scenario (S2) assumes moderate changes in the technology diversions from fired bricks, indicating lower penetration of non-fired bricks in the market, leading to a decrease of 195 GT and 421 GT CO_2 equivalent in 2030 and 2050, respectively (Fig. 3a). However, when all the SLCPs including OC, SO_2 and NOx are considered, the emission reduction observed is 219 GT and 3955 GT CO_2 equivalent in 2030 and 2050, respectively.

The prospective policies (S3) scenario moves towards the cleaner path, resulting in a reduction of emissions compared to S2. In 2030, as much as 555 GT CO_2 equivalent can be reduced, by adapting non-fired bricks into use. The emission reductions achieved could be as high as 1345 GT CO_2 equivalent by 2050, if the dependence on non-fired bricks is increased up to 70% (Fig. 3b). The reduction in emissions changes

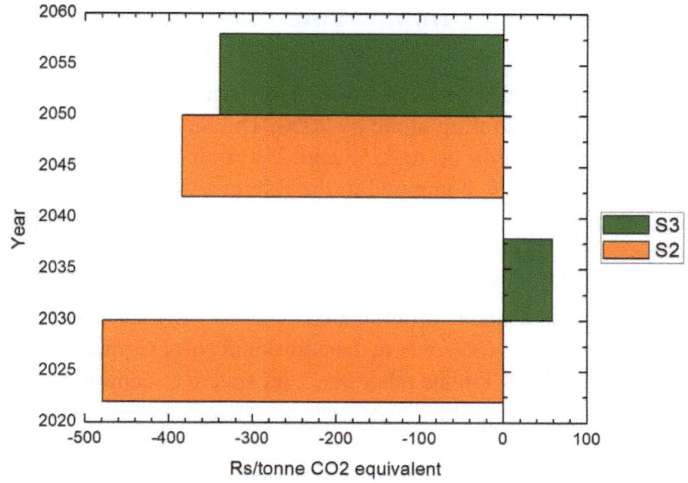

Fig. 4 Mitigation costs for 2030 and 2050 for S2 and S3 by achieving reductions compared to reference scenario in Rs./tonne CO_2 eq.

when negative SLCPs are included, the total emission reductions achieved in 2030 and 2050 amount up to 540 GT and 9227 GT CO_2 equivalent, respectively. The net reductions achieved are higher compared to the S2 scenario, even when compared with only positive SLCPs.

Production costs have been discussed earlier which indicate that operational and capital costs of VSBK are the highest amongst others. Even non-fired bricks have lower capital and operational costs, indicating that promoting the use of non-fired bricks is an economically and environmentally friendly option. As S3 promotes boosted use of non-fired bricks, the cost paid for shifting technology leads to reduced emissions at 300 billion Rs. in 2030 and 500 billion Rs. in 2050. On the other hand, to follow the path indicated under promulgated policies scenario, the costs paid are 50 billion Rs. and 220 billion Rs. in 2030 and 2050, respectively.

Even though the costs incurred in achieving the path under S3 is high, mitigation achieved is comparatively higher leading to mitigation costs of 59 Rs./tonne of CO_2 eq. in 2030 and −339 Rs./tonne of CO_2 eq. in 2050. The mitigation costs of S2, on the other hand, are negative in 2030 and 2050 at −479 Rs./tonne of CO_2 eq. and −384 Rs./tonne of CO_2 eq. due to reduced cost of production compared to S1 (which is indicated in Fig. 4). All the comparisons are based on the reference scenario (S1); where fired bricks are assumed to take up most part of the market demand. Even though, the cost of mitigation for the non-fired bricks is higher under S3, the long-term sustainability of the cement blocks is evident at similar cost in 2050.

5 Aggregate Economic Implications of Brick Industry

SLCP reductions of as much as 1345 GT CO_2 eq. can be achieved by technology shifts of the brick manufacturing alone by 2050. The modest change in technology could also achieve mitigation up to 22% and 25% reductions in the emissions by 2030 and 2050, respectively, compared to the reference scenario. Such a reduction suggests an immediate potential mitigation opportunity to be tapped by minimum efforts. However, the higher efforts could lead to emission reductions up to 60% in 2030 and up to 80% by 2050 under the prospective policy scenario (S3). To achieve reduction, the costs paid under promulgated policies (S2) are lower compared to prospective policy scenario (S3), where the additional costs required by 2030 is 11% and reduces to 8% by 2050. On the other hand, prospective scenario requires a much higher investment of 53% in 2030; however, the investment reduces with time to 20% by 2050.

Such investments are the addition over a period of time, making it possible for the producers to plan their strategy wisely. Moreover, investing 20% additional to the reference scenario may reap up to 80% of emission reductions is an attractive prospective for the cleaner future only from brick production. The mitigation potential of SLCP is huge compared to CO_2, as the recent study by McKinsey suggests that India has potential to mitigate 2.7 Gt CO_2 per year in 2030 which is 54% reduction from the business as usual scenario under the study (McKinsey 2009). The current study shows that the moderate efforts could lead to reductions of 219 Gt CO_2 per year by 2030. Moreover, the energy efficiency improvements are natural to an industry with time; however, they have not been considered as a function of time in this study. Considering the natural rate of efficiency improvement may lead to better emission reductions in the given cost.

It is evident from the above discussion that SLCP mitigation is an important aspect of climate change and mitigation debate with the additional benefits of local air pollution mitigation. Considering the modest mitigation may also lead to huge emission reductions compared to only GHG mitigation. The brick manufacturing industry is one such easy target, wherein the growth can be controlled and mitigations can be achieved. Technological standards should be set by the government in order to mitigate the emissions. To target the mitigation, costs paid by the industry is as low as 59 Rs./tonne CO_2 eq. compared to 1500 Rs./tonne CO_2 eq. required for reduction of GHGs (McKinsey 2009). A policy should be proposed for brick industry to reduce the emissions which will act as a bridge before the long-term actions are taken towards GHG mitigation.

The study has delved into the environmental and economic impact of mitigation from the brick industry for various scenarios in the given timeframe. The analysis suggests that policy should have a short-term and a long-term focus targets, the short term being from 2015 to 2030; 15 years of time span, and long term should extend up to 2050. It would be advisable to slowly reduce emissions by directly converting fired bricks manufacturing to non-fired bricks manufacturing. Such an approach would reap better results in the short run and long run, as non-fired bricks cost less and emissions are also less. The study recommends a mitigation strategy specifically designed for brick manufacturing with two sets of timelines and goals.

The study has comprehensively studied SLCP mitigation opportunities from brick industry in the future with a set of plausible scenarios. However, it is not certain that any of the scenario firmly represent the future of brick industry, rather combination of such scenarios are likely outcome of the brick emissions. The scenarios are developed based on the technologies that exists in the present, future innovation of technologies have not been taken into account. Moreover, the costs assumptions under the study are based on the current market prices, inflation over the future has not been taken into account. Lastly, the sole focus of the study is on SLCPs; GHGs have not been addressed in the study.

6 Summary

Global warming reductions would require actions against GHG mitigations; however, SLCPs can provide an opportunity to slow down the warming in the short run. Brick manufacturing is one such industry which can lead to potential mitigation for SLCPs as well as local air pollutants. The study estimates emissions from two policy scenarios: one promulgated and two prospective. The mitigation potentials are compared with the reference scenario which is an extension of the existing trends of brick production. The study extends to calculate mitigation cost per tonne of CO_2 eq. avoided using cost-effective analysis for which costs are assumed from the surveys performed.

Under brick productions, the major mitigations are a result of technology change from clamps to VSBK, and from fired bricks to non-fired bricks. The least polluting bricks are non-fired bricks, as the pollution from the same comes from energy use for the production process, which amounts to be almost null. On the other hand, the costliest technology is VSBK, due to higher capital cost requirements. Currently, promulgated policies advocate to use fired bricks manufactured from high-efficiency path such as zigzag firing and VSBK; however, the share of fired bricks in the market would remain higher at a total of 70 and 55% in 2030 and 2050, respectively. The prospective policies are expected to promote use of non-fired bricks with a total share of 40% in 2030 and 70% share in 2050. The resulting mitigation potentials are two times compared to S2 in 2030 and 2.5 times in 2050.

An economic analysis using CEA tool has been used to compare the mitigation costs for each scenario. As mentioned earlier, VSBK is the costliest technology amongst all; the overall mitigation cost under S3 is highest at 59 Rs./tonne CO_2 eq. in the short run (2030). The mitigation cost in 2050 reduces to −338 Rs./tonne CO_2 eq. due to proposed increase in non-fired bricks. On the other hand, S2 leads to negative costs due to lower investment requirement with reduced number of blocks required at −479 Rs./tonne CO_2 eq. and −383 Rs./tonne CO_2 eq. in 2030 and 2050, respectively.

From the above discussion, it is evident that in the long run, non-fired bricks are environmentally and economically friendly. However, promoting VSBK as an emission friendly option may not have economically sound effect on the country. The current directive by CPCB suggests that any brick kilns producing bricks using clamps or fixed chimney BTK has to convert the kiln into a natural draft zigzag kiln (CPCB 2017). The study presented here advocates in the short-term to continue with the zigzag firing technology and aim to slowly move towards the non-fired bricks as it is financially more viable. However, there is no substitute for non-fired bricks and must be promoted in the short as well as long run. Similarly, a policy should be designed that reaps maximum benefits by putting in short-term cost of 59 Rs./tonne CO_2 eq. with the potential reductions of 555 GT CO_2 eq. by 2030 by moving away from BTK and clamps. Whereas in the long term, non-fired bricks along with no dependence on clamps and BTK by 2050 should be promoted costing −338 Rs./tonne CO_2 eq. to mitigate 1345 GT CO_2 eq.

References

Antimiani A, Costantini V, Paglialunga E (2015) The sensitivity of climate-economy CGE models to energy-related elasticity parameters: implications for climate policy design. Econ Model 51:38–52. https://doi.org/10.1016/j.econmod.2015.07.015

Bhushan C, Basu D, Kumar Yadav N, Kumar R (2016) National brick mission—a scoping paper. Cent Sci Environ New Delhi:1–28

Busby JW, Shidore S (2017) When decarbonization meets development: the sectoral feasibility of greenhouse gas mitigation in India. Energy Res Soc Sci 23:60–73. https://doi.org/10.1016/j.erss.2016.11.011

CCAC (2015) 5-year strategic plan (2020)

CPCB (2017) Directions of Air (Prevention and Control of Pollution) Act, 1981 for brick kilns

Farmer K, Steininger KW (1999) Reducing CO2-emissions under fiscal retrenchment: a multi-cohort CGE-model for Austria. Environ Resour Econ 13:309–340. https://doi.org/10.1023/A:1008243611666

Kumar S, Maithel S (2016) Introduction to brick kilns & specific energy consumption protocol for brick kilns. https://breathelife2030.org/wp-content/uploads/2016/09/12.pdf

Lalchandani D, Mithel S (2013) Towards cleaner brick kilns in India

Li W, Jia Z (2016) The impact of emission trading scheme and the ratio of free quota: a dynamic recursive CGE model in China. Appl Energy 174:1–14. https://doi.org/10.1016/j.apenergy.2016.04.086

Maithel S (2003) Energy utilisation in brick kilns

Mass R, Grennfelt P (2016) Towards cleaner air. Scientific Assessment Report 2016

McKinsey (2009) Pathways to a low-carbon economy: Version 2 of the global greenhouse gas abatement cost curve

Ministry of Environment Forest and Climate Change Government of India (1986) The environment (protection) rules, 1986

Ministry of Environment Forest and Climate Change Government of India (2015) India first biennial update report to the United Nations framework convention on climate change

Myhre G, Shindell D, Bréon F-M et al (2013) Anthropogenic and natural radiative forcing. Clim Change 2

Rajarathnam U, Athalye V, Ragavan S et al (2014) Assessment of air pollutant emissions from brick kilns. Atmos Environ 98:549–553. https://doi.org/10.1016/j.atmosenv.2014.08.075

Ramanathan R, Parikh JK (1999) Transport sector in India: an analysis in the context of sustainable development. Transp Policy 6:35–45. https://doi.org/10.1016/S0967-070X(98)00030-4

Reddy MS, Venkataraman C (2002) Inventory of aerosol and sulphur dioxide emissions from India: I—fossil fuel combustion. Atmos Environ 36:677–697. https://doi.org/10.1016/S1352-2310(01)00463-0

Sadavarte P, Venkataraman C (2014) Trends in multi-pollutant emissions from a technology-linked inventory for India: I. Industry and transport sectors. Atmos Environ 99:353–364. https://doi.org/10.1016/j.atmosenv.2014.09.081

Sulukan E, Sağlam M, Uyar TS, Kirlidoğ M (2010) Determining optimum energy strategies for Turkey by MARKAL model. J Nav Sci Eng 6:27–38

UNEP (2014) The emissions gap report 2014: a UNEP synthesis report

USEPA (2012) Report to congress on black carbon

Valdés HM (2016) CCAC annual report 2015–16

Venkataraman C, Sagar AD, Habib G et al (2010) The Indian national initiative for advanced biomass cookstoves: the benefits of clean combustion. Energy Sustain Dev 14:63–72. https://doi.org/10.1016/j.esd.2010.04.005

Venkataraman C, Ghosh S, Kandlikar M (2016) Breaking out of the box: India and climate action on short-lived climate pollutants. Environ Sci Technol 50:12527–12529. https://doi.org/10.1021/acs.est.6b05246

Venkataraman C, Brauer M, Tibrewal K et al (2017) Source influence on emission pathways and ambient PM2.5 pollution over India (2015–2050). Atmos Chem Phys. https://doi.org/10.5194/acp-2017-1114

Victor DG, Zaelke D, Ramanathan V (2015) Soot and short-lived pollutants provide political opportunity, pp 1–3

Wen Z, Meng F, Chen M (2014) Estimates of the potential for energy conservation and CO_2 emissions mitigation based on Asian-Pacific integrated model (AIM): the case of the iron and steel industry in China. J Clean Prod 65:120–130. https://doi.org/10.1016/j.jclepro.2013.09.008

Weyant C et al. (2014) Emissions from South Asian brick production. Envir sci tech 48(11):6477–6483

Xu Y, Masui T (2009) Local air pollutant emission reduction and ancillary carbon benefits of SO_2 control policies: application of AIM/CGE model to China. Eur J Oper Res 198:315–325. https://doi.org/10.1016/j.ejor.2008.07.048

Energy Efficiency Policies in India: Implications for Climate Change Mitigation

Manisha Jain, Anand B. Rao and Anand Patwardhan

Abstract The role of energy efficiency in meeting domestic energy challenges and global environmental issues is well acknowledged. The need for policy intervention to maximize the potential of energy efficiency that can be realized is also widely accepted. The impact of the energy efficiency policies determines the extent to which they address the barriers to adoption of efficiency measures. Evaluating the impact of policy instruments is crucial to increase their effectiveness and maximize their energy and emission reduction potential. The data to estimate the impact of a policy intervention are often disaggregated, particularly in developing countries. The policy interventions adopted in India to increase energy efficiency include information programs, regulations, financial incentives, and other market-oriented mechanisms. The standards and labeling program is one the most important energy efficiency policies in India. It has been found that the program has significantly contributed to the energy and emission reduction in the country. A positive response from the consumers to the program has also been reported. In this chapter, the policy interventions in India to improve energy efficiency and impact on climate change mitigation are discussed.

M. Jain
Indira Gandhi Institute of Development Research, Mumbai, India
e-mail: manishajain@igidr.ac.in

A. B. Rao (✉)
Interdisciplinary Program (IDP) in Climate Studies, Indian Institute of Technology Bombay, Mumbai, India
e-mail: a.b.rao@iitb.ac.in

A. B. Rao
Centre for Technology Alternatives for Rural Areas (CTARA), Indian Institute of Technology Bombay, Mumbai, India

A. Patwardhan
School of Public Policy, University of Maryland, College Park, USA
e-mail: apat@umd.edu

C. Venkataraman et al. (eds.), *Climate Change Signals and Response*,
https://doi.org/10.1007/978-981-13-0280-0_18

1 Introduction

As part of the international agreement, India has submitted to reduce the emission intensity of its gross domestic product (GDP) by 33–35% from 2005 levels by 2030. The role of energy efficiency in addressing the problem of carbon emissions and climate change is widely acknowledged. But, many of the policies for promoting energy efficiency implemented globally are derived originally from different rationales. In India, the primary drivers for the energy efficiency policies were meeting the demand–supply gap, enhancing energy access, increasing productivity, and reducing energy dependence. Energy efficiency is now also recognized as an important contributor to mitigating the risk of climate change and environmental degradation associated with energy use. The low adoption of cost-effective energy efficiency measures provided a basis for the policy interventions. The policy interventions adopted in India include information programs, regulations, financial incentives, and other market-oriented mechanisms. The standards and labeling (S&L) program, implemented in 2006, a combination of information and regulation program, is a flagship program to improve efficiency in appliances. The program has led to significant savings in the energy and emission reduction since its launch. The program has also received a positive response from the consumers and the market for appliances.

In this chapter, an overview of the energy efficiency potential and barriers to energy efficiency policies in India is presented. The impact of S&L program on energy and emission reduction and consumer response is also discussed. The chapter is divided into four sections. In Sect. 1, an introduction to the rising energy demand of the country and the global challenge of climate change mitigation is given. In Sect. 2, a review of studies that estimate the potential of energy efficiency and indicate the possible barriers to adoption is presented, followed by a discussion on policy interventions implemented in India. In Sect. 3, the impact of the S&L program in energy savings, emission reduction, and consumer response to labels is presented. The conclusions are summarized in Sect. 4.

1.1 India's Energy Demand

The total consumption of energy from conventional sources in India increased at a compounded annual growth rate (CAGR) of 5.5% from 2006–07 to 2015–16 (MOSPI 2017). The largest consumption of raw coal (61%) and the second largest consumption of natural gas (23%) were in electricity generation. The total electricity consumption has increased at a CAGR of 8.2% from 456 billion units (BUs) in 2006–07 to 1001 BUs in 2015–16. India's total emissions were 2137 million tonnes (MT) of the CO_2 equivalent greenhouse gas (GHG) emissions in 2010. In the overall emissions, the energy sector was the most significant contributor (71%) followed by agriculture (18%), industrial processes and product use (8%) and waste (3%). The land use, land-use change, and forestry offset 12% of the total GHG emissions. The total GHG emissions from energy industries and electricity production were 42 and 38% of the total emissions, respectively.

In 2015–16, the industry sector accounted for the most significant share in the total electricity consumption (44%), followed by domestic (22%), agriculture (18%), and commercial sector (9%). The electricity consumption in industry sector increased at CAGR of 9.5% followed by domestic and commercial at CAGR of 8% during 2006–07 and 2015–16 (MOSPI 2017). As per GOI (2017), urbanisation is expected to go up to 47%, and the population is predicted to go up to 1.6 billion by 2040. These developments will result in the energy demand to increase by 2.7–3.2 times between 2012 and 2040. The energy demand from buildings in India, including residential and commercial, is expected to grow at a CAGR of 7% during this period. As a result of rising standards of living and increasing urbanisation, the penetration of electrical appliances is rising. In 2011–12, television sets were possessed by 50% of the rural households compared to 26% in 2004–05, and by 80% urban households compared to 66% in 2004–05. Refrigerators were possessed by 44% urban households in 2011–12 compared to 32% in 2004–05.

The data on household's monthly electricity consumption show that in the urban and rural areas, the share of households consuming less than 100 kWh per month is 60 and 90%, respectively (Chunekar et al. 2016). The appliance-specific energy consumption is commonly estimated using appliance ownership and usage patterns. The estimates vary across studies due to a difference in assumptions. Room air conditioners and refrigerators are one of the major contributors to the total electricity consumption in households. The sale of room air conditioners in India has increased from a million in 2003–04 to 3.1 million in 2010–11 and is estimated to be 4.3 million in 2016–17. The total number of refrigerators sold in 2010–11 was 8 million. These two appliances have been mainly studied in the literature owing to their significant contribution in electricity consumption and their potential in energy savings. It has been found that in addition to income and urbanisation, an increase in the number of cooling degree days may also contribute to an increase in appliance sales.

1.2 The Challenge of Mitigation

The global mean surface air is warming approximately proportional to cumulative CO_2 emissions. The relationship has been called the transient climate response to cumulative carbon emissions and has been used to link a global quota on cumulative CO_2 emissions to a nominated temperature threshold with a specified probability of success (Friedlingstein et al. 2014). Following the Copenhagen Accord in 2009, the United Nation Framework Convention on Climate Change (UNFCCC) formally decided in 2012 to pursue actions in line with a target to hold global temperature increase to below 2 °C above pre-industrial levels. About two-thirds of the available budget for keeping warming to below 2° have already been emitted (Friedlingstein et al. 2014).

Limiting warming to any level implies that the total amount of CO_2 that can ever be emitted into the atmosphere is finite implying that global CO_2 emissions need to become net zero. Limiting global warming to 2° requires negative CO_2 emissions (Gasser et al. 2015). It has been found that the positive CO_2 emissions are more effective at warming than negative emissions are at subsequently cooling, and the cooling effectiveness of negative CO_2 emissions decreases if applied at higher atmospheric CO_2 concentrations, further emphasizing the need for urgent action.

As part of the Paris Agreement in 2015, countries submitted plans for post-2020 to address a range of issues relating to climate change mitigation and adaptation. In these plans, known as Intended Nationally Determined Contributions (INDCs), countries have committed reduction of GHG emissions and have identified specific actions for the same. Rogelj et al. (2016) find that the INDCs collectively lower GHGs compared to where current policies stand, but still implies median warming of 2.6° to 3.1 °C by 2100. After the Paris Agreement, the INDCs are now nationally determined contributions. Studies have reported that if the current NDCs are found to be inadequate, there is a possibility of enhancement in the target in a potential new climate agreement in 2020 (Friedlingstein et al. 2014). As part of this agreement, India has submitted to reduce the emissions intensity of its GDP by 33 to 35% by 2030 from 2005 level. The mitigation strategies to achieve the target include reducing energy demand by energy efficiency and infrastructure development and meeting energy demand by renewable energy sources (GOI 2015).

In the literature, there is a divergence in view on technology readiness to address the carbon and climate-related problem over the short and the long term (Hoffert et al. 2002; Pacala and Socolow 2004). However, the importance of end-use efficiency in meeting the short-term climate mitigation is well established (Pacala and Socolow 2004; Dietz et al. 2009). The negative emission technologies are at an early stage of development, and hence, it is recommended that conventional mitigation, that is, reduced consumption of fossil fuels should remain a significant part of any climate policy aiming at this target.

2 End-Use Energy Efficiency

Most of the research and policy action relating to end-use efficiency began post-oil crisis in the 1970s. A techno-economic approach is most commonly used to calculate the potential of energy efficiency. The techno-economic studies are based on life-cycle cost analysis of efficiency measures. The costs and benefits of a durable are calculated over its lifetime and discounted at a market rate of interest for the individual. The investment with the lowest life-cycle cost is preferred to all other. The concern on profitable energy-saving measures not undertaken has been extensively studied in the literature indicating the presence of market failures such as inadequate information, indifference to energy cost, low energy cost, high capital cost, etc. Various policy interventions have been made, and the impact of these programs are being studied.

2.1 Estimates of the Potential and Barriers

Several studies have estimated the potential of end-use efficiency in India over the last three decades. In one of the early estimates, Nadel et al. (1991) assessed the technical and achievable conservation potential of end-use efficiency in different scenarios. The savings were estimated from twenty-seven conservation and load management measures using detailed techno-economic approach. The study determined that savings of 104–167 BUs were possible in 2004–05 from 1990 to 91, which is around 20% of the projected consumption. The weighted average capital cost of the measures was found to be ₹8800/kW. Reddy and Parikh (1997) estimated the economic and environmental impact of twelve actions in industries and found that 9.2% of the energy requirement can be reduced annually during 1995–2010 (around 113 BUs in 2010). Shrestha et al. (1998) estimated the economic potential of electricity generation savings from national, utility, and user perspective for appliances used in different sectors such as bulbs, refrigerators, and motors. The study estimated that around 1494 BUs could be saved during the period 1997–2015 which is 10% of the total cumulative electricity generation. The potential of the efficiency improvements in the residential sector from lighting and refrigerators was found to be 7.9% of the cumulative electricity generation.

In more recent studies, McNeil et al. (2008) calculated cost-effectiveness potential of air conditioners, refrigerators, motors, and transformers. The study projected that the electricity consumption of these four products would reach 410 BUs by 2020 and 4.3% of the total electricity consumption (54 BUs) can be saved using cost-effective measures from the base year of 2005. Garg et al. (2011) estimated the potential of energy savings from efficient air conditioners and refrigerators in the residential and commercial sector, and agriculture pumps in the agriculture sector in Gujarat. Assuming penetration rate of efficient technologies, the study estimated saving of 8.8 BUs over a period of 10 years from 2006 to 07. In the most recent study, Parikh and Parikh (2016) estimated the annual energy savings and emission reduction from efficiency improvements in air conditioners, refrigerators, televisions, and ceiling fans. The study estimates a range of electricity savings from these appliances between 52 and 151 BUs in 2030 from 2009 depending on the penetration of efficient technologies without any financing.

The projections in the studies discussed have used stock forecast approach to calculate the aggregate saving in the given period. The savings are estimated based on cost-effectiveness metrics such as the cost of conserved energy or the life-cycle cost. The results are not comparable due to differences in end-use technologies, period, and other assumptions. While most studies calculate the total economic potential assuming a 100% penetration of cost-effective technologies, other studies assume a penetration rate to account for other factors such as barriers to penetration of efficient technologies. The cash flows in energy efficiency investments are spread over a given period, and hence, assumptions on discount rates, duration of project or life of the appliance, and changes in electricity cost are also made. The potential of CO_2 emission reduction is calculated by multiplying the electricity savings in the generation with the emission factor for electricity generation.

The top three barriers identified by Nadel et al. (1991) to energy efficiency in India were lack of information, measure cost exceeding willingness to pay and electricity prices lower than marginal costs and the cost of electricity production. Given the existing efficiency programs, the study made several recommendations and emphasized on promoting high-efficiency technologies in India. Reddy (1991) presented an actor-centric approach to identify a comprehensive list of barriers for different stakeholders such as consumers, equipment manufacturers and providers, generation and distribution utilities, financial institutions and policymakers. The study suggested a combination of measures with a focus on policy assisted and market-oriented mechanisms. The study also emphasised on promoting innovation rather than energy efficiency alone. The study suggested that poor consumers have less capital, less information, and are more risk averse and hence offering financing schemes on efficient alternatives to convert the initial down-payment into a payments stream that coincides in time with the savings stream only to those who are first cost sensitive should be considered. Reddy (1996) offered some intervention mechanisms for the promotion of energy-efficient appliances, such as financial incentives like rebates and tax subsidies, installing energy-efficient equipment in households by government/electricity boards and collecting the payments in monthly installments and educating the consumers.

2.2 *Policy Interventions*

Following the oil crisis in the 1970s, several policy interventions have been made globally to address the barriers to energy efficiency. The first major policy initiative in India to coordinate various activities associated with the efficient use of energy and its conservation was the enactment of the Energy Conservation (EC) Act in October 2001. The EC Act provided creation of the Bureau of Energy Efficiency (BEE) which was established in March 2002. BEE works under the Ministry of Power and is responsible for spearheading the improvement of energy efficiency in the economy through various regulatory and promotional instruments. Among the responsibilities of BEE are the planning, management, and implementation of appliance standards and labeling (S&L), as well as Energy Conservation Building Codes. The Indian electricity sector is governed by the Electricity Act 2003, under the premise of which the Government of India notified the National Electricity Policy (NEP) in 2005. The policy mandated the BEE to set standards on energy conservation with a voluntary and self-regulating approach, to begin with, followed by a more regulatory approach. The Integrated Energy Policy Report (IEPR) released in August 2006 entrusted distribution utilities to undertake energy efficiency and demand-side management (DSM) programs.

In 2008, the National Action Plan on Climate Change (NAPCC) was introduced to address the growing issue of climate change. The plan identifies energy efficiency as an essential tool for addressing climate change issues. One of the eight missions of the plan is the National Mission for Enhanced Energy Efficiency (NMEEE). The four pillars of the NMEEE include a market-based mechanism to improve energy efficiency in energy-intensive industries, accelerating the shift to energy-efficient appliances in designated sectors, and to make the products more affordable, the creation of mechanisms that would help finance demand-side management program and the development of financial instruments to promote energy efficiency. The policy regime before the EC Act and till the formulation of the NAPCC is presented in detail by Balachandra et al. (2010).

The three most widely used policy interventions in developing countries identified in the literature are information programs, regulations, and financial incentives (Kelly 2012). Information programs such as appliance labeling address the barrier of inadequate information to consumers, regulation such as efficiency standards on appliances limit consumer choices and address barriers such as insufficient information and indifference to energy costs, and financial incentives aim to lower the high capital cost of efficient technology and push the market toward higher efficiency.

In the industry sector, BEE implemented the Perform, Achieve, and Trade (PAT) scheme in 2012 under NMEEE. PAT scheme is a regulatory instrument linked with market mechanism by way of certification of energy savings. The identified energy-intensive industries, called as designated consumers, were given unit-specific targets. The targets were designed to reflect relative responsibility by giving less target for more efficient and more for less efficient. It covered 478 designated consumers in eight energy-intensive industrial sectors accounting one-third of total energy consumption in the country. A decline of 4–5% in the specific energy consumption of designated consumers in 2015 from 2012 has been reported.

The first large-scale program implemented for appliance efficiency was the S&L program in 2006. It is a combination of information and regulatory intervention with the objective to inform consumers about actual and relative electricity consumptions of appliances and remove least efficient models from the market. In 2010, the Government of India initiated the Bachat Lamp Yojana (BLY) to transform the lighting market from incandescent lamps to energy efficiency compact fluorescent lamps (CFLs). In 2015, the BLY scheme was replaced by National LED program, renamed in 2016 as Unnat Jyoti by Affordable LEDs for All (UJALA) to promote energy-efficient LED lamps. The UJALA scheme has been extended to energy-efficient tube-lights and fans and is now renamed as Unnat Jeevan by Affordable LEDs and Appliances for All. These programs aim to reduce the high cost barrier to adoption of efficient technologies. The scheme operates in conjunction with the S&L program as the energy-efficient ceiling fans are identified by their labels. The impact of these programs on energy and emission reduction has been reported in the relevant policy documents. The savings are estimated primarily using engineering approach. However, study on the impact of the programs on consumers has not been undertaken. In the following section, a detailed analysis of the S&L program in India and its impact on energy and emission reduction and consumer repose is presented.

3 Standards and Labeling Program in India

BEE launched the S&L program with the objective to provide information on the electricity consumption of appliances to the consumer and remove the least efficient models from the market. The S&L program was developed using consumer research and stakeholder interactions to integrate consumer preferences in the design process (Dethman et al. 2000). The label design research was initiated in 1999, and the program was launched in May 2006 under the name of the National Energy Labeling Program. The program is implemented by the Bureau of Energy Efficiency (BEE) under the provisions of the EC Act 2001 (CLASP 2007). Under this program, a comparative star-labeling system has been established where higher numbers of stars imply lower electricity consumption. A total of twenty-one electrical appliances have been covered by 2016–17 (BEE 2017). The EC Act 2001 was amended in 2010 which provided a process for the phasing out of energy-inefficient products for which mandatory labeling had been introduced. There are now six appliances on which labeling have been made mandatory. For the appliances under mandatory labeling, the lowest level of efficiency is the minimum energy performance standard (MEPS). The labels are also mandatory for tubular fluorescent lamps, distribution transformers, electric geysers, and television.

3.1 Energy Savings and Emission Reduction

Few studies have evaluated the impact and the potential of the S&L program in energy savings and emission reduction. GOI (2014) shows that the estimated savings in electricity consumption from labeling of four key appliances—air conditioner, refrigerator, ceiling fan, and television—are 136.8 BU in 2030. The estimated savings are from the use of star-rated appliances as compared to non-rated appliances in 2030 from 2009 to 10. The penetration rates of the star-rated appliances are assumed based on the cost-effectiveness of higher star rating. The contribution of air conditioners and refrigerators in the estimated saving is 23% and 54%, respectively. Dhingra et al. (2016) report the cumulative energy saving and emission reduction achieved from the S&L program in 2012. The total energy saved from eight product categories is estimated to be 6 BUs and GHG reduction of 5.5 Mt of CO_2 in 2012. The contribution of air conditioners and refrigerators in the savings and emission reduction is 22 and 61%, respectively. The cumulative energy savings estimated by BEE (2018) in 2016–17 since 2011–12 are 81 BUs from all appliances covered under the program.

The S&L program was implemented in 2006, and the air conditioners and frost-free refrigerators were the first two appliances on which the labeling was made mandatory in 2010. The data on the share of different star rating in the total sales are available on the Web site of BEE from 2011–12 to 2017–18 (BEE 2018). The impact of labeling program is seen on the sales of the labeled products since the

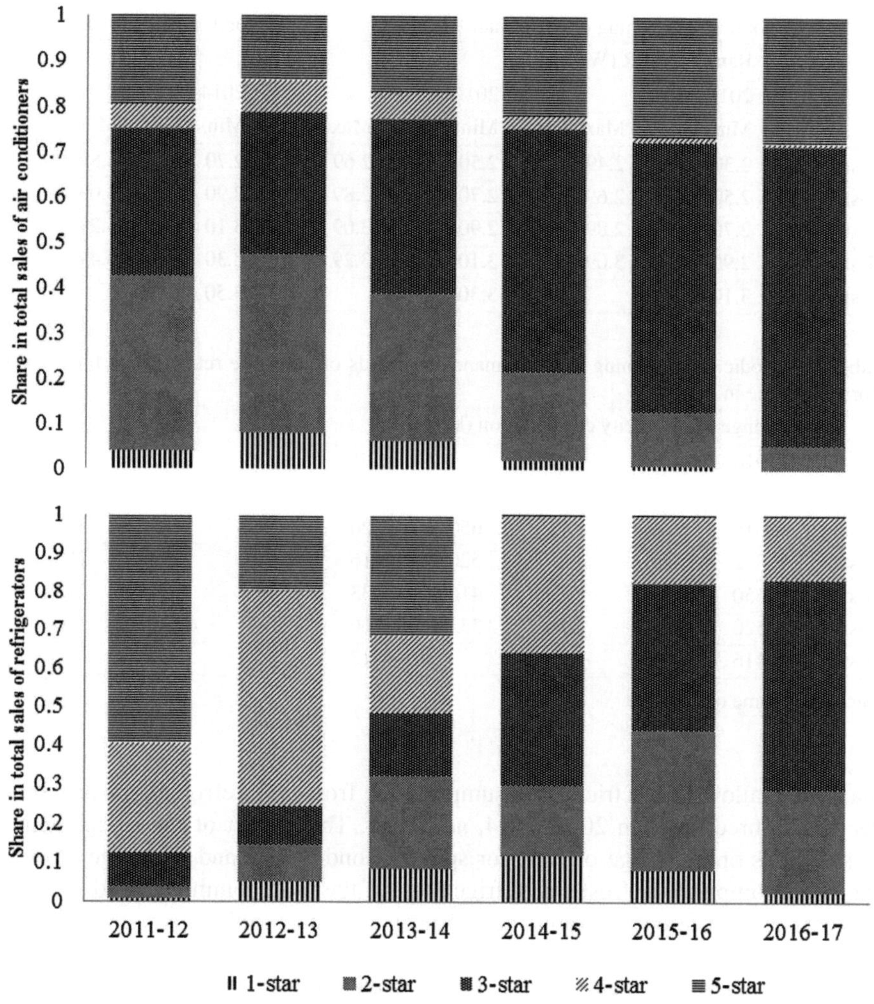

Fig. 1 Share of star rating in total sales of air conditioners and refrigerators in India

launch of the program. As shown in Fig. 1, the share of higher star rating model has been increasing since the launch of the program.

The star rating on air conditioners is given based on a range of energy efficiency ratio (EER) calculated using cooling output (W) and power input (W). The lowest level of allowed EER corresponds to the lower value of the range of 1-star rating. The star rating in refrigerators is given based on a range of electricity consumption calculated using total adjusted volume, constant multiplier (kWh/litre/year), and constant fixed allowance (kWh/year). The maximum allowed electricity consumption corresponds to the lower value of the range of 1-star rating. The minimum allowed EER for room air conditioners have been increased two times in 2012 and 2014. The

Table 1 Periodic strengthening of performance standards of split air conditioners in India

	Range of EER (W/W)					
	2010–11		2012–13		2014–	
	Min	Max	Min	Max	Min	Max
1-star	2.30	2.49	2.50	2.69	2.70	2.89
2-star	2.50	2.69	2.70	2.89	2.90	3.09
3-star	2.70	2.89	2.90	3.09	3.10	3.29
4-star	2.90	3.09	3.10	3.29	3.30	3.49
5-star	3.10		3.30		3.50	

Table 2 Periodic strengthening of performance standards of frost-free refrigerators for a given storage volume in India[a]

	Range of electricity consumption (kWh/y)							
	2010–11		2012–13		2014–15		2016–	
	Min	Max	Min	Max	Min	Max	Min	Max
1-star	1015	812	812	650	520	416	416	333
2-star	812	650	650	520	416	333	333	266
3-star	650	520	520	416	333	266	266	213
4-star	520	416	416	333	266	213	213	170
5-star	416		333		213		170	

[a] Storage volume of 250 litre

maximum allowed electricity consumption of frost-free refrigerators has been decreased three times in 2012, 2014, and 2016. The impact of the strengthening of standards on the range of EER for split air conditioners and the range of electricity consumption for frost-free refrigerators of the most commonly sold size of a refrigerator in different star rating categories is shown in Tables 1 and 2.

The impact of the strengthening of standards on the share of the star rating in the two appliances is also seen in Fig. 1. In the air conditioner sales, the share of 3-star and 5-star has increased steadily despite an increase in efficiency standards in 2012 and 2014. In the refrigerator sales, after an increase of standards in 2014, the market is dominated by 2-star and 3-star refrigerators. This shows that the impact of the strengthening of the standards is not the same in the two appliances. Particularly, this indicates that after strengthening of standards, the air conditioner market continued moving toward higher star rating, but the refrigerator market shifted from higher star rating models to lower star rating models.

The impact on the electricity consumption of the two appliances due to the program can be calculated using data on the share of star rating in sales and allowed electricity consumption in each star rating. The maximum allowed electricity consumption in each star rating is used to calculate the maximum weighted electricity consumption in the two appliances. The annual electricity consumption of refrigerators under test conditions is given on the star labels in kWh/year. The information

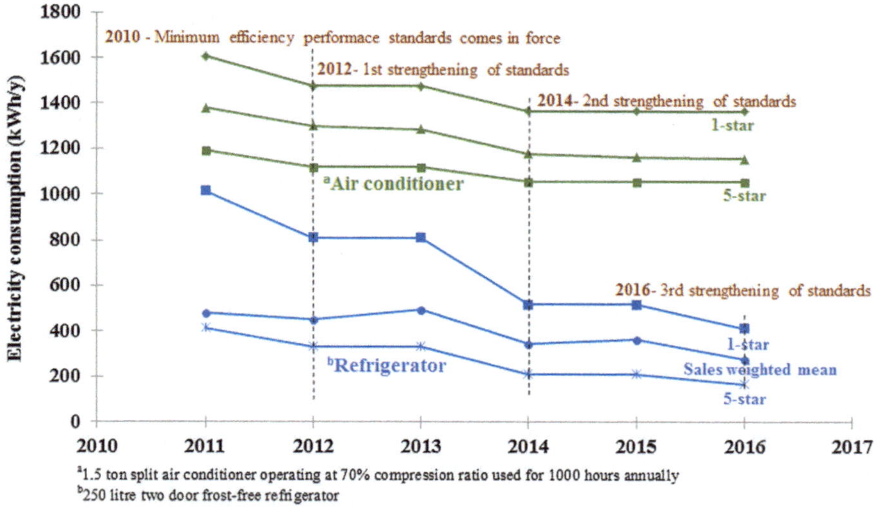

Fig. 2 Star rating range and sales-weighted electricity consumption of air conditioners and refrigerators in India

on the labels relating to the electricity consumption of air conditioners is the EER. The annual electricity consumption of air conditioners can be calculated by assuming usage and compression ratio. The cooling capacity of a 1.5-ton air conditioner is 5275 W (18,000 Btu/hr). The ratio of cooling capacity and EER gives the power input of the air conditioners. The annual electricity consumption in different star ratings is calculated assuming compression ratio of 70% and usage of 1000 h. The range of electricity consumption between 1-star and 5-star rating and the sales-weighted electricity consumption using the maximum allowed electricity consumption is shown in Fig. 2.

The electricity consumption of the air conditioners has declined at an average annual rate of 3% during 2011–2017. The percentage of reduction in energy consumption in refrigerators in the period 2011–2017 is 8%. The electricity consumption is calculated for the most commonly sold sizes of the two appliances for comparison purpose. The sales-weighted electricity consumption is estimated using the maximum allowed electricity consumption in each star rating band and hence is only indicative of the change in electricity consumption. The maximum allowed electricity consumption of air conditioners and refrigerators decreased at annual rate of 2.3 and 12%.

A rough estimate of the realized energy savings from the program can be made using the growth rate in sales of appliances and the change in the rate of efficiency improvements due to the program. The sales of room air conditioners have increased from a million in 2003–04 to 3.4 million in 2010–11 and are estimated to be 4.3 million in 2016–17 suggesting a CAGR of 10% over the last thirteen years. The sale of refrigerators has increased from 8.4 million in 2010–11 to 11.4 million in 2016–17 growing annually at 7%. It has been reported in the literature that in the absence of S&L program the rate of efficiency improvements will be in the range of 0.5–1%. Assuming 1% annual reduction in energy consumption in the absence of the S&L program in India, the savings due to 3 and 8% reduction in electricity consumption of air conditioners and refrigerators, respectively, are calculated. The cumulative energy saving in 2016–17 from 2011 is 3.5 BUs due to air conditioners and 4.3 BUs due to refrigerators. The cumulative energy saving from the two appliances from 2011 to 2017 is approximately 7.8 BUs.

The electricity savings at the generating plant is calculated by taking into account the transmission and distribution losses (T&D). The emission reduction is calculated by multiplying the saving at the generating end by the emission factor for electricity generation. In India, the CO_2 emission factor for all grids in 2013–14 was 0.83 t CO_2/MWh, and the T&D losses in 2015–16 were 22%. The cumulative emission reduction due to energy efficiency improvements in air conditioners and refrigerators is 7.8 MT of CO_2 equivalent.

The potential cumulative energy reduction can be calculated assuming the growth rate of annual sales of appliances and reduction in energy consumption due to energy efficiency. If the energy consumption of the two appliances continues to decrease at 3% annually, the energy consumption of average appliance in 2035 will be equal to the most energy-efficient appliance in India in 2017–18. If the sales of appliances continue to grow at the current rate and the energy consumption decrease at 3% each, the emission reduction from the two appliances will be 156 Mt CO_2 equivalent in 2035. The emission reduction can be accelerated by boosting the rate of efficiency improvements in appliances.

3.2 Consumer Response

The rate of energy efficiency improvements in appliances and hence the potential of emission reduction depends critically on consumer willingness to pay (WTP) for higher efficiency. Based on a country-level survey, Dhingra et al. (2016) show that consumers have responded positively to the S&L program. The study reports that energy efficiency is not a top priority for the consumer in the purchase decision as compared to other factors such as product life and brand name. The consumer WTP for a labeled product is also estimated using direct elicitation method. It is found that most consumers are willing to pay 10% price premium for a labeled product.

In a recent study by the authors, the consumer WTP for the presence of labels and higher star rating is estimated using a discrete choice experiment. In a stated preference survey in April 2015, the choices of a random sample of 302 consumers in Mumbai suburban district were observed in a hypothetical purchase of air conditioners and refrigerators. The findings of the study on consumer preference for labels in air conditioner purchase can be found in Jain et al. (2018). It is found that consumer WTP for the presence of a label on air conditioners is 36% of the price of the air conditioner. It is also found that the consumer preference for star rating levels varies in the sample such that 69 and 78% of consumers place a positive value on the 3-star and 5-star, respectively. The mean of the value placed on the 3-star and 5-star is 12 and 25% of the price of air conditioner, respectively. The consumer WTP for higher star rating is also found to be varying in the sample. The distribution of WTP is such that 62% of the consumers have placed a higher value on the 5-star over 3-star. The mean WTP for the 5-star as compared to 3-star is 12% of the mean price of the air conditioner. The differentiated response to the star rating levels in refrigerators is also reported by the authors in Jain (2018). It is found that consumers differentiate between star rating levels in air conditioners but not in refrigerators. In a between-group experiment design, the impact of additional information on annual operating cost on consumer WTP is also estimated. The impact of the additional information is not significant in air conditioner purchase. However, the consumers use the annual operating cost information in refrigerator purchase to differentiate between star rating levels.

Empirical research on consumer response to policy intervention can provide useful input to improving the effectiveness of energy efficiency policies. For example, as found by the authors, the difference in consumer response to labels in air conditioners and refrigerators shows that even large appliances cannot be considered together in policy design. Energy efficiency programs are sometimes designed in conjunction with labeling programs such as information programs and financial incentives. The choice between alternative policy instruments based on cost-effectiveness can be facilitated by research on consumer response. For example, it is found that in general, consumers have high WTP for efficient air conditioners and may not require financial incentives for adoption of higher-rating models. Consumers are also found to have high WTP for efficient refrigerators. However, the reported indifference to star rating levels could be due to factors such as lack of information on operating cost. Hence, provision of information on operating cost can be considered to increase the rate of penetration of high star rating refrigerators.

4 Conclusions

The mitigation strategies identified in India's NDC to achieve the emission reduction target rely upon technology policies. Ensuring efficient outcome of these policies is hence critical. Energy efficiency is playing an important role in the reduction of emissions intensity of the Indian economy. The improvements in energy efficiency in appliances have been facilitated by policy interventions such as information programs, regulations, and financial incentives. The potential of energy efficiency can be realized by monitoring the impact of the existing programs and designing new programs to complement the existing programs. Development of database of key parameters relating to the program to facilitate empirical research can contribute in overcoming the barriers to program evaluation. Improving the effectiveness of policy interventions by timely evaluation can enhance the contribution of energy efficiency to energy savings and emission reduction.

References

Balachandra P, Ravindranath D, Ravindranath N (2010) Energy efficiency in India: assessing the policy regimes and their impacts. Energy Policy 38(11):6428–6438

BEE (2017) Official website of the Bureau of Energy Efficiency, Ministry of Power, Government of India. URL: http://220.156.189.26:8080/beeLabel/index.jsp. Accessed on 15/01/2017

BEE (2018) BEE star label energy savings, Bureau of Energy Efficiency. URL: https://www.beestarlabel.com/Home/EnergySavings. Accessed on 01/01/2018

Chunekar A, Sapekshya V, Shantanu D (2016) Residential electricity consumption in India: what do we know? Prayas (energy group)

CLASP (2007) India labeling program impacts: case study. Technical report on international institute for energy conservation

Dethman L, Unninayar I, Tribble M (2000) Transforming appliance markets in India: consumer research leads the way. In: Proceedings of the 2000 ACEEE summer study on energy efficiency in buildings—consumer behavior and non-energy effects, California

Dhingra N, Walia A, Mukherjee PK (2016) Measuring the impact of India's standard and labeling program. In: Proceedings on the international energy policies and programmes evaluation conference, Amsterdam

Dietz T, Gardner GT, Gilligan J, Stern PC, Vandenbergh MP (2009) Household actions can provide a behavioral wedge to rapidly reduce US carbon emissions. Proc Natl Acad Sci 106(44):18,452–18,456

Friedlingstein P, Andrew RM, Rogelj J, Peters G, Canadell JG, Knutti R, Luderer G, Raupach M, Schaeffer M, Van Vuuren D (2014) Persistent growth of CO_2 emissions and implications for reaching climate targets. Nat Geosci 7(10):709–715

Garg A, Maheshwari J, Mahapatra D, Kumar S (2011) Economic and environmental implications of demand-side management options. Energy Policy 39(6):3076–3085

Gasser T, Guivarch C, Tachiiri K, Jones C, Ciais P (2015) Negative emissions physically needed to keep global warming below 2°C. Nat Commun 6

GoI (2014) The final report of the expert group on low carbon strategies for inclusive growth. Planning Commission, Government of India

GoI (2015) India's Intended Nationally Determined Contribution: working towards climate justice. Government of India

GoI (2017) Draft national energy policy. NITI Aayog, Government of India

Hoffert MI, Caldeira K, Benford G, Criswell DR, Green C, Herzog H, Jain AK, Kheshgi HS, Lackner KS, Lewis JS (2002) Advanced technology paths to global climate stability: energy for a greenhouse planet. Science 298(5595): 981–987

Jain M (2018) Consumer choices in the purchase and utilisation of energy intensive appliances. PhD thesis, Indian Institute of Technology Bombay

Jain M, Rao AB, Patwardhan A (2018) Consumer preference for labels in the purchase decisions of air conditioners in India. Energy Sustain Dev 42:24–31

Kelly G (2012) Sustainability at home: policy measures for energy-efficient appliances. Renew Sustain Energy Rev 16(9):6851–6860

McNeil MA, Iyer M, Meyers S, Letscher VE, McMahon JE (2008) Potential benefits from improved energy efficiency of key electrical products: the case of India. Energy Policy 36(9):3467–3476

MOSPI (2017) Energy statistics 2017. Central Statistics Office, National Statistical Organisation, Ministry of Statistics and Programme Implementation, Government of India

Nadel S, Kothari V, Gopinath S (1991) Opportunities for improving end-use electricity efficiency in India. American Council for an Energy-Efficient Economy

Pacala S, Socolow R (2004) Stabilization wedges: solving the climate problem for the next 50 years with current technologies. Science 305(5686):968–972

Parikh KS, Parikh JK (2016) Realizing potential savings of energy and emissions from efficient household appliances in India. Energy Policy 97:102–111

Reddy AKN (1991) Barriers to improvements in energy efficiency. Energy policy, pp 953–961

Reddy BS (1996) Consumer discount rates and energy carrier choices in urban households. Int J Energy Res 20:187–195

Reddy BS, Parikh JK (1997) Economic and environmental impacts of demand side management programmes. Energy Policy 25(3):349–356

Rogelj J, Den Elzen M, Hohne N, Fransen T, Fekete H, Winkler H, Schaeffer R, Sha F, Riahi K, Meinshausen M (2016) Paris agreement climate proposals need a boost to keep warming well below 2°C. Nature 534(7609):631–639

Shrestha RM, Natarajan B, Chakaravarti K, Shrestha R (1998) Environmental and power generation implications of efficient electrical appliances for India. Energy 23(12):1065–1072

MIX
Papier aus verantwortungsvollen Quellen
Paper from responsible sources
FSC® C105338

If you have any concerns about our products,
you can contact us on
ProductSafety@springernature.com

In case Publisher is established outside the EU,
the EU authorized representative is:
Springer Nature Customer Service Center GmbH
Europaplatz 3, 69115 Heidelberg, Germany

Printed by Libri Plureos GmbH
in Hamburg, Germany